Trends in Development of Accelerated Testing for Automotive and Aerospace Engineering

Trends in Development of Accelerated Testing for Automotive and Aerospace Engineering

Lev M. Klyatis
Professor Emeritus
Habilitated Dr.-Ing., Dr. of Technical Sciences, PhD.

ACADEMIC PRESS
An imprint of Elsevier

ELSEVIER

Academic Press is an imprint of Elsevier
125 London Wall, London EC2Y 5AS, United Kingdom
525 B Street, Suite 1650, San Diego, CA 92101, United States
50 Hampshire Street, 5th Floor, Cambridge, MA 02139, United States
The Boulevard, Langford Lane, Kidlington, Oxford OX5 1GB, United Kingdom

Notices
Knowledge and best practice in this field are constantly changing. As new research and experience broaden our understanding, changes in research methods, professional practices, or medical treatment may become necessary.

Practitioners and researchers must always rely on their own experience and knowledge in evaluating and using any information, methods, compounds, or experiments described herein. In using such information or methods they should be mindful of their own safety and the safety of others, including parties for whom they have a professional responsibility.

To the fullest extent of the law, neither the Publisher nor the authors, contributors, or editors, assume any liability for any injury and/or damage to persons or property as a matter of products liability, negligence or otherwise, or from any use or operation of any methods, products, instructions, or ideas contained in the material herein.

Library of Congress Cataloging-in-Publication Data
A catalog record for this book is available from the Library of Congress

British Library Cataloguing-in-Publication Data
A catalogue record for this book is available from the British Library

ISBN: 978-0-12-818841-5

For information on all Academic Press publications visit our website at
https://www.elsevier.com/books-and-journals

Language Editor: Edward L. Anderson
Publisher: Matthew Deans
Acquisitions Editor: Carrie Bolger
Editorial Project Manager: Ana Claudia Garcia
Production Project Manager: Paul Prasad Chandramohan
Cover Designer: Matthew Limbert

Typeset by TNQ Technologies

Working together
to grow libraries in
developing countries
www.elsevier.com • www.bookaid.org

Dedication

To my wife Nellya Klyatis

L. M. K.

Contents

About the Author

Lev M. Klyatis

Lev M. Klyatis is a senior adviser at SoHaR, Incorporated. He holds three doctoral degrees: Engineering Technology, PhD; Engineering Technology, Dr. of Technical Sciences (a high-level East European doctoral degree); and Engineering, Habilitated Dr.-Ing. (a high-level West European doctoral degree).

His major scientific/technical expertise has been in the development of a new direction for the successful prediction of product efficiency during any given time, including accelerated reliability and durability testing technology, and accurate physical simulation of real-world conditions. This direction is based on new ideas, and unique approaches will lead to future developments for improving society. This new direction was founded on the integration of field/flight inputs, safety aspects, and human factors, improvement in the engineering culture and accelerated product development. Developing a new methodology for reducing complaints and recalls, he formalized concepts on how to avoid the negative aspects of accelerated testing and the misconceptions that are prevalent in today's engineering. His approach has been verified in various industries, primarily those involved in automotive, farm machinery, aerospace, and aircraft industries. He has shared these methods working as a consultant to Ford, DaimlerChrysler, Nissan, Toyota, Jatco Ltd, Thermo King, Black and Decker, NASA Research Centers, Carl Schenk (Germany), and many others.

He was qualified as a full professor by the USSR's Highest Examination Board and was employed as a full professor at the Moscow University of Agricultural Engineers. He has selected to be on the US–USSR Trade and Economic Council, the United Nations Economic Commission for Europe, the International Electrotechnical Commission (IEC), and the International Standardization Organization (ISO). He also served as an expert for the United States, and an expert at the International Standardization Organization and

International Electrotechnical Commission (ISO/IEC) Joint Study Group in Safety Aspects of Risk Assessment. He was the research leader and chairman of the State Enterprise TESTMASH, Moscow, Russia, and principal engineer of a state test center.

He is presently a member of the World Quality Council, the Elmer A. Sperry Board of Award, the SAE International G-41 (former G-11) Reliability Committee, the Integrated Design and Manufacturing Committee (IDM) of SAE International World Congresses; a Session Chairman for SAE World Congresses in Detroit since 2012; and a member of the Governing Board of the SAE Metropolitan Section. He has also been a seminar instructor for the American Society for Quality (ASQ).

Lev Klyatis is the author of over 300 publications, including 12 books, and holds more than 30 patents worldwide. Dr. Klyatis frequently speaks at technical and scientific events that are held around the world.

Preface

If you analyze all engineering areas, including automotive and aerospace, over the last dozen years, you will discover that the number of recalls, complaints, and life cycle cost are increasing. Safety, reliability, durability, and maintainability are increasingly becoming serious issues for manufacturers, users, and even third parties who become collateral damage to these defects.

Why is this happening?

The basic reasons for these increasing recalls are largely safety- and reliability-related issues. While it is true that these are the apparent reasons for the recalls, they are not the fundamental reasons. The fundamental cause is the failure to predict the product's proper operation during research, design, and manufacturing. And in today's world one of the basic components of prediction is the use of accelerated testing.

This leads to the question "Why are these prediction and testing results failing?"

The answer is that while the products have become more complicated, and while the technical progress in technological applications is more rapid than ever, the technical progress in the testing area is increasing far more slowly. Ideally, more complicated and accurate technology needs to be accompanied by corresponding advances in accurate testing. And, this more complicated testing demands more accurate simulation of field/flight conditions during research, design, and manufacturing.

But, this is not happening. Simply comparing the technology of today's cars or trucks as compared with those of 80–100 years ago, one sees dramatic technical progress. But comparing testing technology of 80–100 years ago with that of today will show that mostly similar vibration, corrosion, environmental (temperature + humidity) factors, and other testing technology being used. Most progress over this time relates to progress in the system of control and reflects largely the progress in electronics. Separate influences simulation remains the basic approach for testing as it was many years ago.

So, as a result the outcomes of testing (most notably accelerated testing) do not provide accurate initial information for successful prediction, especially for the long-term prediction of complicated products.

This book details how to overcome this obstacle and how to improve the product's accelerated testing. Implementing the concepts presented in this book leads to increased quality, reliability, safety, durability, maintainability and decreased recalls and complaints, life cycle cost, and other efficiency components.

In the quest for this improvement, caution is required as changes in accelerated testing can be positive or negative. Changes which adopt the negative trends will lead to degradation of the testing process and can be a degrading factor in the engineering development of the product.

Therefore, this book focuses on how to avoid those negative trends, and on how to implement the positive trends in testing to influence improvement to the product's effectiveness. Of course, using only this book cannot fully inform one on how and what needs to be done. But, in combination with the previous books of this author [1−4], it will provide the knowledge needed to improve accelerated testing and prediction.

Consider one practical example. The professional society SAE International, which is involved in all aspects of mobility engineering, organizes an annual SAE World Congress in Detroit. Tens of thousands of professionals from automotive, aerospace, and other industries from many countries attend this event, and most of these attendees are from world-class industrial companies. At this congress, there is a session entitled "Trends in Development of Accelerated Reliability and Durability Testing," which has been chaired by this author since 2012.

During his years as the chair, 2012−2019, most presentations, especially including unaccepted paper proposals, were related to the negative trends in the development of accelerated testing (a description of negative trends can be found in Chapter 4). Unfortunately, this is reflective of the common situation that can be found in automotive and aerospace engineering.

There are two fundamental concepts in accelerated testing of automotive and aerospace engineering:

1. Most laboratory or proving ground testing are accelerated testing because it obtains results faster than in normal field conditions. For the same reason, intensive field or flight testing is also accelerated testing.
2. The effectiveness of the product is highly dependent on the effectiveness on the testing. Many examples of this can be seen in real life.

There are several concepts prevalent as to what constitutes accelerated testing. For example, in Ref. [5], accelerated testing is presented as "Historically accelerated testing providing and analyzing began from involving times-to-failure data (of a product, system, or component) obtained under normal operating conditions in order to quantify the life characteristics of the product. In many situations, and for many reasons, such life data (for times-to-failure data) is very difficult, if not impossible, to obtain."

The reasons this is difficult include the long life of many products, the small time period available from product design through release, and the challenge of testing products that are used continuously.

Given these difficulties coupled with the need to better understand a product's failure modes and life characteristics, reliability practitioners have attempted to devise methods to force these products to fail more quickly than they would under conditions of normal use.

So, they attempted to accelerate failures. And, as a result, over the years, the term "Accelerated Life Testing (ALT)" often has been used to describe all such practices.

Traditionally, a variety of methods that served different purposes have been termed "Accelerated Life Testing." While this accelerated testing is designed for the single purpose of qualifying the life characteristics of the product under normal use conditions, this testing is frequently performed in specific, and usually artificial conditions.

A variety of approaches to Accelerated Life Testing (ALT) are widely used in quality, reliability, durability, maintainability, supportability, and for other objectives. ALT is a test method where the test's stresses are greater than that experienced in normal field/flight stress operations in order to compress the time to failure and to uncover problems with the test subjects.

Accelerated testing is widely used in many areas of research, design, manufacturing, and usage, and not just in engineering.

ALT is now the primary source for obtaining initial information for predicting a product's reliability, safety, durability, and maintainability. This book demonstrates how such testing often fails to provide relevant information. This occurs because the product's degradation process in testing conditions differs substantially from the product's degradation process under field conditions.

Consequently, testing data such as the mean time to failure is skewed, resulting in actual product problems with recalls, reliability, and maintainability that were not predicted.

In 2009−2010, Toyota's global recalls jumped to 9 million cars and trucks. A similar situation has happened to other automakers, as well as to companies in other areas of industry. This leads to losses in the billions of dollars. For example, in Ref. [6], Steven Ewing writes that automotive recalls cost $22 billion in 2016, a 26% increase over the previous year; according to *Automotive News*, "A staggering 53.1 million vehicles were recalled in the US in 2016, citing a study by AlixPartners, this resulted in $22.1 billion in claims and warranty accruals by automakers and suppliers, a 26 percent increase over the previous year."

The inaccurate simulation of real-world conditions for testing results in inaccurate prediction of the product's reliability, safety, durability, and quality during design and manufacturing and these societal and financial losses.

The prediction based on data from inaccurate ALT where the long-term degradation (failure) process differs substantially from the product's random degradation process under actual field conditions.

Traditionally, a variety of stress methods that are designed to serve different purposes have been termed "accelerated life testing." ALT should involve acceleration of failures with the single purpose of quantification of the life characteristics of the product under normal use conditions. But, much of what being called ALT is based on traditional approach, where only part of the field influences are simulated, and they are not integrated with factors such as safety and human factors, as well as not taking into account the real nonstationary random process of field influences. A detailed analysis of this can be seen in this author's book [1].

Test engineering needs to become smarter. This author agrees with Dr. Leuridan who wrote in Ref. [7] that it will "require fundamental, innovative, new approach as part of the redesign of the end-to-end development process."

Although written about aircraft products, this relates equally to automotive, aerospace, and other areas of engineering.

In addition, in order to meet society's expectations, the next generation of these products needs to be far more economical and ecological. Successful product development programs will have to deal with new materials, new technologies, and greater complexity, while still staying within budget and with the need for shorter time to market.

While traditional test processes are extremely important during the design, development, assembly, and validation stages, they will need a major structural overhaul. Testing time needs to be shorter, tests need to address a wider range of conditions and technologies, and testing methodology must be able to cope with increasingly complex systems.

This will require a fundamental, innovative, new *approach* as part of the redesign and development process. This book will explain how this testing can be performed quickly, efficiently, and accurately.

This will require more intelligent testing systems, more consistency and ease of use, and more results in a shorter time.

Growing product complexity also necessitates enhanced computing power. New testing equipment must enable the collection of more data through more channels, for networked systems, for software that shares data seamlessly between different applications, with built-in scalability and customization possibilities, and for data that can be more easily interpreted [7] and used for optimization of the product.

Classical test strategies need to be upgraded and adopted, and robust controllers will become a major contributor toward reaching these higher performance targets. This is why today's development programs intensively use technologies such as MIL-SIL-HIL, allowing more realistic test configurations. And the test planning needs to be developed to fully support this enhanced testing process. Full system testing requires robust procedures to analyze the data and the correlation of testing with physical interactions to fully understand the impact on the final product's performance characteristics.

Another important consideration is that new design concepts and technologies require more testing in order to fully understand the physical phenomena involved and to control risks. Models should be used to carefully plan the validation test campaign for new designs.

Traditional development testing can serve as a prototype for validation. But, when using this process, the simulation of real-world conditions is usually inaccurate. This is a significant problem, which professionals need to resolve during the development of the testing.

Automotive and aerospace engineering in particular needs answers that address all of these challenges. The important problems that need to be evaluated are model-based test strategies, updating development strategies through feedback from test data, and accelerated testing of subsystems and systems in real-world conditions. In accomplishing this, it is very important that interacted and integrated accurate simulation of real-world conditions be included. This is a very important need in accelerated testing development in both automotive and aerospace engineering. Field/flight testing must be based on accurate simulation of real-world conditions. And, while test subjects are becoming more complicated and more expensive, the accurate accelerated testing must also be more complicated and have much higher corresponding costs. But too often we hear that this accelerated testing development is too expensive, especially in comparison with traditional ALT.

As was mentioned in the author's previous publications, Phillip Coule, the former director of the Operational Test and Evaluation Office (Pentagon) said in the US Senate [2] that "...if, during the design and manufacturing of the complicated apparatus such as satellite, one tries to save a few pennies in testing, the end results may be a huge loss of thousands of dollars due to faulty products which have to be replaced because of this mistake."

This observation equally relates to other products. Or, in more generic language, if during the design and manufacturing of any products, including automotive and aerospace, one tries to save money in testing and making it easier by using incorrect simulation of real-world conditions, the end result will be in huge loss, including reliability, safety, quality, durability, maintainability, supportability, life cycle cost, profit (for both producer and user), and other components of effectiveness.

This relates not just to engineering areas but to many other areas of people's activity (physics, chemistry, mathematic, medical, pharmaceutical, teaching, and others, as well as to many areas of the arts and sciences).

The final result of poor prediction will be decreased value to society and increased costs, and not just to the engineering culture but also to the society.

For these reasons, the concepts of accelerated testing are presented in this book, and this author's other publications. Solutions in successful prediction of product efficiency are important not only for the present generation but also for future generations.

Reducing the negative trends and wider implementation of the positive trends in the development of accelerated testing in automotive and aerospace engineering will lead to the development of greater product effectiveness and will result in products that are more beneficial to the society. As will be seen in Chapter 2, the current result of this failure is multibillion dollar losses just in the automotive industry.

This book has been written to be a useful tool for implementing these methods particularly in the marine, electronic, electrotechnical, and other areas of engineering.

The author has briefly considered some of these aspects in the trends in the development of accelerated testing in his previous publications [8—12] and others; but this book completely considers the problems of moving to effective accelerated testing in automotive and aerospace engineering.

Bibliography

[1] Klyatis LM. Accelerated reliability and durability testing technology. Wiley; 2012.

[2] Klyatis LM, Anderson EL. Reliability prediction and testing textbook. Wiley; 2018.

[3] Klyatis LM. Accelerated quality and reliability solutions. Elsevier; 2006.

[4] Klyatis LM. Successful prediction of product performance (quality, reliability, safety, durability, maintainability, recalls, life-sycle cost, and other components. SAE International; 2016.

[5] Pantelis Vassilion and Adamantios Mettas. Understanding accelerated life testing analysis. 2003 Annual reliability and maintainability symposium (RAMS) tutorial notes.

[6] S. Ewing. Automotive recalls cost $22 billion in 2016. That's a 26 percents increase over the previous year. January 31, 2018. Road show. Car industry.

[7] Jan L. On the right road. Aerospace Testing International. SHOWCASE 2013.

[8] Klyatis L. About trends in the strategy of development accelerated reliability and durability testing technology. SAE paper 2012-01-0206. SAE 2012 World Congress. Detropit.

[9] L. Klyatis. Development of accelerated reliability/durability testing standardization as a component of trends in development accelerated reliability/durability testing (ART/ADT). SAE 2013 World Congress and Exhibition. Paper 2013-01-0151. Detroit, MI, April 16—18, 2013.

[10] Klyatis LM. Principles of accelerated reliability testing. 1998 IEEE Workshop on accelerated stress testing proceedings. Pasadena, CA, September 22—24, pp. 1—4.

[11] Klyatis L. A new approach to physical simulation and accelerated reliability testing in avionics. Development Forum. Aerospace Testing Expo2006 North America. Anaheim, California, November 14—16, 2006.

[12] Klyatis L. Why current types of accelerated stress testing cannot help to accurately predict reliability and durability? SAE 2011 World Congress and Exhibition. Paper 2011-01-0800. Also in book Reliability and Robust Design in automotive engineering (in the book SP-2306). Detroit, MI, April 12—14, 2011.

Chapter 1

Terms and definitions

Abstract

Too often, even professionals in the areas of testing and reliability use mistaken or incorrect terms and definitions. As a result, frequently, there are misunderstandings of the type of testing being provided, and what the testing's results actually prove. Frequently, it can be seen even in published literature, vibration testing being referred to as "durability testing," or proven ground testing being referred to as "reliability" or "durability" testing, or fatigue testing being referred "durability testing," etc. The end result of this confusing terminology is testing and product evaluations that are inaccurate or disappointing in the prediction of the product's performance. The basic reason is not the failure of the testing but of not having a clear understanding of the proper definitions to be used in the testing. Very often, this results in the laboratory, proving ground, or field/flight testing being different than the real-world, long-term results. This mistaken terminology can dramatically influence the quality, reliability, safety, durability, and other components of a product's effectiveness in real-world operation.

Recognizing this, this author includes here, and in some of his other books, a special chapter entitled "Terms and Definitions."

The following terms and definitions are largely based on the following sources:

- Standards [1—11]
- Published literature [12—16]

Correct terms and definitions

Accelerated testing - Testing in which the deterioration of the test subject is accelerated as compared to actual service in the real world. This testing can be related to most laboratory testing, proving grounds testing, intensive field testing, or research in the laboratory.

Accelerated Reliability and Durability Testing (ART/ADT) testing in which:

A. The testing consists of an integrated combination of laboratory testing and periodic field testing. The laboratory testing must provide a simultaneous

Trends in Development of Accelerated Testing for Automotive and Aerospace Engineering.
https://doi.org/10.1016/B978-0-12-818841-5.00001-5
1

combination of a whole complex of multi-environmental testing, mechanical testing, electrical/electronic testing, and other types of real-world testing. The periodic field testing accounts for those factors that cannot be accurately simulated in the laboratory, such as the stability of the product's technological processes during usage, how the operator's and management's operational and maintenance practices influence the test subject's reliability and durability, and the interaction of real field influences on the product's safety, and others. Accurate simulation of field conditions requires accurate simulation of all of the integrated field input influences, safety, and human factors.

B. The testing produces physics-of-degradation mechanism that meets the stated criteria for similarity to the failure mechanisms that occur in normal service conditions (e.g., a similar percentage of metallic or plastic deformation, or creep rate, etc.).

C. The reliability indicator measurements (times to failure, degradation metrics, etc.) that have a high correlation with the corresponding real-world measurements in normal service conditions using stated criteria.

NOTE 1 Accelerated reliability testing or accelerated durability testing, or durability testing that provide useful information for the accurate prediction of reliability and durability because they are based on accurate simulation of real-world conditions.

NOTE 2 ART/ADT is identical to reliability testing if reliability testing is used for accurate reliability and durability prediction during the service life, warranty period, or other defined cycle.

NOTE 3 Accelerated reliability and durability testing, as in any type of accelerated testing, is related to the applied stress. Higher stress level results in a higher acceleration coefficient (the ratio of time to failures in the field as compared to the time to failures during ART). A lower correlation between field results and ART results means there will be a less accurate prediction.

NOTE 4 Accelerated reliability testing (ART) and accelerated durability testing (ADT) (or durability testing) have the same basic objective—the accurate simulation of the field conditions. The only difference is in the indices of these types of testing and the length of the testing. For reliability testing, it is usually the mean time to failures (MTTF), time between failures, and other parameters of interest, while for durability, it is the amount of time without failure or the out of service time.

NOTE 5 Accelerated reliability testing can be for different defined lengths of time, i.e., warranty period, 1 year, 2 years, service life, and others.

Acceleration coefficient The acceleration coefficient is the ratio of the determined time for the product's timeline in service (years, hours, cycles, etc.) as the numerator versus the time of testing as the denominator.

NOTE Thus, an accelerator coefficient of 1 would indicate testing time equal to real-world time, and accelerated testing would yield accelerated co-efficients greater than 1.

Accurate simulation of real-world conditions A. Simulation in which all important groups of real-world influences (multi-environmental, mechanical, electrical, and other groups) act simultaneously and in mutual combination.

B. Requires comprehensive simulation of the above influencing factors, combined with safety and human factors.

Accurate simulation of the field/flight input influences

Is simulation for which all field/flight influences act simultaneously and in mutual combination, and are accurately simulated.

Accurate physical simulation of real-world conditions Simulation which provides the same physical state of degradation (deformation, cracking, corrosion, etc.) or failures, during ART/ADT of the product that differs from those that occur in the real world by no more than a specified allowable difference (generally achieved by using an acceleration coefficient).

NOTE Misleading in the accurate physical simulation of real-world input influences leads to misleading in using output variables (vibration, loading, output voltage, etc.) that one are often uses instead of product degradation (deformation, corrosion, cracking, etc.).

Allowable stress Maximum permissible stress that can be permitted in a structural part for a given operating environment that will prevent rupture, collapse, detrimental deformation, or unacceptable crack growth.

Accurate physical simulation. Occurs when the physical state of degradation in the laboratory differs from those in the field/flight by no more than the allowable limit of divergence.

Assessment A systematic method of obtaining evidence from tests, examinations, questionnaires, surveys, and collateral sources that are used to draw inferences about the characteristics of people, objects, or programs for a specific purpose.

Classification accuracy A measure of the degree to which neither false positive nor false-negative categorizations and diagnoses occur when a test is used to classify an individual or event.

Climatic categories. Defined categories of specific types of world climates in which materials or products are designed to withstand during operation, storage, and transit.

Cumulative effects. The collective consequences of stresses during the life cycle of the product.

Combination of tests. A combination of tests designed to represent the effects of the real world more realistically than one, or a series of single tests produces. Combined testing is encouraged when these conditions may be expected in the real world.

NOTE Combined testing includes two or more input influences.

Common Cause Failure: Failures of multiple items occurring from a single cause, which is common to all of them.

Common Mode Failure: Failures of multiple similar items that fail in the same fashion.

Common Mode Fault: Faults of multiple items, which exhibit the same fault mode.

Correlation The tendency for two measures or variables, such as height, or weight, or other, to vary together or be related for individuals in a group.

Compliance Test A test used to show whether or not a characteristic or a property of an item complies with the stated requirements.

Development The process by which the capability to adequately implement technology or design is established before manufacture.

NOTE The process may include the building of various partial or complete models of the products and assessment of their performance.

Durability 1. The ability of the item (material, detail, unit, or whole vehicle) to perform a required function under given conditions of use and maintenance until a limiting state is reached. OR 2. Durability is the measure of its useful life.

NOTE Durability is a special case of reliability.

Durability testing (DT) or accelerated durability testing (ADT) Is testing in which: The primary output of the measurement is of the product's useful life (service life), which may be expressed in terms of operating hours, years, miles, or other volumes of work. These measurements must also have a high degree of correlation with corresponding real-world measurements in normal usage conditions using the stated criteria.

NOTE 1 The result of durability testing (accelerated durability testing) or accelerated reliability testing (ART) is normally used for:

A. Obtaining information that allows for the successful predicting of reliability, durability, safety, recalls, etc., for the warranty period, service life, or other specified periods of products performance; **B.** Successful prediction and prevention of problem with a product.

C. ART and ADT (DT) differ only in the associated metrics and the test length. For ART, the associated metrics include MTTF, MTBF, among others. For DT (ADT), the primary metric is service life, which may be expressed in terms of operating hours, years, miles, etc.

NOTE 2. A. Durability testing is different from discrete vibration testing and proving ground testing because durability testing requires the interaction of the multi-environmental, mechanical, electrical, and other necessary testing with the corresponding testing equipment and interaction with other system components.

B. Vibration testing is only a part of proving ground testing or mechanical testing.

C. As a result of the often misleading terminology in the literature for the above types of testing, SAE International G-11 Division, Reliability

Committee prepared and approved testing definitions which are included in standard JA 1009/A Reliability Testing Glossary and Terms.

Endurance Test A test carried out over a time interval to investigate how the properties of an item are affected by the application of stated stresses and by their time duration or repeated application.

Environmental analysis. Technical activities covering the analytical description of the effects that various environmental factors, such as temperature, humidity, and their rates of changes have on materiel, subsystems, and component effectiveness.

Evaluation and predicting Evaluation is ascertaining or fixing the value or worth of parameters in test conditions. If the testing is in the laboratory or proving grounds, the resulting evaluation relates to laboratory or proving grounds conditions only but not to real-world conditions.

Predicting is to ascertain or fix the value or worth the product of product performance parameters in actual conditions of future use. If the corresponding methodology plus the testing provided in the laboratory coupled with the field/flight testing is accurate, the predicting relates to what will happen in future actual field/flight conditions.

Failure mechanism The physical, chemical, or other processes that lead to failure.

Fatigue testing Testing for determining the failure or degradation behavior of materials, which are subject to fluctuating loads. A specified mean load (which may be zero) and an alternating load are applied to a specimen, and the number of cycles required to produce failure (fatigue life) is recorded.

Field testing (sometimes referred to a real world test). A test administered to check the adequacy of (sometimes referred to a real-world test) testing procedures in the actual normal service, generally, including test administration, test responses, test scoring, and test reporting.

HALT (Highly Accelerated Life Test) An acronym for Highly Accelerated Life Test. It is designed to expose products to extreme levels of thermal, vibration, and product-specific stresses in order to precipitate failures as quickly as possible. Because of the extreme stresses (mostly thermal and vibration) it does not accurately simulate real field conditions.

HASS (Highly Accelerated Stress Screen) An acronym for Highly Accelerated Stress Screen. HASS tests are less extreme versions of HALT that are conducted on a manufactured product before shipping to the customer. The HASS test uses information that has been obtained from the HALT testing to create the stress screen parameters. As with HALT, it also does not accurately simulate real field conditions.

Human factors (in general) The scientific discipline that is concerned with understanding the interactions between humans and other elements of a system.

Human factors (in engineering) The scientific discipline dedicated to improving the human-machine interface and human performance through the application of the knowledge of human capabilities, strengths, weaknesses, and characteristics.

The interactions between units and details A. Units and details are components of entire machines and interact within machines. If one provides separate testing of them, one ignores these interactions and cannot obtain, as a result of such testing, accurate evaluation of the product's field performance.

B. To achieve successful performance prediction, including units and details, it is necessary to accurately simulate these interactions during the design and manufacturing of units and details, as well as to accurately simulate the real-world influences of the combination. For example, if performing transmission or engine reliability/durability testing in the laboratory, it must take into account vibration and climatic influences produced by the vehicle and its operating environment.

Laboratory test A compliance test made under prescribed and controlled conditions, which may or not simulate field conditions.

Life cycle The time envelope of a product, consisting of four basic phases: research and development, production or construction, operation, and maintenance. In other words, the life cycle consists of the timeline, including disposal, design, manufacturing, and usage.

Model A physical or abstract representation of relevant aspects of an item or process that is developed as a basis for calculations, predictions, or other product assessment.

Multi-environmental complex of field input influences. The combination of the various independent.

Input influences environmental factors that affect a product during its life cycle. For example, the operating environment may include temperature, humidity, pollution, radiation, wind, snow, rain, and rates of fluctuations. Some basic input influences combine to form a multifaceted complex. For example, chemical pollution and mechanical pollution combine in the pollution complex. These factors are often interconnected, and interact simultaneously in combination with each other.

Output variables The direct results of the action of the input influences the product. Output variables can be loading, tension, output voltage, vibration, and others. The output variables have the potential to produce product degradation (deformation, crack, corrosion, vibration, overheating) and failures of the product.

Quality is the ability of the product or service to satisfy the user's needs.

NOTE 1 As in many instances, the user's needs can change over time; this implies a need for the periodic review of requirements to maintain quality.

NOTE 2 Needs are usually translated into characteristics with specified criteria. Needs may include, for example, aspects of performance, usability, dependability (reliability, availability, maintainability), safety, environment, economics, and esthetics.

Prediction What is expected to happen at some time in the future. A prediction is often, but not always, based upon prior experience or knowledge.

Successful prediction of a product's performance in engineering A high degree of correlation between the predicted performance and actual performance of a product. Generally, it consists of two integrated components— employing the advanced methodology and using accelerated reliability and durability testing (ART/ADT).

Real-world (or field/flight) testing. A test administration used to check the product's performance in actual normal service conditions Generally includes testing administration, testing response, testing scoring, and test reporting.

Reliability The ability of an item to perform a required function under given conditions for a given time interval.

NOTE 1 It is generally assumed that the item is in a condition to perform this required function at the beginning of the time interval.

NOTE 2 The term "reliability" is also used to denote the nonqualified ability of an item to perform a required function under stated conditions for a specified period of time.

Reliability growth A condition characterized by a progressive improvement of the product's reliability performance when measured over time.

Reliability testing Testing in actual normal service use that provides initial information for the evaluation of the measurement of reliability indicators during the time of the testing.

NOTE 1 If used for accurate reliability prediction during any specified time (service life, warranty period, or other), it is identical to accelerated reliability or durability testing.

NOTE 2 Often incorrect or misleading. For example, some companies may test automobiles for 160,000 miles in several months and claim that they can predict the vehicle's reliability over its service life. This is not incorrect, because it does not take into account the multi-environmental influences that are related to the time of exposure during the vehicle's service life.

Service life. Time from the release of the product by the manufacturer through its retirement from service and including any disposition cost.

Life cycle cost The total cost of a system or product from "need identification" until disposal. This consists of acquisition cost, ownership cost, and disposal costs.

Software Programs, procedures, rules, and any associated documentation pertaining to the operation of a computer system.

Types of stress testing Classified as constant stress, step stress, cycling stress, and random stress.

Random stress A stress whose type, frequency, amplitude, duration, and magnitude are selected through random processes. Frequently used in an attempt to simulate real-world conditions.

Random process. A process represented by an ensemble of time history records that have properties described in terms of parameters estimated from statistical computations at selected times. In this publication, it will be assumed that one or more sample records from the random process are related to a repeatable experiment that completely describes the phenomenon under consideration.

Stationary random process. A process represented by stationary random process is an ensemble of time history records that have statistical properties that are not a function of time, and hence, are invariant with respect to time translations. The stationary random process may be ergodic or nonergodic.

Non-stationary random process. A process that is an ensemble of time history records that cannot be defined to be stationary. In general, the statistical properties of a nonstationary process are a function of time and not invariant with respect to time translations. In this standard, if either the mean (first moment) estimate or mean-square (second moment) estimate or both from a random process ensemble vary with time over the ensemble, the random process is considered nonstationary. If the ensemble has a deterministic component that varies in time, the ensemble may or may not be considered nonstationary depending on whether the random part of the ensemble is nonstationary or stationary.

Systems Engineering A discipline concerned with the architecture, design, and integrated elements that, when taken together, comprise a system. Systems engineering is based on an integrated and interdisciplinary approach, where components interact with and influence each other. In addition to the technological systems, the systems considered include human and organizational systems where they incorporate critical human factors with other interacting factors that directly affect achieving the enterprise objectives.

Systems of Systems Unit that are composed of components that are systems in their own right (designed separately and capable of independent action) that work together to achieve the shared goals.

Test development The process through which a test is improved, planned, constructed, evaluated, and modified, including consideration of content, format, administration, scoring, item properties, scaling, and technical quality for its intended purpose.

Test development system A generic name for one or more programs that allow a user to the author, and edit test items (i.e., questions, choices, correct answer, scoring scenarios, and outcomes) and to maintain testing definitions (i.e., how items are delivered with a test).

Testing Techniques Methods that are used in order to obtain a structured and efficient testing protocol to achieve the testing objectives.

Validation Confirmation by the examination and the provision of objective evidence that the particular requirements for a specifically intended objective are fulfilled.

NOTE 1 In the design and development phase, validation concerns the process of examining a product to determine if it is in conformity with the user's needs.

NOTE 2 While final validation is normally performed on the final product under normal operating conditions, it may be necessary to be performed at earlier stages.

Whole-life Costs (WLC): Whole-life cost is the total expense of owning an asset over its entire life, from purchase to disposal, as determined by financial analysis. It is also known as a "life-cycle" cost, which includes design and building cost, purchase and installation cost, operating costs, maintenance, associated financing costs, depreciation, and disposal costs. Whole-life cost also takes into account certain costs that are usually overlooked, such as those related to environmental and social impact factors.

References

[1]   SAE International standard JA 1009-1 reliability testing. Glossary. (third draft).
[2]   SAE International standard. ARP 56 38 RMS terms and definitions.
[3]   Available from European cooperation for space standardization (ECSS). ECSS Secretariat, ESA ESTEC.
[4]   ECSS-Q-30B.European cooperation for space standardization (ECSS). Space product assurance. Dependability.
[5]   ECSS-Q-30. Glossary of terms.
[6]   ISO 9000. Quality management systems − fundamentals and vocabulary. 2000.
[7]   IEC 60050-191. Quality vocabulary − Part 3: availability, reliability and maintainability terms − section 3.2, glossary of International terms. 1990.
[8]   IEC 60050-191. International electrotechnical vocabulary − chapter 191:dependability and quality in service.
[9]   MIL-STD-280. Definitions of item levels, item exchangeability, models, and related terms.
[10] MIL-STD-721C definitions of terms for reliability and maintainability.
[11] MIL-HDBK-781 reliability test methods, plans and environments for engineering development, qualification, and production.
[12] Chan HA, Paul Parker T, Felkins C, Antony O. Accelerated stress testing. IEEE Press; 2000.
[13] Klyatis Lev M. Accelerated reliability and durability testing technology. John Wiley & Sons, Inc.; 2012.
[14] Klyatis LM, Klsyatis EL. Accelerated quality and reliability solutions. UK: Elsevier; 2006.
[15] Nelson W. Accelerated testing. New York, NY: John Wiley & Sons; 1990.
[16] Toolkit R. Commercial practices edition. Reliability Analysis Center; 1993.

Chapter 2

Analysis of the current status of accelerated testing

Abstract

This chapter will describe four general directions of accelerated testing development. Specifically:

- field/flight accelerated testing;
- accelerated testing using computer/software simulation;
- laboratory and proving ground testing using physical simulation of field conditions;
- accelerated reliability and durability testing.

Each will be discussed in some detail in this chapter.

2.1 Introduction

Accelerated testing was actually developed hundreds of years ago. There are different approaches to accelerated testing, and the effectiveness of the selected approach to this testing can be positive or negative. The major positive outcome sought through accelerated testing is shorter (as compared to testing in normal usage) time from design to market, increased quality, reliability, safety, maintainability, supportability, profit, and decreased life-cycle cost, recalls, etc.

Negative outcomes resulting from accelerated testing lead to the opposite results. The time from design to market, although initially appearing to be shorter, in the long term leads to longer times and greater investment in product development with increased expenses (and life cycle cost) and can also have a negative impact on the company's reputation. Therefore, the recognition and understanding of the trends in accelerated testing development, both—positive and negative—are very important for anyone who is involved in product development and production, and these will ultimately have impacts on the user.

Trends in Development of Accelerated Testing for Automotive and Aerospace Engineering.
https://doi.org/10.1016/B978-0-12-818841-5.00002-7

Many of the accelerated testing approaches are negative, and therefore not effective, because they are based on utilizing very high stresses in testing, and these very high stresses cannot provide the proper information necessary for the accurate prediction of the product's performance in the real world during the product's service life (or any other specified time).

The past experience in accelerated testing methods (ATM) in the aerospace industry can be found by researching the available literature. Many of these publications on ATM are related to the area of space exploration, which dates from about 1960 [2.1a]. One particularly interesting chapter in this reference [2.1a] is titled "DEVELOPING AN ACCELERATED TESTING TECH-NOLOGY ROAD MAP." A key observation in this chapter was the acknowledgment that scientific endeavors in space activities, as well as other programs, such as weather or communications, have upstaged the importance of the analysis of associated engineering and hardware problems. But utilizing this approach only postpones addressing the engineering related decisions, thereby leaving very limited time for the design, testing, and performance evaluation of the mechanical components, and as future space missions envision longer missions, greater efforts will need to be devoted to solving the space hardware problems.

It was further written in this referenced publication [2.1a] that: "One promising solution would be to develop a more rational approach to the technique of accelerated testing. By carefully increasing the severity of the test conditions, it should be possible to hasten early failures without altering the is reasonable well understood that the operating conditions used in these tests meet two requirements:

a) The severity of the test conditions can be adjusted to magnify the mode of the failure.
b) The new test conditions will not be severe enough to cause the component to undergo a different mode of failure, i.e., a transition from mild wear to galling and seizure."

But these two requirements are not always possible to obtain in practice. Consider, as an example, Fig. 2.1, which depicts the lubrication regimes from dry friction through full hydrodynamic lubrication. In this example, three discrete regimes are presented.

- The dry friction regime;
- The boundary lubrication regime;
- The full hydrodynamic lubrication regime.

In the thin-film regime, there is a transition from hydrodynamic fluid film lubrication to a mixed film region, where surfaces are only partially supported on an oil film, and there is significant asperity contact taking place through the film. It should be noted that most of the slow-speed, lubricated contact mechanisms that are used in space vehicles operate under these conditions.

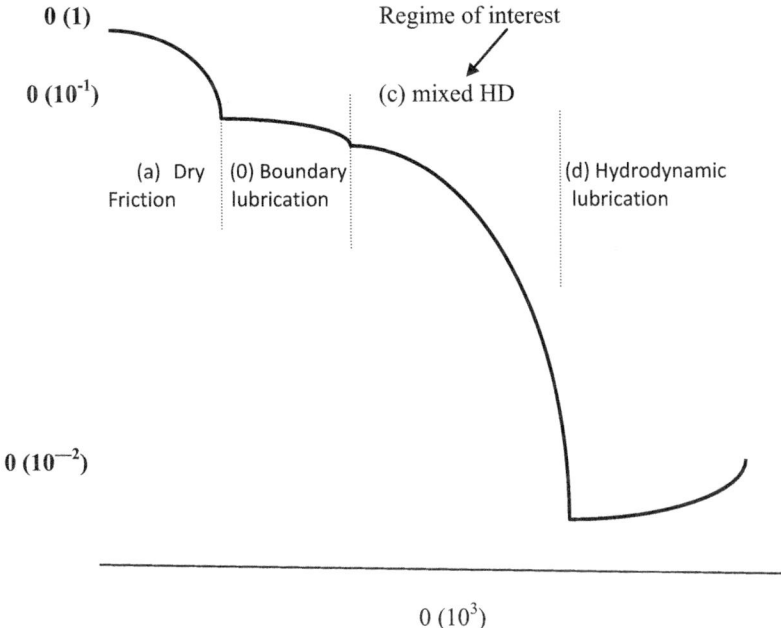

FIGURE 2.1   Tribological regimes.

The authors [2.1a] wrote that it is difficult to stay within the regime brackets or even to support the claim that the test conditions might favor a certain operating regime unless means are available to monitor the performance and demonstrate the failure mode.

In this example, various accelerated testing techniques were tried, including varying speed, load, temperature, surface roughness, lubricant starvation, etc. The effects of changing lubricant type and retainer material were also investigated. Although, much time and effort was expended on these tests, the result did not lead to clean solutions. The authors [2.1a] revealed that these tests did not yield the kind of predictive capabilities that would make it possible to accurately predict impending failures before they occurred.

In this case, the problem was not with the concept of accelerated testing with the limited capabilities of the instrumentation that was used. These authors wrote of the need for nondestructive sensor techniques that can indicate an impending problem without shutting down the test or before experiencing a catastrophic failure. But, since the release of this paper, many of these needed techniques have already been established. As noted in the Technical Background section of reference [2.1a], the methods of monitoring motor current, speed, temperature, load, and torque have all been used in DMA studies. The authors then wrote: "However, the techniques being used do not reflect the advances that have been made in sensor technology (circa 1995—author). The truth is, there is more sophisticated measuring, monitoring, and controlling capabilities in the engine of a today's automobile that there is in the mechanical assembly of a multi-million dollar satellite. This is where the emphasis must be placed in accelerated testing is to become a realistic tool for predicting life. While the immediate goal is concerned with ground tests, the ultimate goal is to be able to monitor performance in both ground and orbit locations with the some instrumentation, built-in as an integral part of the mechanical assembly."

Citations such as the above example have happened and continue to happen. The above has happened because there is no wide implementation of this author's strategic approach, which is described in his previous books [2.10,2.11,2.13], in the aerospace, as well as other industries.

2.1.1 International standard in accelerated testing

Presently there is only one international standard in accelerated testing, namely Standard IEC 62506.

Methods for Products Accelerated Testing [2.1].

The basic contents of this standard are:

3. Terms, definitions, symbols, and abbreviations
4. General description of accelerated test methods
    4.1 Cumulative damage model
    4.2 Classification, methods, and types of test acceleration
        4.2.1 General
        4.2.2 Qualitative accelerated tests
        4.2.3 Quantitative accelerated tests
        4.2.4 Quantitative time and event compresses tests
5. Accelerated test models
    5.1 Qualitative accelerated tests
        5.1.1 Highly accelerated limit tests (HALT)
        5.1.2 Highly accelerated stress test (HAST)
        5.1.3 Highly accelerated stress screening/audit (HASS/HASA)
        5.1.4 Engineering aspects of HALT and HASS
    5.2 Quantitative accelerated test methods
        5.2.1 Purpose of quantitative accelerated testing
        5.2.2 Physical basis for the quantitative accelerated type B test methods
        5.2.3 Type C tests, time (C_1) and event (C_2) compression
    5.3 Failure mechanisms and test design
    5.4 Determination of stress levels, profiles, and combinations in use and test stress modeling
        5.4.1 General
        5.4.2 Step-by-step procedure

5.5 Multiple stress accelerated methodology—type B tests

5.6 Single and multiple stress acceleration for type B tests

    5.6.1 Single stress acceleration methodology

    5.6.2 Stress models with stress varying as a function of time—type B tests

    5.6.3 Stress models that depend on repetition of stress applications. Fatigue models

    5.6.4 Other acceleration models—Time and event compression

5.7 Acceleration of quantitative reliability tests

    5.7.1 Reliability requirements, goals, and use profile

    5.7.2 Reliability demonstration of life tests

    5.7.3 Testing of components for a reliability measure

    5.7.4 Reliability measures for components and systems/items

5.8 Accelerated reliability compliance or evaluation tests

5.9 Accelerated reliability growth testing

5.10 Guidelines for accelerated testing

    5.10.1 Accelerated testing for multiple stresses and the known use profile

    5.10.2 Level of accelerated stresses

    5.10.3 Accelerated reliability and verification tests

6. Accelerated testing strategy in product development

    6.1 Accelerated testing sampling plan

    6.2 General discussion about test stresses and durations

    6.3 Testing components for multiple stresses

    6.4 Accelerated testing of assemblies

    6.5 Accelerated stresses of systems

    6.6 Analysis of test results

7. Limitations of accelerated testing methodology

This standard also includes 7 Annexes, 17 figures, and tables.

Description of this standard:

IEC 62506 provides guidance on the application of various accelerated test techniques for the measurement or improvement of product reliability. The identification of potential failure modes that could be experienced in the use of a product or item and their mitigation is instrumental in ensuring the dependability of the product or item. The purpose of these methods is to identify potential design weaknesses or to provide information on the item's dependability, or to achieve the desired reliability/availability improvement, within a compressed or accelerated period of time. This standard addresses accelerated testing of both nonrepairable and repairable systems. It can be used for probability ratio sequential tests, fixed duration tests, and reliability improvement/growth tests, where the measure of reliability may differ from the standard probability of failure occurrence. This standard also extends to present accelerated testing or production screening methods that would identify weaknesses that could compromise product dependability that is introduced into the product by manufacturing error.

As the above part, which is only a portion of the table of contents, includes 49 chapters and subchapters but consists of only 49 pages in total, one can understand that it is so short that it cannot describe the full essence of the mentioned approaches (methods and equipment) for accelerated testing.

2.1.2 Recalls as source material for providing official information about product defects

Where is the world going with respect to accelerated testing development, and how is it proceeding? The answer to these questions is interesting with regard to the wide aspects of engineering (research, design, manufacturing, usage, service, and others).

Recalls are the powerful source material for providing official information about product defects.

A measure of the product's overall quality, at is the final results of quality, reliability, durability, safety, and other product's performance attributes can be:

- Recalls data (numbers and cost) and death/injury numbers as a result of road (flight) incidents, which are attributable to product defects;
- Functional or other reported incidents of the product's diminished or compromised effectiveness.

Of these two, recalls are the most credible indicator of the product deficiencies, because there are government reporting requirements that companies must comply with when their product is the subject of a recall. Some information about recalls and how the dynamic of this factor is changing over the years is presented in the following.

A product recall is the process of retrieving and replacing defective goods for consumers. When a company issues a recall, the company or manufacturer absorbs the cost of replacing and fixing the defective products. For large companies, the costs of repairing faulty merchandise can accumulate to multibillion dollars in losses.

We can read from Ref. [2.1A]:

"The 10 Biggest Recalls in 2018

1. 1.6 million 2015—18 Ford F-150 ....
2. 1.3 million 2014—18 Ford Fusion, Lincoln MKZ ....
3. 1.3 million 2012—18 Ford Focus ....
4. 807,000 2010—14 Toyota Prius, Prius v ....
5. 507,600 2010—13 Kia Forte, Optima, Optima Hybrid, Sedona ....

6. 504,000 2013–16 Ford Escape, Fusion ....
7. 343,000 2012–17 Audi A4, A4 Allroad, A5, A6, Q5 ....
8. 240,000 2017–18 Chrysler Pacifica."

In Ref. [2.1B] we read: "A new study highlighting the flurry of U.S. automotive recalls asserts that automakers and suppliers remain focused on innovation and cost cuts while vehicle quality takes a hit. The study, "The Auto Industry's Growing Recall Problem—and How to Fix It," shows that automakers and suppliers paid almost $11.8 billion in claims and recorded $10.3 billion in warranty accruals for U.S. recalls in 2016. That $22.1 billion total is an estimated 26% increase over the previous year. Michael Held, enterprise improvement director at the global consulting firm AlixPartners, which led the study, told *Automotive News* that automakers and suppliers totaled an estimated $17.5 billion on claims and warranty accruals in 2015.

The number of vehicles recalled in the U.S. in 2016 rose 4.5% to 53.1 million, from 50.8 million in 2015 –making 2016 the highest year on record, the study says. Nearly half of those recalled vehicles were attributed to Takata Corp.'s defective airbag inflators or General Motors' faulty ignition switches, which combined for 23 million.

In addition, suppliers' share of total recall costs has tripled from 5% to 7% from 2007 to 2013, to 15% to 20% since 2013, according to the study, while "the frequency that suppliers are named in recall notices has doubled.'

So, recalls are costing companies more and more money, and these numbers do not even include the financial impact to the consumer in bringing vehicles to the dealers and their loss time waiting for their vehicles to be repaired.

The study, "The Auto Industry's Growing Recall Problem—and How to Fix It" (PDF), suggests that automakers and suppliers have cut as much as 50% of quality control spending in recent years, directly leading to the larger number of recalls." [2.1C].

But, as has been shown in our research, the words "quality control" are only a commonly used but an inaccurate term. In fact, the essential need is to use more of the positive trends in accelerated testing and to eliminate the negative trends in the development of accelerated testing (AT) that leads to decreasing quality, reliability, safety, and other performance components.

Rather than cutting "quality control," the organizations need to focus on the costs of recalls resulting from the negative trends in not aggressively developing accelerated testing for their product. And one needs to take into account the influencing negative trends in the development of accelerated testing on the cost of recalls.

In the above case, an important information was published on April 14, 2019 [2.1D] in the article "After second recall, Toyota Prius electrical system is still overheating" about the problem with Toyota Prius safety recall." The article described many examples, especially an important one, when one customer (Felo) had to tow his vehicle to the dealer, who gave him (Felo) the bad news that it would cost $3000 to replace the show-box sized unit.

One of Toyota's largest dealer in Southern California, Roger Hogan, told the manufacturer in 2017 that he was seeing inverter failures on vehicles that had received the software modification. At the time, he was refusing to resell about 100 used Priuses that he had taken on trade-in, asserting he did not believe they were safe. Hogan filed a defect petition with the NHTSA in December 2017, asking for a safety investigation, and filed a lawsuit against Toyota that alleged the recall was a sham. A trial in Orange County Superior Court is set for next month.

There is more information that the car's low quality, reliability, and safety led to multibillion-dollar losses. For example, in the same article, one can read more examples of dangerous situations in this area. But companies have not cardinally improved this situation, because they do not use accelerated reliability and durability testing technology, which is more expensive than the simple testing, which is used now. We can read from the same article:

"Even though Toyota says defective inverters are safe as long as they enter limp-home, it has extended the warranty on 2010 to 2014 models. Owners with failed inverters can get a free replacement for up to 15 years with no mileage limit.

In its defect information report filed with MHTSA on October 4, Toyota described the warranty as intended 'to support increased customer satisfaction' and describes it as a 'separate program' from the recall. Hogan, however, contends Toyota is trying avoid a multibillion-dollar replacement program for the defective inverters.

'Unfortunately, there's been a trend of manufacturers trying to turn safety recalls into what sound like performance-related technical bulletins,' Levine said. 'It's deceptive and it's dangerous.'

Even though Hogan petitioned the MHTSA for a defect investigation in December 2017, the agency never initiated a formal process."

As was provided in Ref. [2.2], there is statistical data representing the number of cars that were recalled for safety and other reasons in the United States between January 2015 and March 2016, by automotive manufacturers.

During that period, GM's recalls had affected 10.6 million vehicles. But for the industry, around 51.26 million vehicles were recalled with 2015 being a big year for recalls: General Motors, Fiat Chrysler Automobiles, and the automotive supplier company, Takata, were the major culprits, while used car dealers were the most negatively affected. Daimler, Ferrari, Ford, GM, Honda, Subaru, and child seat manufacturer Graco have kicked off the year 2016 by issuing new recalls.

Fig. 2.2 depicts the number of vehicles (in millions) affected by major manufacturers' car recalls in the United States between January 2015 and March 2016.

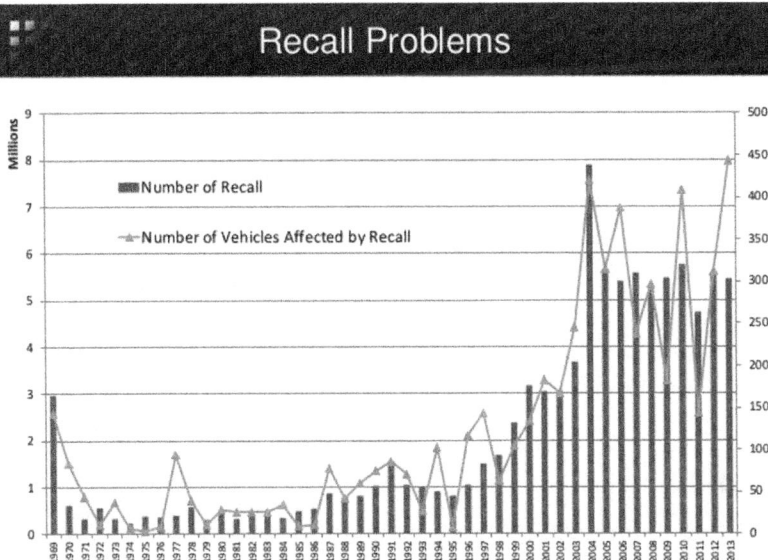

FIGURE 2.2 Number of recalls of automotive vehicle from 1969 to 2013 [K. Suzuki, Espec Corp., 2015].

Another measure of the impact of defective products is the total civil penalties levied by the Consumer Product Safety Commission (CPSC) in the United States from 2006 to 2018 (in millions of U.S. dollars), which is explained in Ref [2.3]. In the fiscal year 2015, the CPSC levied civil penalties against companies totaling 5.18 million U.S. dollars. An even more troubling statistic is the fatalities estimates from 1950 to 2010, which are attributable to insufficient safety, quality, reliability, durability, and other performance components in the automotive area [2.6].

These trends are indicative of the need for the development and implementation of accelerated testing to improve the quality, reliability, durability, life cycle cost, profit, and other components of product and technology performance in the automotive and aerospace engineering fields. The official documentation demonstrating this need is products recall statistics.

If we look at the product recall statistics for the last dozen years, the number of recalls is increasing, and this metric is neither stabilizing or diminishing, which should be the desired goal. This is demonstrated by the information in Fig. 2.2 showing the increasing number of recalls from 1975 to 2014 [2.4].

The number of vehicles affected by major manufacturers' car recalls in the United States between January 2015 and March 2016 (in Millions) is:

1. Honda—**13.58**
2. GM—**10.6**

3. Toyota—**8.42**
4. Fiat Chrysler—**6.87**
5. Ford—**5.41**
6. Mitsubishi—**4.89**
7. Nissan—**3.99**

Similar situations are present in other countries, and with the worldwide business supply chain, some defects are global in nature, such as that with the Takata airbag. Australia is ordering automakers to recall more than two million vehicles fitted with potentially deadly airbags [2.6]. As was written in Ref. [2.5]:

"The Australian government said Wednesday that the cars contain airbags made by Takata (TKTDQ), the Japanese company at the heart of a scandal that has led to tens of millions of vehicle recalls around the world in recent years. A defect can cause the airbags to explode and blast shrapnel into drivers.

Voluntary recalls in Australia have previously been announced by big car brands including BMW, Chevrolet, Honda (HMC), Nissan (NSANF), and Toyota (TM). But the government said those steps haven't been enough to deal with the danger, leaving around 2.3 million vehicles with the defective airbags still on the country's roads."

" … The voluntary recall process has not been effective in some cases, and some manufacturers have not taken satisfactory action to address the serious safety risk which arises after the airbags are more than 6 years old," said Michael Sukkar, an assistant Australian Treasury minister [2.5].

The new compulsory recall includes models made by major carmakers, such as Ford (F), Mercedes Benz, Tesla (TSLA), and Volkswagen (VLKAF).

The total number of affected cars in Australia is four million, or nearly one in five passenger vehicles on the country's roads. All the faulty Takata airbags have to be replaced by the end of 2020, according to the government.

"The company has pleaded guilty to corporate criminal charges and agreed **to pay a $1 billion fine in the U.S.** It filed for bankruptcy last year, and much of its operations are being taken over by Key Safety Systems, a Chinese-owned company based in Michigan.

In another example, Ford is recalling 1.4 million vehicles because the steering wheels can become loose and even come off while driving [2.2]. Ford said the problem is that a steering wheel bolt could come loose, which could cause the steering wheel to potentially detach."

"More than 60 million vehicles have been recalled in the United States, double the previous annual record in 2004 [2.3]. In all, there have been about 700 recall announcements—an average of two a day—affecting the equivalent of one in five vehicles on the road.

The eight largest automakers have each recalled more vehicles in the United States this year than they have on average since 1966, when data collection began, with G.M., Honda, and Chrysler each setting corporate records, the review by The Times found."

"Toyota overhauled its safety practices a few years ago after a spate of recalls for unintended acceleration resulted in a criminal penalty in March of $1.2 billion, the largest ever for a carmaker in the United States. Still the company this year has had to take further steps to improve its recall rates on vehicles with defective airbags made by the supplier Takata.

While G.M. has spent billions of dollars fixing recalled cars and setting up a fund to compensate ignition-switch accident victims and their families, the company has had to take extreme measures to restore trust in its products and management."

"And even with its internal changes, the company's reputation for quality and safety will take years to rebuild. "G.M. has changed on the surface, but it has yet to walk the walk," Mr. Blumenthal (Senator) It is unclear how much bipartisan support will emerge in Congress for new safety laws that, among other things, could hold auto executives criminally liable for deliberately concealing safety defects. A U.S. Senate panel on Tuesday will ask automakers and regulators why tens of millions of vehicles with faulty Takata air bag inflators remain on the road years after deaths prompted the largest auto safety recall in history [2.5]."

Nearly 30 million vehicles remain unrepaired in the recall impacting 19 automakers. At least 22 deaths and hundreds of injuries worldwide are linked to Takata inflators that can explode, unleashing metal shrapnel inside cars and trucks. The defect led Takata to file for bankruptcy protection in June.

The recall process "may play out for another 10—15 years," Senator Jerry Moran, who heads the Senate Commerce Committee panel that deals with consumer protection and other issues, said in his written opening statement for the hearing.

"… Takata pleaded guilty in 2017 single felony count of wire fraud to resolve a U.S. Justice Department investigation and agreed to a $1 billion settlement."

And yet another example, after years of litigation and probes, Toyota announced a $1.1 billion settlement Wednesday to resolve lawsuits alleging "unintended acceleration" in some Toyota and Lexus models [2.6].

Toyota will create a fund for retrofitting 3.2 million Toyota and Lexus cars with technology that will make it easier to stop them in a panic situation, as part of the settlement in a U.S. District Court case that sought class-action status. Owners of models that cannot be retrofitted will receive cash pay-outs. And those who sold their vehicles in late 2009 and all of 2010 will be eligible for compensation due to lowered resale value due to the issue.

When journalists from the *Wall Street Journal*, *The New York Times*, and other newspapers and magazines write that recalls are increasing over the years, and that the impacts are decreasing safety and reliability, we may think that the writers who may not be professionals in engineering, do not understand the essential correct reasons for the increasing recalls. But professionals from the engineering community are writing much the same in their publications, so we must conclude that the problem is relevant to the engineering community. It means that engineering professionals and societies are not recognizing that safety and quality (or reliability) issues are the results and not the reason for recalls.

In fact, recalls are the result of insufficient prediction of the product's performance (where product performance include quality, reliability, safety, maintainability, life cycle cost, profit, and other components) [2.10].

And some further examples:

"Automaker BMW says it is expanding a recall to cover 1.6 million vehicles worldwide due to possible fluid leaks that could result in a fire. BMW said Tuesday that in some diesel vehicles coolant could leak from the exhaust gas recirculation module, part of the emissions reduction system. The leaks could combine with soot at high temperatures and lead to a fire" [2.7].

"Hyundai and Kia have soared to the top of J.D. Power's quality rankings in recent years and earned respectable marks from *Consumer Reports*. But a brewing crisis surrounding hundreds of reports of noncollision fires in their vehicles could put the rewards of those accolades at risk and rekindle memories of their early years in the U.S. when the brands struggled with quality. The South Korean stablemates are facing pressure from the Center for Auto Safety, which renewed calls this month for them to recall almost three million crossovers and sedans for potential fires that could erupt while people are driving. The group is asking for recalls of all 2011−14 Kia Sorento, Kia Optima, Hyundai Sonata and Hyundai Santa Fe models, as well as all 2010−15 Kia Souls" [2.8].

Ref [2.8a-1] presents NHTSA's findings on recalls by decade and potentially affected vehicles. It demonstrates: "The number of vehicles affected by safety defect vehicle recalls has increased dramatically since 2011."

Below information for 2018 [2.8a-1]:

Subaru Extends JDM Recall Over New Cases Of Inspection Cheating:

Fresh from announcing the first quarterly loss in many years, Subaru acknowledged that the inspection scandal is much wider than we thought in the first instance. Even Tomomi Nakamura, the chief executive officer of the Japanese automaker, said that the lack of respect for regulations led to things going out of control.

Fast-forward to the shareholders meeting from June 2018, and that's when Yasuyuki Yoshinaga stepped down as CEO of Subaru's automotive division. Developments in the investigation are ongoing, and so far, Subaru announced that it would recall an additional 100,000 vehicles after discovering that the inspection of the braking system was not performed in compliance with quality control regulations.

The head honcho ensured that this is the final recall connected to the inspection scandal, and Subaru will pony up 6.5 billion yen in costs related to fixing these vehicles. Converted to U.S. dollars, the sum translates to no less than $57 million.

But this is not the only trouble that Subaru has to face. Last month, the automaker announced that it would recall 400,000 vehicles to fix a flaw with the valve springs of the engine found in the Impreza and Forester. To make matters worse, Subaru is also preparing for a hike in tariffs on Japanese imports, which would raise the price of several popular models in the United States of America, including the Forester.

Ford Recalls 1.5 Million Focus Models in North America [2.8a-2]:

"Since the beginning of 2018, millions of models, including the GT sports car, the F-150 pickup truck the Focus Electric had to be sent back to the shop for this or that problem.

The Focus, this time the version of it powered by an internal combustion engine, is at the center of the latest such announcement, made on Thursday by Ford.

The company says 1.5 million cars wearing the Focus nameplate have to be recalled due to a malfunctioning canister purge valve that could leave motorists stranded without fuel in the middle of nowhere."

BMW EGR Fire Recall Grows to 1.6 Million Cars Globally [2.8a-3]:

"Back in August, BMW said it is recalling around 100,000 cars in South Korea on fears the vehicles can catch fire. That number grew the same month with the addition of 300,000 cars possibly affected in Europe.

Now, due to further examination of engines with a similar technical setup, BMW says it will recall around 1.6 million vehicles worldwide.

The problem which led to this massive recall was first discovered in 2016. It has to do with the exhaust gas recirculation module (EGR) and a malfunction which, in certain conditions, may cause engine fires."

Toyota Recalls Millions Of Hybrid Vehicles [2.8a-4]:

"After recalling 1.03 million hybrid vehicles over wire harness and software issues, Toyota has announced one more recall that encompasses 2.43 million examples of the breed. Older Prius and Auris Hybrid models are affected, built between October 2008 and November 2014."

Shanghai GM Recalls More Than 3.3 Million Vehicles In China [2.8a-5]:

"2018 hasn't been better with motorists in the Middle Kingdom either, with General Motors leading the ranking. Starting October 12th, General Motors will call back more than 3.3 million vehicles in China over an issue with the suspension system. Under extreme operating conditions, the suspension arms of vehicles manufactured between 2013 and 2018 could deform, leading to the loss of control."

Aviation also is experiencing problems with quality and reliability of their product related to adequate or proper accelerated testing. While recalls are not normally seen in the aviation industry, extensive accident or incident investigations typically result in voluntary changes by the involved parties or the issuance of an "airworthiness directive" or similar legal document that mandates corrective actions.

Over the last several years, several crashes of commercial airlines have occurred involving systems designed to prevent human error, but in some situations compromising the pilot's ability to control the aircraft.

Examples include SAS flight 751 where a system called the "Automatic Trust Restoration" (ATR) was a factor preventing the pilots from controlling the engine throttles; and Air France Flight AF447 where a contributing factor in the accident was the pilot tubes, which were believed to have become iced resulting in the loss of accurate airspeed and altitude information. The pilot tubes were known to have a problem with icing and had been replaced by several other airlines. A more recent example is the Lion Air Flight 610, which crashed in October 2018. While at the writing of this book, the crash is still under investigation, and the final report is not yet issued, some facts that had already come to light indicate an instance of an automated control designed to prevent pilot error accidents multifunctioning and affecting the pilot's ability to control the aircraft [2.8a]. JAKARTA, Indonesia—The "black box" data recorder from a crashed Lion Air jet shows its last four flights all had an airspeed indicator problem, investigators said Monday, after distraught relatives of victims confronted the airline's cofounder at a meeting organized by officials [2.8a].

National Transportation Safety Committee chairman Soerjanto Tjahjono said the problem was similar on each of the four flights, including the fatal flight in October 29 (2018) in which the plane plunged into the Java Sea minutes after takeoff from Jakarta, killing all 189 people on board.

Problems with the plane's previous flight, from Denpasar on Bali to Jakarta, were widely reported and "when we opened the black box, yes, indeed the technical problem was the airspeed or the speed of the plane," Tjahjono told a news conference. "Data from the black box showed that two flights before Denpasar-Jakarta also experienced the same problem," he said.

"Rumors circulating on social media are so great and here we want to clarify that in the black box there were four flights that experienced problems with the airspeed indicator."

Relatives questioned why the 2-month-old **Boeing 737 MAX 8 plane** had been cleared to fly after suffering problems on its Bali to Jakarta flight on October 28 that included a rapid descent after takeoff that terrified passengers [2.8a]."

It was further written in Ref. [2.8b]: "The final moments of Lion Air Flight 610 as it hurtled soon after dawn from a calm Indonesian sky into the waters of the Java Sea would have been terrifying but swift."

The single-aisle Boeing aircraft, assembled in Washington State and delivered to Lion Air less than 3 months ago, appears to have plummeted nose-first into the water, its advanced jet engines racing the plane toward the waves at as much as 400 m.p.h. in less than a minute.

While investigators have not yet concluded what caused Flight 610 to plunge into the sea, they know that in the days before the crash, the plane had experienced repeated problems in some of the same systems that could have led the aircraft to go into a nosedive.

"... the Federal Aviation Administration of the United States warned that erroneous data processed in the new, best-selling Max 8 jet could cause the plane to abruptly nose-dive. Investigators examining. Flight 610 are trying to determine if that is what happened."

A Critical Sensor.

The downed airplane may have received false input from an angle of attack sensor, which measures the angle at which oncoming wind crosses the plane.

A high angle of attack indicates to the pilot that the plane is flying nose-up and may need to be adjusted. A plane that is not adjusted may lose speed and risk stalling.

During the 2 days before Flight 610 began its final journey, there were repeated indications that pilots were being fed faulty data—perhaps from instruments measuring the speed and a key angle of the plane—that would have compromised their ability to fly safely.

"... The recommended response issued by Boeing and the F.A.A. this week would not be a pilot's natural reaction. The flight crew is instructed to switch off the electricity powering stabilizers in the tail of the aircraft that are propelling the downward pitch of the nose.

But without specific training on this anomaly, what pilot would think to turn off part of the plane? When flight crews learn how to helm a new model of aircraft, they typically study the differences between older and newer models. Aviation experts worry that pilots at hard-driving carriers like Lion Air may not be given adequate time for such training.

In addition, differences between models sometimes manifest themselves only after months or years of operation. The Boeing Max 8 went into service just last year. Even though Captain Suneja was an experienced aviator for his age, he would not have had time to fully familiarize himself with the latest version of Boeing's workhorse jet" [2.8b].

In the reference "Boeing's troubled jet is costing $ 1 billion to fix so far" David Koenig wrote [2.8e]. "Boeing is already estimating a $ 1 billion increase in costs related to its troubled 737 Max and has pulled its forecast of 2019 earnings because of uncertainty surrounding the jetliner, which remains grounded after two crashes that killed 346 people.

The $ 1 billion figure is a conservative starting point. It covers increased production costs over the next few years but does not include the company's spending to fix software implicated in the crashes, additional pilot training, payments to airlines for grounded jets, or compensation for families of the dead passengers."

From the article published in the *Wall Street Journal* "Boeing Sees More 737 Costs," April 2019 ([2.8f]), we read that "... More than370 MAX planes had been delivered to customers forcing carriers to cancel flights and reconfigure schedules ahead of the busy summer travel season."

And, "… Investigators have cut some $27 billion off Boeing's market value since a 737 MAX operated by Ethiopian Airlines crashed last month, leaving the company valued at about $212 billion." [2.8f].

The recalls result are connected with accelerated testing, especially negative aspects of trends in development accelerated testing.

During design, manufacturing, and usage, one needs to provide a complex analysis of factors that influence a product's efficiency, and as can be seen in Fig. 2.3, the accelerated testing level is playing an important role in these processes. One can find a detailed description of these processes in this author's book [2.11]. This book [2.11] explores in greater detail the problems that also relate to trends in the development of accelerated testing in the automotive and aerospace engineering areas. It also discusses the real reasons for recalls, how these trends are influencing the engineering effectiveness of the product.

A fundamental requirement for increasing design and manufacturing effectiveness, especially for new products, is the accurate simulation of the real-world conditions for testing the product. If the simulation does not adequately reflect real-world operational conditions, the results of testing may be very different from the real-world results.

Too often, managements think that by saving money in the testing protocols, the organization will decrease the cost of the product and improve the

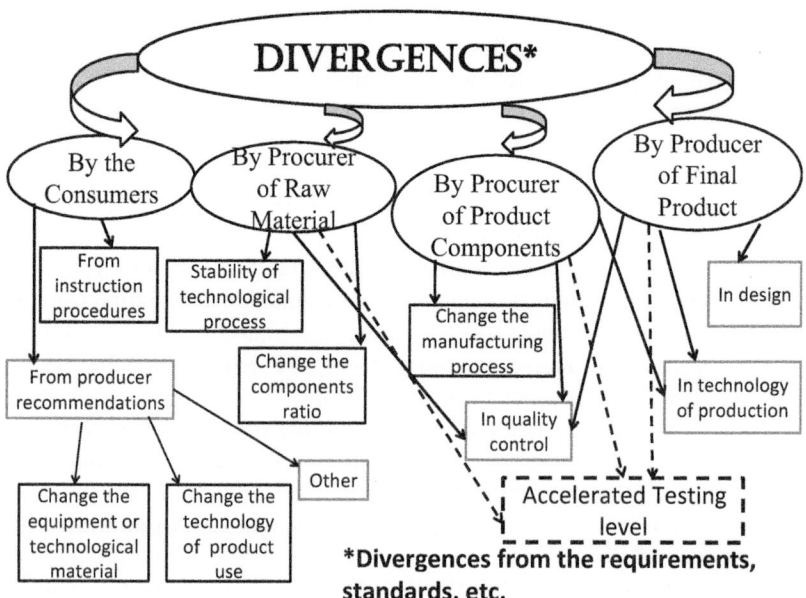

FIGURE 2.3 Diagram of the complex analysis of factors that influence product efficiency (for producers and consumers) and whose analysis is necessary for accelerated testing during design, manufacture, and usage.

organization's profitability. The fallacy of this approach is that the cost of accurate testing is one of many components of the cost of producing the product and assuring its performance. As has been shown, the failure to accurately simulate the real-world experience while saving money for the testing process usually leads to increasing expenses and losses throughout the product's life cycle, including during design, manufacturing, and usage. Simplified testing processes with an inaccurate simulation of the real-world conditions or with a simulation where the conditions are tested separately, frequently leads to future safety, quality, and economic problems. This is because real-world conditions are interacted and interconnected. If these interactions are not part of the simulation, they do not accurately simulate the real-world results.

Similar situations can even be found by companies that specialize in testing. For example, in the note about MIRA's durability testing [2.12] it was stated that among the Proving Ground Durability Circuits & Features was that MIRA's proving ground is used extensively for accelerated durability testing on the whole vehicle and utilizes these traditional durability surfaces:

- Belgian Pave;
- Corrugations;
- Resonance Road;
- Stone Road.

While one might think that this would provide real-world simulation, there may be many other proving ground surfaces and features needed to provide a track equivalent to the real-world operating conditions [2.13].

But the real answer was contained in the statement: "... With recent issues in the performance of many of the systems, it is imperative to focus on approaches that minimize risk to the public. Simulation and test track testing can help reduce this risk. The industry is still skeptical to simulation, especially sensor simulation, as an appropriate tool to replace field or real-world testing. However, simulation provides a number of benefits to field testing, it saves time and money, it is ideally suited for the evolution of dangerous scenarios, and it provides exact ground truth."

If we accept this citation as true, how do we overcome industry's skepticism concerning simulation? Of course, one should not use the Monte Carlo model for simulation as a correct model. As will be demonstrated in the references in the next chapters, simulation **must be accurate for successful testing and high-quality prediction.**

But the above-cited reference does not provide the roadmap for developing an **accurate** simulation of real-world testing, and, from the Test Expo Show Preview [2.57], as well as the contents of this Expo and other Expos, we do not see the trends in the development of the basic concepts of accelerated testing in the automotive and aerospace engineering fields, especially as related to new design and technology. The question remains, "Where are the strategic solutions in accelerated testing that are necessary to reduce or prevent recalls, economic and technical problems?"

In this book, the author will try to demonstrate the way to solving this problem.

2.2 Basic general directions of accelerated testing development

The literature and the industry provide many different approaches and types of accelerated testing (AT) related to the fields of automotive and aerospace engineering. This author has classified these approaches into four general directions of AT development, as shown in Fig. 2.4.

Let us briefly describe each of them.

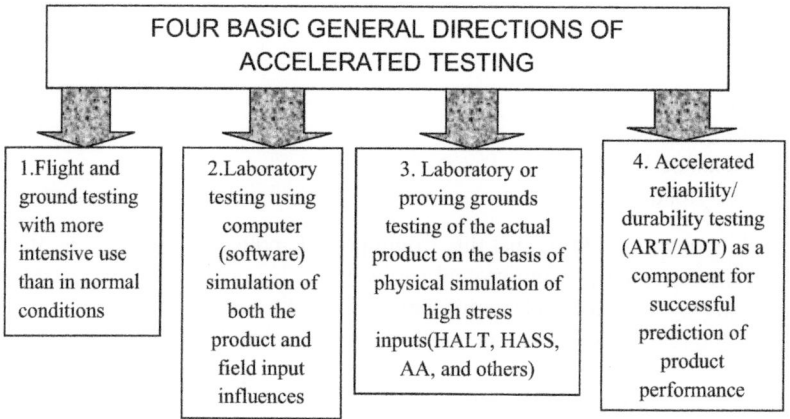

FIGURE 2.4 Four general directions for accelerated testing development.

2.2.1 The first general direction (field/flight accelerated testing)

The first general direction is field/flight testing with more intensive usage than that to be experienced in normal use. For example, an automobile is typically in use for only about 5—6 h per day. By using that same automobile 18 to 20 or more hours per day, the testing is assumed to represent a true accelerated testing and should provide enhanced durability research into an automobile's parameters of interest. However, this approach also results in a shorter nonoperating interval than that normally experienced.

This type of testing does provide results that are more accelerated than those that would be evident under normal field operating conditions.

In Ref. [2.13] the author considers three types of field testing, although there is some overlap between them.

1. Evaluation of the conceptual design and product prototypes in connection with research and development engineering.
2. Evaluation of products during the process of product release, working with different departments.
3. Evaluation of products during or after market introduction, primarily for the purpose of publishing a report in a magazine or other public or trade media.

In the first of these three types, the evaluation of prototypes requires a high level of knowledge and experience with nondestructive metal detectors, the ability to communicate findings to the engineers, and a keen appreciation for the need for confidentiality. In general, those who do such evaluation receive neither money nor "freebies"—they do it out of love for the interest. The people who are often best at this, do field testing for more than one company, which places a high premium on their professional reputation for protecting trade secrets. Persons without brand loyalty are usually preferred for this kind of field testing because this normally results in more objective reporting of findings to many company's departments.

In the second of these three types, evaluation of products for engineering feedback during the process of product release the requirements are for knowledge and experience with metal detectors, particularly with those which are somewhat similar to the ones involved in the tested product, the ability to communicate findings to engineers, and the ability to refrain from divulging information without explicit permission. The person doing the evaluation is not normally paid but is often allowed to keep the unit they tested, and with the factory providing upgrades to production specifications as may become necessary.

Usually, once the product is fully released, the field tester is given permission to openly discuss what they know about the released product, but they are encouraged to refrain from discussing issues, which arose during field testing, especially ones that are not representative of the actual released product.

The person doing the testing is, preferably, without brand loyalty, or has some loyalty to the company on behalf of whom the field testing is done, but not so much that objectivity in reporting to the engineering department might be compromised.

In the third of these three types, evaluation of products during the process of product release on behalf of marketing department the requirements are knowledge and experience with metal detectors, especially of competitors' units in the same general category, and the ability to communicate clearly in public settings such as Internet forums and in the competition's claims of the desirable characteristics of their product. The field tester must use good judgment in releasing information about the product, which is representative of what the product is or will be while refraining from commenting on issues that are still not fully determined. Because this activity is primarily a marketing activity, some brand loyalty on the part of the person doing the field testing is usually expected, and said person will often have their own business interests involved, such as a dealership or metal detecting. It is customary for the person doing the field testing to be allowed to keep the unit they test, including all updating of the unit at the factory's expense if necessary to bring it up to full product release specifications. Marketing departments may make deals involving other forms of compensation.

This evaluation of products during or after introduction is primarily for the purpose of publishing a "field test," for instance, in a magazine. It requires someone who is knowledgeable enough to do the kind of evaluation appropriate for that product, is skilled in the art of the written or video medium, where the field test will be published, and has the necessary connections to get the work published.

This kind of testing is usually arranged through an organization's marketing department. One example excerpted from the report of the U.S. Department of Energy, INL/EXT 06−01,262 [2.15], which stated: "A total of four Honda Civic hybrid electric vehicles (HEVs) have entered fleet and accelerated reliability testing since May 2002 in two fleets in Arizona. Two of the vehicles were driven 25,000 miles each (fleet testing), and the other two were driven approximately 160,000 miles each (accelerated reliability testing). One HEV reached 161,000 miles in February 2005, and the other 164,000 miles in April 2005. These two vehicles will have their fuel efficiencies retested on dynamometers (with and without air conditioning), and their batteries will be capacity tested. Fact sheets and maintenance logs for these vehicles give detailed information such as miles driven, fuel economy, operations and maintenance requirements, operating costs, life-cycle costs, and any unique driving issues ..." [2.15].

Another example is the following, which was written in the final report cited above:

"... One of the field evaluation tasks of the Program is the accelerated reliability testing of commercially available electric vehicles. These vehicles are operated with the goal of driving each test vehicle 25,000 miles within 1 year. Since the normal fleet vehicle is only driven approximately 6000 miles per year, accelerated reliability testing allows an accelerated life-cycle analysis of vehicles. Driving is done on public roads in a random manner that simulates normal operation."

The report in Ref. [2.16] summarizes the accelerated reliability testing of three nickel-metal hydride (NiMH) equipped, Toyota RAV4 electric vehicles by the Field Operation Program and its testing partner, Southern California Edison Company (SCE). The three vehicles were assigned to SCE's Electric Vehicle Technical Center located in Pomona, California. The report adds "... To accumulate 25,000 miles within 1 year of testing, SCE assigned the vehicle employees with long commutes that lived within the vehicles' maximum range. Occasionally, the normal drivers did not use their vehicles because of vacation or business Travel." In that case, SCE attempted to find other personnel to continue the test.

While this is a useful example of work in many areas of accelerated testing, practice shows that this type of field testing cannot be used for accurate reliability, durability, and maintainability prediction. There are several reasons for this, including:

1. The term "accelerated reliability testing" in the above reports does not correspond to an accurate definition of the term [2.13].

   Many years of field testing of several specimens are required to obtain accurate initial information needed for accurate quality, reliability, and maintainability prediction during a given period.
2. The methodology and equipment that can accomplish this objective and do it much faster and at a lower cost can be found in Ref. [2.13].
3. Companies usually change their design and manufacture every several years, and not always on a regular basis. In such a situation, the test results of the previous model have only relative usefulness, and may not be directly applicable.

As will be shown later in the book, field testing can only provide partial initial information for identifying problems related to an integrated system of quality, reliability, and maintainability during the service life or warranty period. Experience has shown that even after describing its field testing, and testing of the following models, Toyota still had many problems in the areas of reliability and safety. These problems ultimately led to faster than predicted degradation, product failures, complaints, and recalls.

One more example is also taken from Toyota's practices. In the report entitled "Hybrid Electric Vehicle End-of-Life Testing on the Honda Insight" [2.16] it was written that: "Two model year 2004 Toyota Prius hybrid electric vehicles (HEVs) entered accelerated reliability testing in one fleet in Arizona during November 2003. Each vehicle will be driven 160,000 miles. After reaching 160,000 miles each, the two Prius HEVs will have their fuel efficiencies retested on dynamometers (with and without air conditioning), and their batteries will be capacity tested. Each sheets and maintenance logs for these vehicles give detail information such as miles driven, fuel economy, operations and maintenance requirements, operating costs, life-cycle costs, and any unique driving issues ..."

In fact, this accelerated field testing is conducted by professional drivers for short periods of time (maximum 2—3 years). But this testing cannot provide the necessary information for accurate prediction of reliability, life-cycle costs, and maintenance requirement during real service life, since it does not take into account the following interactions experienced during the service life of the vehicle:

- the corrosion process and other output parameters, as well as input influences which act during service life;
- the influences of the customers' reliability on the automotive reliability, because it was operated by professional drivers during the above testing;

- the influences presented by many other real-life situations.

Mercedes-Benz calls similar testing "durability testing." For example, the test program for the new Mercedes-Benz C-Class stated in Ref. [2.18] states:

"... For the real-life test work, that involved 280 vehicles expending a wide range of climatic and topographical conditions. Particularly significant testing was carried out in Finland, Germany, Dubai, and Namibia. The program included tough 'Heide' endurance testing for newly developed cars, equivalent to 300,000 km (186,000 miles) of everyday driving by a typical Mercedes customers. Every kilometer of this endurance test is around 150 times more intensive than normal driving on the road, according to Mercedes. Data gathered is used to control test rigs for chassis durability testing ...."

Of course, again, this testing is improperly called "durability testing" for the same reasons listed in the above examples of what Toyota referred to as "accelerated reliability testing."

There are other examples of such field testing. A similar situation existed with the Ford Otosan's 2007 durability testing [2.19]. In an article published about LMS Supporting Ford Otosan in Developing Accelerated Durability Testing in 2007, it was stated that "Ford Otosan and LMS engineers developed a compressed durability testing cycle for Ford's durability testing cycle for the Ford Otosan's new Cargo truck. LMS engineers performed dedicated data collection, applied extensive load data processing techniques, and developed a 6 to 8 week test track sequence and 4-week accelerated rig test scenario that matched the fatigue damage generated by 1.2 million km of road driving" [2.19].

Similar situations can be seen relating to accelerated flight testing. Consider the testing involving a single component, the Flight Attendant Panels (FAPs). FAPs have evolved from a simple communications device for the flight attendant to communicate with the pilot and passengers, to a sophisticated interactive control device providing reporting and control of many of the aircraft's passenger comfort and convenience systems. So much so that in Ref. [2.20] it was written that verification of the design of FAPs requires an interdisciplinary approach. Two companies involved in testing TestPlant and Vector have combined their domain tools for efficient testing of the entire system. Ever since sophisticated FAPs were first introduced in the Airbus A320 series, the number of functions checked and monitored and which they control has grown regularly. This is coupled with improved touchscreen technologies that are continually making the human-machine interface more efficient and convenient. In modern aircraft, the crew uses FAPs to control and monitor many cabin functions, among them lighting announcements, door status indication, smoke detection, and temperatures. The units are also used for functions relevant to maintenance, for instance to add entries to the digital cabin logbook which is used to log faults. Furthermore, the FAB indicates safety information, such as smoke detection or emergency signals. Easy efficient and reliable operation of FAPs through graphic user interfaces plays an important role in the airlines' satisfaction with the units.

In Ref. [2.20] it was also written that quality assurance measures must be adapted to meet the increasing performance of software-based user interfaces. Complex logic must be applied, especially in the operation of embedded systems, which pose new challenges in the development processes. Trends such as extending functionally by adding new software components or increasing flexibly by adapting the user interface can further increase this complexity. Consequently, the test phase must assume growing importance in the development of these types of user interfaces.

To properly address this increasing importance, it is necessary to validate functions in the early phases of development. Often the target hardware is either unavailable or incomplete at the time testing must be started. Frequently, the validation must be performed in a purely virtualized environment or on an isolated subsystem using the remaining bus simulation.

Desirable user interfaces are characterized by those visualizing the underlying cabin applications in a way that is both clear and easy to understand—despite their complexity—and by offering intuitive and simple operation (usability). The challenge is to design FAP systems to be as smart as smartphones. This leads to FAP designs that are intuitive and highly responsive touch-based infotainment user.

Flight test methods may also be found in the following documents:

- FAA AC 23-8C—Flight Test Guide for Certification of Part 23 Airplanes
- FAA AC 23-15A—Small Airplane Certification Compliance Program
- FAA AC 90-89A—Amateur-built Aircraft And Ultralight Flight Testing Handbook
- C. Edward Lan and Jan Roskam: Airplane Aerodynamics and Performance, Roskam Aviation Co.
- Russel M. Herrington et al.: Flight Test Engineering Handbook, Air Force Technical Report no. 6273, NTIS No. AD 636.392 National Technical Information Service, Springfield, VA.
- National Advisory Committee for Aeronautics (Technical Note 2098)—The Effects of Stability of Spin-recovery Tail Parachutes on the Behavior of Airplanes in Gliding Flight and In Spins. By Stanley H. Scher and John W. Draper, Langley Aeronautical Laboratory, Langley Air Force Base, Va.
- NASA Technical Memorandum 80237—A Spin-Recovery Parachute System for Light General-Aviation Airplanes by Charles F. Bradshaw, Langley Research Center, Hampton, Virginia
- CS-23 Book 2, Flight Test Guide
- USNTPS-FTM-No. 103 U.S. NAVAL TEST PILOT SCHOOL—FLIGHT TEST MANUAL—FIXED WING STABILITY AND CONTROL, Theory and Flight Test Techniques. Naval Air Warfare Center Aircraft Division, Patuxent River, MD January 1997

One more example is from the Airbus flight testing, which can be found in Ref. [2.22]:

"The Airbus A 350 jetliner has been the manufacturer first test of an evolved development program that could see improved aircraft entering service more quickly," says flight and integration tests senior vice president Fernando Alonso. Having seen ever-lengthening certification periods, Airbus decided to introduce accelerated test procedures that also aimed to increase customer satisfaction from the outset through the delivery of more nature machines.

In the 1980s, flight test programs had grown longer, from a year through 14—16 months to 20 months, according to Alonso. While regulations were becoming more sophisticated and technical, no development served to act as a catalyst to change certification procedures or pace. It was to introduce flight test involvement "upstream to be more involved earlier; there was no justification not to change things." One objective in the development of the A 350, which completed certification flying in mid- August is to try to reduce the lead time while also improving safety. Accordingly, as the giant A 380 double-decker quadjet entered service in 2007, the company established a new flight and integration tests center (F&ITC) that has seen pilots also participating in systems testing.

Aiming to "excel in testing and test for excellence," Alonso wanted to involve pilots as far upstream as possible in a search for, as it were, flight development holism. The philosophy has become "Look at the aircraft as if you are the first customer, not the designer. Remember, it is going to be flown by people to carry people," says the Airbus official.

The trend toward longer certification periods will be reversed by beginning the test earlier, the A 350 campaign—about 15 months long—contrasted with the 20 months devoted to the A 380 quadjet (Table 2.1). The relatively greater speed with which the A 350 approval program has been completed is very quickly it might become possible to certificate future new aircraft.

TABLE 2.1 Certification time for different models.

Model	Certification time
A 300	1974—17 months
A 330	1998—12 months (1200 flight hours)
A340	1992—14 months (2000 flight hours)
A 330	1993—12 months (1800 flight hours)
A 380	2007 (20 months (2500 flight hours)
A 350	2014—15 months (2500 flight hours)

Before 2002 when Airbus was already heavily engaged in the development of the A 380, launched 2 years earlier following long project studies, the manufacturer has separate flight and ground testing sections with the drawing office, including both design and testing of systems. By 2007, when the A 380 entered service, the test organization involved two distinct sections, one looking after systems and cabin testing and the other overseeing development and production testing of new aircraft."

The Airbus flight and integration tests management decided that to reduce the certification periods, one must start testing earlier, on the ground [2.22]. But while much can be done with test rigs with adequate investment and more representative equipment, there is also much that you can see only in the air.

One example of the evolving technology of testing and measuring aero-dynamics was the development of a program called "wings of change." While historically wing surfaces carried myriad static-pressure test points linked by bundles of tubes to a blowing system, multiple sensors, a plenum chamber, metallic "gloves" and a leak detector; on the new A 350 many multipressure sensors were taped to the wing surface and connected by a single cable.

Aerospace has been a leading applicant of digital technology, and the past quarter-century has seen exponential growth in the volume of test information that can be captured for analysis under "big data project." Since the A 320 single-aisle twinjet first flew in 1987. The number of parameters recorded has increased by a factor of 50, leading to the opportunity to improve maturity through the analysis of the data collected (see Table 2.2).

Before the first flight, virtual testing enables Airbus to capture better knowledge of the aircraft and to optimize the flight tests. On the A 350, the flight testers were first engaged in the systems integration tested, involving many simulators and rigs, in an effort to check "as much as possible, as soon as possible, and to reduce flight-test time to a minimum."

One obvious advantage of the integrated operation is that knowledge can be shared more readily among all parties. Under previous practice, there had been occasions when it became clear that things discovered in the air had previously been uncovered by ground test engineers without information reaching flight test personnel.

TABLE 2.2 Airbus big data project.

Model	First flight	Parameters monitored	Data archived
A 320	1987	12,000	8.5 TB
A 340	1991	14,000	12.8 TB
A 380	2005	320,000	57.0 TB
A 350	2013	670,000	53.0 TB

In general, this stimulated motives to improve procedures and results through better communication between test disciplines.

2.2.2 The second general direction (accelerated testing based on computer/software simulation)

This direction is virtual testing and based on computer (software) simulation or analytical/statistical methods. A computer simulation is a computer program that attempts to simulate an abstract model of a particular system.

Computer simulations have become a useful part of the mathematical modeling of many systems in engineering, and to gain insight into the operation of those systems.

Computer simulation and statistical analysis

For example, DRI [2.24] uses a wide variety of computer simulation and statistical analysis methods to perform research and development projects for their clients. Computer simulation methods include multibody and Finite Element approaches for crash, vehicle dynamics, and other topics. Statistical analysis can be applied to simulation results, as well as to their state, national, and international accident databases.

Computer simulation software

- LS-DYNA
- MADYMO
- ATB
- Nastran
- CarSim, TruckSim, BikeSim

Statistical analysis software

- SPSS
- R
- Matlab

Commonly used databases

- FARS
- Hurt
- CPSC
- NASS/CDS
- NASS/GES
- NASS/PCDS (Pedestrian Crash Data System)
- MCCS (US Motorcycle Crash Causation Study)

Computer-based simulations

Vehicle and driver-vehicle software simulations

Another example Dynamic Research, Inc. (DRI) [2.24], based in California, who has developed and applied on behalf of its clients a range of computer simulations for vehicle dynamics and control analysis.

Versions are available for a wide range of vehicles, including:

- Cars
- Trucks and utility vehicles
- Articulated vehicles
- Motorcycles
- All-terrain vehicles
- Buses
- Aircraft

Human response, control modeling, and simulation applications

DRI's human response and control models and simulations are also applicable to a range of different areas:

- Driver (pilot, rider, etc.) active control of vehicle motions (e.g., steering, throttle, braking in a variety of on-road and off-road tasks, maneuvers and conditions, for predictive modeling of driver behavior and handling performance;
- Human body active and passive response as related to vehicle control (e.g., human limb impedance, body-active control of small vehicles, etc.)
- Human body biomechanical response to impacts and large amplitude motions, for purposes of vehicle crash and rollover simulations.
- Human injury potential models, probabilistic models based on measured biomechanical forces and motions, and expressed in normalized injury cost terms, as are useful in injury risk-benefit analyses
- Human comfort rating models, statistically derived empirical models of juries of human subjects, for predictive quantification of comfort due to measurable physical variables, in the areas of vehicle ride, handling noise, and vibration.

In Ref. [2.25] the company wrote: "Under relentless pressure to shorten development cycles and reduce costs, auto manufacturers are increasingly exploring ways to use analysis tools to perform meaningful virtual evaluations of vehicle designs early on, well before the availability of physical prototypes. The ability to obtain accurate predicted - or calculated - loads in this fashion augments the effectiveness of component physical testing, enables early identification and elimination of design flaws, reduces the need for rework and reliance on prototypes, and streamlines design validation."

The Virtual Proving Ground method involves "driving" a vehicle model over a digitized road. But, the company further stated, while having the advantage of relying entirely on virtual models, this approach yields inaccurate loads that are difficult to validate through physical testing. A semi-analytical method, on the other hand, uses spindle loads acquired from an existing vehicle to excite the vehicle model. This delivers better results than the Virtual Proving Ground but establishes boundary conditions not entirely appropriate for the vehicle model, so calculated loads still lack accuracy and are difficult to validate. A third method, Virtual Testing, overcomes these problems by integrating a model of an actual physical test system into the simulation to excite the vehicle model.

Virtual Testing is the simulation of a physical test, using finite element analysis tools, multibody dynamic analysis tools, and RPC iteration techniques to derive accurate loads, motion, and damage information of a vehicle system very early in the development process. MTS, Inc has written and claims that there are several advantages of this approach [2.25]. First, because it is easier to model the constraints of a physical test system than proving ground surfaces or tires, virtual testing establishes far more effective boundary conditions than the other methods. Second, stated the company, virtual testing leverages an arsenal of proven, well-established physical testing tools and techniques, which have demonstrated utility in the analysis realm. Third, the incorporation of a modeled physical test system greatly streamlines the validation of results through subsequent physical testing and provides an opportunity to improve physical test setups and fixtures designs. By spanning the analytical and physical test disciplines, virtual testing requires advanced knowledge of both CAE tools and physical testing, the development of a process to link RPC Pro software and analysis models, and preferably some degree of virtual testing exposure and experience.

For years, MTS Systems Corporation has focused on refining the virtual testing approach, conducting a variety of demonstration projects with customers like Hyundai Motor Company (HMC) and Thermo King to evaluate various methodologies. These projects included virtual tests of full vehicles and subsystems, all yielding correlation with either measured road load data (RLD) or actual physical testing. The basic methodology gleaned from these experiences comprises the following steps:

1) Connect test rig models with specimen models;
2) Couple the models with RPC Pro;
3) Reproduce road load data on the virtual test rig;
4) Extract calculated loads;
5) Create the physical component/subsystem test.

A recent demonstration project with the SAIC Motor Corporation Limited provides an example of this methodology in practice:

1. Connect test rig models with specimen models: At the outset of the project, MTS built a collection of test rig models in ADAMS, Simulink, and ADAMS-Simulink cosimulation formats. The MTS test rigs that were modeled included a variety of Model 329 Spindle-coupled Road Simulators, a Model 353.20 Multiaxial Simulation Table (MAST) System, and a TestLine component test system. Select test rig components were also modeled, including a FlexTest digital controller, MTS actuators and servovalves, and a transformation that converts actuator displacement, acceleration, and force into DOF displacement, acceleration, and force. Due to project time constraints and the relatively slow simulation speeds of the more complex cosimulation models, the ADAMS/Car 329 model was coupled with SAIC full vehicle, front suspension, and rear suspension models, and ultimately used for the bulk of the virtual testing [2.25].

2. Couple the models with RPC Pro [2.25]: A virtual test server was developed to connect RPC Pro software and the ADAMS/Car 329 model. During testing, RPC Pro employs this virtual server to send drive files to the ADAMS model, initiate the ADAMS simulation, and copy the response file from the ADAMS folder to the RPC Pro working directory. Additionally, a Matlab Interface Tool already existing in RPC Pro was used to couple RPC Pro with the cosimulation models.

3. Reproduce road load data on the virtual test rig: The RPC iteration technique was then used to reproduce road load (spindle force) data collected from the proving ground for a variety of select events and maneuvers. Interestingly, initial RPC iterations on the virtual 329 rigs exhibited no convergence, suggesting flaws in the SAIC vehicle model. Subsequent analysis of the model did indeed reveal deficiencies, which were corrected in short order. With the model improved, the RPC iterations finally converged, showing a correlation between the desired and achieved signals and RMS errors across all channels for both full and partial vehicle virtual simulations.

4. Extract calculated loads: Upon RPC iteration convergence, calculated loads for any mechanical component or subsystem could then be readily obtained from the vehicle model as functions of time.

MTS' ongoing exploration of virtual testing demonstrates that it is a means for evaluating component loads prior to the development of physical parts or prototypes. Once these calculated loads are obtained, the same RPC Pro tools and techniques used to obtain them can then be used to create and conduct the physical testing. The potential impacts of effective virtual testing on the vehicle development process are considerable. As MTS has written "… the role of physical testing will evolve: while testing will continue to be necessary to achieve final validation of a vehicle design, it will increasingly be tailored to validate the vehicle model."

Extensive aerospace applications experience

Building on more than four decades of experience of working closely with the world's leading aerospace manufacturers, MTS is one resource that provides the technology and readily adapts to new test requirements to enhance overall testing productivity.

The above processes are designed to address the scope of aerospace testing requirements, from structural and subassembly evaluation to materials characterization. Whether the primary focus is certification testing for a new airframe design, completing flight-by-flight spectrums, or carefully evaluating the strength of individual components, the MTS approach can be helpful.

If the need is for test products for the wireless world, Interlab is a source for fast and reliable test procedures [2.26].

Wireless communications product development and market access processes need to be accompanied by time-crucial and reliable test procedures.

The company 7 layers [2.26] has responded to the development of the Interlab Test Solutions, which is based on conformance and interoperability test procedures.

Interlab Test Solutions are especially suited to test procedures that demand the synchronization of diverse test equipment. They can be used for product development purposes as well as for conformance test procedures, depending on the status of the involved test equipment and test cases for both automotive and aerospace engineering.

Choices of test solutions for wireless test laboratories:

- Interlab Test Solution Bluetooth RF
- Interlab Test Solution DEVICE/UICC
- Interlab Test Solution TTY
- Climatic Chamber Control
- Interlab Test Engines
- Interlab Feature Explorer

Main features of Interlab Test Solutions:

- One graphical user interface means reduced complexity of test procedures, irrespective of the different test equipment involved
- Fast and straightforward configuration of the test set-up, including test equipment configuration and OUT description
- Direct access to all tests combined with intensive filtering, grouping, sorting, and duplicating functionalities

- Automatic availability of final results by combining verdicts of all involved test equipment
- Optional usage of remote control OUT interfaces for fully automated test procedures
- Suitable for R&D and conformance purposes depending on their validation status

The test solutions are available with further services on demand, including set-up services, support, and maintenance, on-site services, etc.

Virtual testing users need to understand that while this works well during predesign procedures. But this testing is not presently ready for design testing, and especially manufacturing testing, because it does not accurately simulate the real-world conditions, nor all of the product's components interaction.

2.2.3 The third general direction (laboratory and proving ground testing with physical simulation of field conditions)

This direction is the oldest in accelerated testing. But while the actual vehicle is studied or tested with a simulation of field/flight input influences by special equipment (vibration test equipment, test chambers, proving ground, etc.), usually, the level of the product loading is higher than that encountered in normal usage. But, this direction of testing does provide a physical simulation of the field input influences on the actual test subject, as well as some components of safety problems and human factors.

Unfortunately, too often, such testing wrongly simulates field input influences separately, such as temperature, humidity, radiation, pollution, or only several from the many field input influences. In such cases, this type of testing cannot offer accurate initial information for providing successful prediction, and as a result, accelerated development of quality, reliability, durability, or other performance components may be less than predicted.

For example, when utilizing this type of testing, in critical vibration testing, accurate, reliable controls with advanced safety features to protect valuable flight hardware is essential, as written in Ref. [2.27].

Safety-controlled vibration testing is required for spacecraft and satellites prior to launch. This testing is designed to qualify the structure against the various static and dynamic loads that are to be experienced during launch. Advanced safety features have been developed and proven to reduce or eliminate testing anomalies and to even protect the test article against unusual external events such as a power outage.

Some of these classifications of accelerated testing methods for aerospace are presented in Ref. [2.27] and will be briefly reviewed below.

There are a variety of test methods that are used to simulate the various load sources encountered in aerospace applications.

Random vibration testing is used to verify strength and structural life by introducing random vibration through the mechanical interface. Random tests are typically performed in the frequency range of 20–2000 Hz.

Sinusoidal vibration testing includes testing at low levels to verify the natural low frequency and at higher frequency levels to verify the strength of structures. Responses are monitored, and input forces are reduced or limited as necessary to ensure that the target responses or member loads are not exceeded. This testing, as well as others detailed below, are not based on accurate simulation of field conditions that have a more complicated random character.

Pyrotechnic shock tests are used to verify resistance to high-frequency shock waves caused by stage separations. This can include the introduction of high-energy vibration at frequencies up to 10 kHz.

Sine burst testing involves short-duration constant-amplitude sine excitation. This is a quasi-static load to validate the strength design for flight and can be used as an alternative in static loads or centrifuge testing.

The Data Physics Vector single-shaker controller and Matrix multishaker controller offer a feature set that addresses all requirements specific to spacecraft and satellite testing.

Swept sine testing is among the more difficult tests required for the qualification of spacecraft and satellites. The primary objective of swept sine testing of spacecraft and satellites is to verify the strength of the primary and secondary structures. Qualification tests consist of one sweep through the required frequency 5–100 Hz, although some tests will be required up to 150 Hz. This sweep is performed in three orthogonal axes. The amplitudes of the sweep are defined by the launch vehicle's characteristics.

Sweep sine testing control requires tracking filters to accurately measure the amplitude and phase of the sinusoidal signal. Tracking-filter type (fixed or proportional bandwidth) and bandwidth are user selectable to optimize the tracking filter for the frequency range and sweep parameters. Swept sine testing of spacecraft and satellites is typically done with high sweep rates, up to four octaves per minute, to minimize the number of cycles at resonance frequencies. The combination of fast sweep rate and low-frequency, highly damped resonances is particularly challenging for a vibration control system.

Reducing the vibration amplitude over specific frequency ranges is often used to prevent excessive loading of the spacecraft or satellite structure. The levels are typically determined by coupled loads analysis using mathematical models of the launch vehicle and spacecraft.

The protection of valuable spacecraft and satellites is a paramount concern. To provide more useful vibration qualification, several new safety features have been incorporated in the Data Physics sine vibration controller that was used for testing of the James Webb Space Telescope at NASA Goddard Space Flight Center's Large Vibration Test Facility (LVTF).

An accelerated test plan considering an economic approach was described in Ref. [2.28]. It was introduced as a general framework to develop plans of accelerated testing with a specific objective, such as controlling the cost of the testing. The test plans are developed by considering the prior knowledge of reliability, including the reliability function and its scale and shape parameters, and the appropriate model to characterize the accelerated life. This information is used in Bayesian inference to optimize the test plan. In this analysis, prior knowledge is used to reduce the uncertainty of the reliability of the new product. The proposed methodology consists of defining the accelerated testing plan while considering an objective function based on economic value, using Bayesian interference for optimizing the test plan, and using the uncertainty of the parameters to obtain a robust testing plan.

In considering the above method, as well as other information on proving ground testing, it should be remembered that while one can use these approaches for fatigue testing, they are not appropriate as predictors for reliability testing, which is a much more complicated process than fatigue analysis.

The above-described analysis also focuses on two factors: the cost linked to testing activities and the cost associated with the operation of the product.

The authors [2.28] attempted to develop their test plan by extending their approach to include the theoretical formulation of the various degrees of freedom with respect to the parameters. To complete this development, the authors improved the algorithm of optimization. The proposed method was illustrated by a numerical example based on a sample problem.

Finally, the authors introduced a general framework to provide an accelerated test plan with the inclusion of a cost objective. The cost objective function is developed in a theoretical formulation with the test plan parameters. Then, this framework is compared with the results obtained from a Genetic algorithm.

Frequently, the Genetic algorithm is a discrete stochastic process that can be considered as a Markov process. Several results can be derived from this process, which enables easily verifying the optimization's efficiency.

As a further example, in Ref. [2.29] the authors wrote a report containing a review of the various existing prediction and accelerated testing methods, some components of which will be reviewed below.

The paper discussed the goal of building longer life unmanned satellites and space probes, which has created a demand for meaningful accelerated test methods to simulate long-term service in space. This is particularly necessary for tribological components, such as the lubrication of bearings and gears. There is an essential need for light-weight, low torque, durable mechanisms that can operate efficiently in a hard vacuum environment.

For being realistic, the ground tests must not only reflect the physical conditions, such as loads, speeds, or temperatures but must also simulate the multienvironmental factors, particularly the hard vacuum conditions that will be encountered in space.

The ultimate goal is to incorporate the sensors into the ground test equipment, and then into the satellite so that actual performance could continue to be monitored in space.

Historical experience with accelerated test methods

There was some previous work documented in the area. During the early 1960s, a number of research organizations did extensive bench testing on small mechanical components, such as instrument size ball bearings, slip rings, and gears in a vacuum environment.

In this testing, the temperature is one of the many factors used for accelerated failures. One obvious effect of increasing temperature is to increase the pressure in the vacuum chamber, especially if liquid lubricants are being used, or if the test pieces are outgassing. Temperature can also be used to adjust the viscosity of a liquid lubricant, and thus, simulate the performance of lighter oil, but this requires very precise instrumentation to measure and control the temperature.

In accelerated life testing, a lead-lubricated bearing was run at 100 rpm with periodic slowdowns to 16 rpm so that the torque spectrum could be measured. The test was run for 1×10^8 revolutions, almost 700 days at 100 rpm without developing excessive torque noise. That was the equivalent of more than six times the required life.

Developing an accelerated testing technology road map

In theory, by carefully increasing the severity of the test conditions, it should be possible to hasten early failures without altering the actual mode of the failure. This concept of accelerated testing presupposes that the failure mode is reasonably well understood and that the operating conditions used in these tests meet two requirements [2.29]:

1. The severity of the test conditions can be adjusted to magnify the mode of the failure.
2. The new test conditions will not be severe enough to cause the component to undergo a different mode to failure, i.e., a transition from mild wear to galling and seizure.

For studies of the DMA bearings, various accelerated testing techniques were used, including speed, load, temperature, surface roughness, lubricant starvation, etc. The effect of lubricant type and retainer material was also investigated. Although much time and effort were invested in these programs, the results did not lead to clean solutions. In this analysis, there were no results that would make it possible to predict impending failures before they occurred.

Frequently, the problem is not with the concept of accelerated testing but is with the limited capabilities of the instrumentation that is being used. There is a need for nondestructive sensor techniques that can indicate an impending problem without shutting down the test or experiencing a catastrophic failure.

TABLE 2.3 Operating characteristics and typical sensors.

Characteristics	Typical sensors used
Sound and vibration	Accelerometers and ccapacitance probes
Temperature	Thermocouples and thermistors
Speed	Magnetic or optical pickups
Torque	Strain gages or piezoelectric sensors
Sliding or rolling	Electrical contact or sound

Many of these techniques are already available. Methods of monitoring motors current, speed, temperature, load, and torque have all been used in DMA studies. **However, the techniques being used do not reflect the advances that have been made in sensor technology.** The truth is, there is more sophisticated measuring, monitoring, and controlling capabilities in the engine of today's automobile than there is in the mechanical assembly of a multimillion-dollar satellite.

This is where the emphasis must be placed if accelerated testing is to become a realistic tool for evaluation. While the immediate goal is concerned with ground tests, the ultimate goal is to be able to monitor performance in both the ground and the orbit locations with the same investment, which needs to be built-in as an integral part of the mechanical assembly.

The first step in outlining a test program directed at a problem is to decide what is to be monitored and measured. The operating characteristics of interest are included in the following (Table 2.3).

Horiba and MIRA presents [2.12] some of their wealth of vehicle accelerated test and development experience covering all vehicle classes, on either MIRA's extensive proving ground or on public roads.

All of their activities are fully risk assessed. As part of the test activities, regular vehicle inspections are undertaken, periodic vehicle measurements made, daily checklists completed, servicing and updates completed, instrumentation fitted and logged, and vehicle audits completed. All of which can be uploaded to a secure web portal for instant customer access.

Of course, proving ground testing cannot account for every eventuality encountered on the road.

In 1889, the Swedish chemist Svante Arrhenius developed a mathematical model for the influence of temperature on chemical reactions. Because chemical reactions are responsible for some failures, the Arrhenius equation has been adopted to model the acceleration of testing. Until now, many professionals used the Arrhenius equation and assumed an exponential distribution for time-to-failure distribution. Even many engineering books continue to use this exponential distribution. But now Lall et al., claim [2.30] that using this distribution in engineering, including in consensus standards,"... have been proven inaccurate, misleading, and damaging to cost-effective and reliable design, manufacture, testing, and support."

In Chapter 4, the author demonstrated that this mathematical model is very old, and the equation is inaccurate when used to simulate real-life conditions. Therefore, it should not be used in reliability testing, including accelerated testing of automotive and aerospace engineering products.

We can also often see in the literature that professionals are using HALT and HAAS for accelerated reliability testing (ART) or accelerated durability testing (ADT). However, this also is not appropriate because as Gregg Hobbs, the author and inventor of HALT and HASS, wrote in his book [2.31]:

"Design engineers involved in quality assurance and students of reliability engineering will benefit from this unique resource detailing the technical aspects of accelerated reliability engineering. Features include:

- Coverage of the physics of failure and useful testing equipment enabling those new to the area to grasp the concepts behind HALT and HASS;
- Overview of the HALT technique demonstrating how to find design and process defects quickly using accelerated stress methodology during the design phase of the project;
- Examination of detection screens and modulated excitation used to detect flaws exposed in HALT;
- Description of how to set up a HASS profile and how to minimize costs whilst retaining efficiency;
- Applications of HALT and HASS and analysis of common mistakes highlighting the pitfalls to avoid when implementing the methods."

From this, we can see that Gregg Hobbs did not consider his approach a means for accurate simulation of the real-life conditions necessary for ART or ADT. But many people after him wrongly used his approach as a method for these types of testing.

2.2.4 The fourth general direction (accelerated reliability/ durability testing)

Development of accelerated reliability testing (ART) and accelerated durability testing (ADT) [2.13] is the most useful trend in the development of accelerated testing in automotive and aerospace engineering. This direction provides the greatest opportunity for improving technical progress and advancing technology in civilization.

It is important to remember that this direction does not relate to accelerated life testing (ALT), which is basically the traditional approach that cannot successfully predict any product's performance during its service life.

Instead, this author has developed this new direction of accelerated testing of engineering products, which he has called "accelerated reliability and durability testing technology" (ART/ADT). This direction is very different from traditional accelerated life testing. In comparison to ALT, ART/ADT has much more robust testing requirements than that employed in ALT. Therefore, using ART/ADT provides accurate initial information on the product's successful performance prediction during any given time. ALT cannot offer this capability. As demonstrated in technical papers presented at the SAE 2012–18 World Congresses in Detroit, this direction is particularly useful for successful new product prediction. The basis of ART/ADT technology is described in detail in Ref. [2.13], as well as in other publications.

The following are the basic concepts of this new accelerated testing direction:

- Provide accurate simulation of the field conditions (quantity and quality of influences, the character of change during the time of each influence, and the dynamics of changes of all influences), including safety, and human factors using the given criteria.
- Conduct simulation testing 24 h a day, every day, but not including
  - Idle time (breaks or interruptions) or
  - Time operating at minimum loading does not contribute failure
- Conduct accurate simulation of each group of input influences (multi-environmental, electrical, electronic, mechanical, and other) in simultaneous combination. For example, the input multienvironmental group is a complex simultaneous combination of pollution, radiation, temperature, humidity, air fluctuation, air pressure, and other environmental factors.
- Utilize a complex system for modeling each of the interacting types of field influences. For example, pollution is a complex system that consists of chemical air pollution + mechanical (dust) air pollution, and both types of pollution must be simulated simultaneously.
- Simulate a whole range of each type of field influences with their characteristics. For example, when simulating temperature, one must simulate the whole range of temperatures from the minimum to the maximum, the rate of change of temperature, the characteristics of the speed of change and the real random characteristics of this temperature if it is changing randomly;
- Use the physics-of-degradation process for an accurate simulation of the field conditions;
- Treat the system as interconnected using systems of systems approach;
- Consider how the interactions of test subject components act within the system;
- Conduct laboratory testing in combination with periodical field testing, as a basic component of ART/ADT;
- Provide periodical field testing at regular intervals after the given hours of laboratory testing;
- Reproduce the complete range of field operational schedules and maintenance or repair actions;
- Maintain a proper balance between field and laboratory conditions;

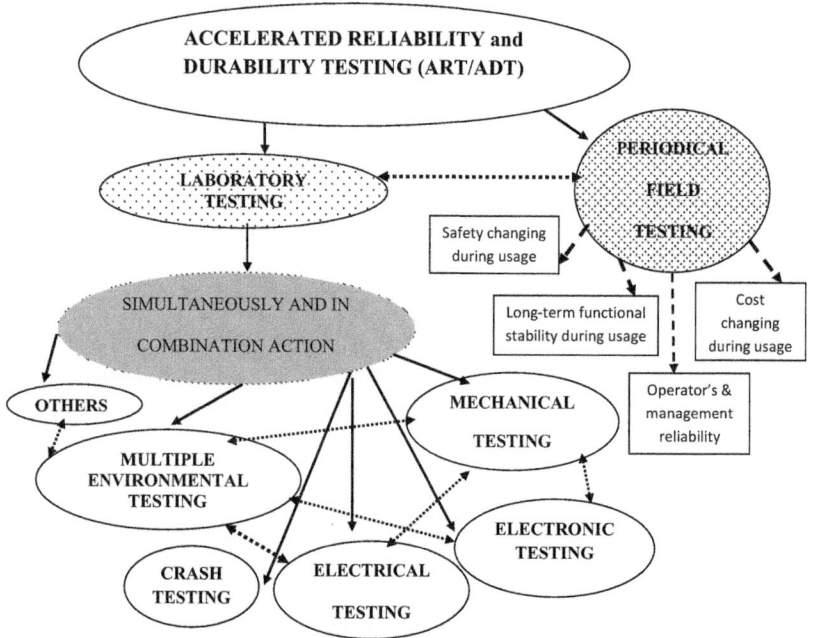

FIGURE 2.5 Scheme of accelerated reliability and durability testing (ART/ADT).

- Correct the simulation system after an analysis of the degradation and failures in the field and during ART/ADT.

This basic scheme of ART/ADT can be seen in Fig. 2.5.

So how does one make practical comparisons between the parameters of real-world degradation mechanisms and those encountered in accelerated reliability/durability testing? This can best be done by comparing the loading processes (output variables) for each sensor. An example from the author's experience follows.

In this example, the results of the analysis of a common drive shaft's (sensor 6) subject to random statistical loadings characteristics (three statistical parameters: mathematical expectation X, standard deviation σ^2, and time of correlation τ_k) are shown in Table 2.4. The resultant solution is that the regimen during ART/ADT corresponds to the real world by a difference no more than 10%). One can see this in Figs. 8.5—8.7 (Chapter 8 of this book).

For successful accelerated testing, one needs to provide a simulation of the other relevant factors experienced in real-world conditions. In order to accomplish this, one needs to duplicate the complex analysis of factors combinations that influence the test subject's real-world exposure and to do this efficiently.

ART/ADT technology is also based on the accurate simulation of all the interacted components (units and details) of the whole vehicle. Fig. 2.6 visually depicts this interaction.

TABLE 2.4 Statistical characteristics of loading processes for components of car trailer.

Studied component	The name of parameter	Real world	ART/ADT Number of regimens	
			1	2
Common drive shaft	Mathematical expectation			
	$\overline{X}$ $(H \bullet M) \times 10^{-1}$	20.3	19.3	21.2
	Standard deviation σ^2 $(H \bullet M)^2 \times 10^{-2}$	181.1	186.0	191.0
	Time of correlation $T_k \bullet c$	0.09	0.10	0.08

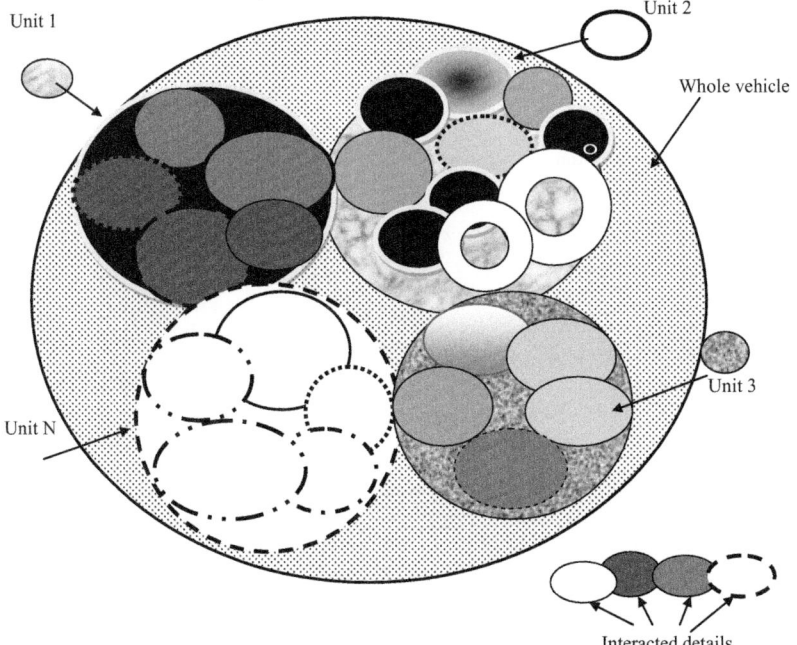

FIGURE 2.6 Vehicle as a combination of interacted units and details.

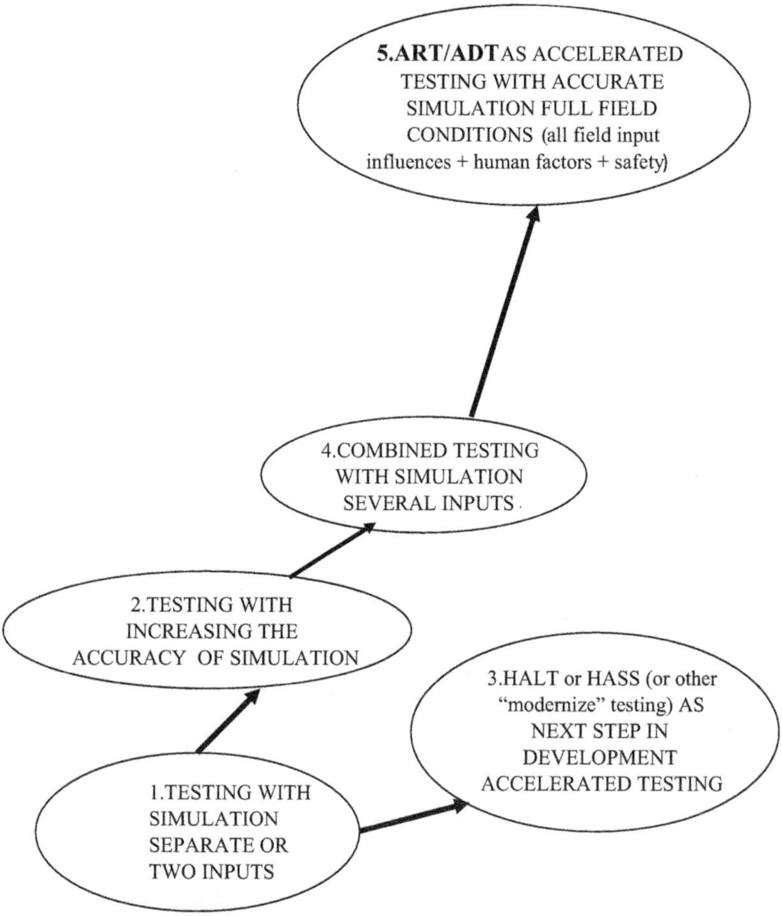

5.ART/ADTAS ACCELERATED TESTING WITH ACCURATE SIMULATION FULL FIELD CONDITIONS (all field input influences + human factors + safety)

4.COMBINED TESTING WITH SIMULATION SEVERAL INPUTS

2.TESTING WITH INCREASING THE ACCURACY OF SIMULATION

3.HALT or HASS (or other "modernize" testing) AS NEXT STEP IN DEVELOPMENT ACCELERATED TESTING

1.TESTING WITH SIMULATION SEPARATE OR TWO INPUTS

FIGURE 2.7 The path from testing with separate simulation or several input influences to ART/ADT.

FIGURE 2.8 Horiba's shaker for the automotive industry. *The author received this photo from Horiba Instruments, Inc.*

So, when ART/ADT is properly performed, it accounts for the interactions and accurate simulation of the different components under field conditions, including the whole vehicle or its details (depending on the test subject).

For those interested in further information concerning this technology, they can find details on ART/ADT in Ref. [2.2]. They can also find information about the results obtained by the implementation of accelerated reliability and durability testing in Ref. [2.10].

Fig. 2.7 depicts how ART/ADT's higher level of accelerated testing, which is based on accurate simulation of field conditions, leads to successful prediction of product performance with all associated components (see Fig. 2.8 and [2.53]).

2.3 Classifications of accelerated testing

There are various approaches to the classifications of accelerated testing seen in the literature. Let us consider some of the prevalent approaches. The following listing is one such example of the approaches to the classification of different types of testing [2.32]. Each type of these tests has been called accelerated testing, and each provides different information about the product and its failure mechanisms.

Generally, in this classification system, accelerated testing is divided into three types:

1. Qualitative testing;
2. ESS (Environmental Stress Screening) and Burn-in;
3. Quantitative accelerated life testing.

2.3.1 Qualitative testing

Qualitative testing is testing that only yields failure information or failure modes. Qualitative testing has been referred to by many names, including:

- Elephant testing;
- Torture testing;
- HALT (highly accelerated life testing);
- Shake and Bake testing.

Qualitative testing is generally performed on small samples with the specimens subjected to a single severe level of stress, to a number of stresses, or to time-varying stress (i.e., stress cycling, cold to hot, etc.).

If the specimen survives, it passes the testing. If the product fails, the appropriate testing actions will be taken to improve the product's design in order to eliminate the cause(s) of failure. Qualitative testing is used primarily to reveal probable failure modes. However, if this testing is not properly designed, it may cause the product to fail due to failure modes that would never have been encountered in real life. In general, qualitative testing is not designed to yield life data that can be used in subsequent analysis or for "accelerated life testing analysis." In general, qualitative testing does not quantify the life (or reliability) characteristics of the product under normal use conditions.

Other limitations with this type of testing are:

- It does not predict product reliability under normal use conditions.
- It is difficult to determine an accurate acceleration coefficient.

2.3.2 ESS (Environmental Stress Screening) and burn-in

Environmental Stress Screening is the process involving the application of environmental stimuli to products (usually electronic or electromechanical products) at an accelerated basis. The stimuli in ESS testing can include thermal cycling, random vibration, electrical stresses, etc. The goal of ESS is to expose, identify, and eliminate latent defects that cannot be detected by visual inspection or electrical testing, but which will cause failures in the field. Generally, ESS is performed on the entire population and does not involve sampling.

Burn-in is a screening process that subjects the items to an elevated stress, often temperature for a specified period of time.

Benefits: provides a means for precipitating defects that contribute to infant mortality. This is especially beneficial in new products;

It can reduce the failure rate in subsequent system operation, especially as compared to when there is no burn-in testing.

This being said, a recent Bellcore study for a natural product using three types of burn-in and no burn-in has revealed no significant difference in the failure rates for burn-in versus no burn-in. Since device failure rates during infant mortality are generally much higher than experienced during the steady-state period, an equipment manufacturer can use a burn-in screening tool to identify defects in the manufacturing process that would otherwise occur in the field. It implies a capital investment for the burn-in testing chambers, as well as the ongoing costs for the burn-in testing. Several industry studies are underway to assess the net economic value of burn-in.

Burn-in can be regarded as a special case of ESS. According to military standards, Burn-in is testing performed for the purpose of screening or eliminating marginal devices. Marginal devices are those with inherent defects or defects resulting from manufacturing aberrations, which will result in time and stress-dependent failures. In these cases, burn-in is performed on the entire population.

2.3.3 Quantitative accelerated life testing

Quantitative accelerated life testing, unlike the other qualitative testing methods, consists of quantitative testing designed to quantify the life characteristics of the product under normal use conditions, thereby trying to provide reliability information. Reliability information can include the determination of the probability of failure of the product under use conditions, mean life under use conditions, and projected results and warranty costs. It can also be used to assist in the performance of risk assessments, design comparison, etc.

Accelerated life testing can take the form of "Usage Rate Acceleration" or "Overstress Acceleration" [2.32].

For all life testing, some time-to-failure information for the product is required since the failure of the product is the event that one wants to understand.

Two methods of acceleration are demonstrated in Ref. [2.32] "Usage Rate Acceleration" and "Overstress Acceleration," both of which have been devised to obtain times-to-failure data at an accelerated pace. For products that do not operate continuously, one can accelerate the time it takes to induce failures by continuously testing these products. Some call this "Usage Rate Acceleration" [2.32]. For products for which "Usage Rate Acceleration" is impractical, one can apply stress(es) at levels that exceed the levels that a product will experience under normal use conditions and use the times-to-failure data obtained in this manner to extrapolate life data to actual use conditions. This is called "Overstress Acceleration."

Parameters commonly used as function of stress for different life distributions [2.32]:

- Exponential (mean of failure rate);
- Weibull (scale parameter);
- Lognormal (Ln-Mean or Median).

Another classification of possible accelerated life testing (ALT) approaches was published by F. Schenkelberg [2.33]. The author considered six approaches that provide a broad range of approaches offering different benefits and limitations.

Time compression approach

The time compression approach simply operates the item more than it actually runs during normal use. A classic example is a toaster. A typically family may use the toaster at a rate of twice a day during the preparation of breakfast. This would equate to the total number of use cycles of 730 cycles per year (twice a day for 365 days per year). A time compression ALT would simply require cycling the toaster 730 times to replicate each a year's use. If each cycle is 20 min, the testing would only require 10.2 days, assuming the testing was continuous for 24 h per day, and would not require any special chambers or equipment. In this case, the acceleration factor Eq. (2.1) becomes:

$$AF = \frac{\text{Use Duration}}{\text{Test Duration}} = \frac{720}{20} = 36 \qquad (2.1)$$

In this case, with an assumed product use of two cycles per day, and each annual cycle running on average 720 min, the lab can achieve an AF of 36.

This approach only works for items that are not in use to such an extent that there is little time for compression, as described in Ref. [2.33]. For example, if the product were a taxi that is in use 18 h per day, it would not benefit testing significantly as running the vehicle the remaining 6 h per day only, provide a 1.33 acceleration factor. While this does provide some acceleration, it would still take about 274 days to replicate 1 year's use.

Build a stress to life relationship approach

The second common ALT approach is to apply higher than expected stress (such as temperature, voltage, etc.) to the product in such a way that the failure mechanism's progression of interest is accelerated, With this approach the difficulty is to properly relate the use factors and test stresses in such a way that the results from the higher testing stress provide meaningful predictions of that which will be experienced in actual use conditions.

If there is no existing acceleration model, the ALT design should first produce data to determine the acceleration model. For example, if you are using temperature as the accelerant and are using the Arrhenius rate equation as the form of the model (unknown activation energy), one approach could be to use three higher temperatures. The plot of the time to failure relative to each temperature may provide a relationship between temperature and life, which may then permit the extrapolation of expected life at normal use temperature.

The three downsides to this approach are

First, the necessity to select appropriately high stresses, but stresses that are not so high as to cause failure mechanisms that would not exist in real-world use.

Second, higher stresses often require special equipment and set up [2.33]. A major assumption in this approach is that the higher stresses selected will actually cause the failure mechanism to accelerate in a fashion that is similar to that which would occur when the item is under real use conditions. Failure analysis is critical to validate this assumption.

Third, it generally takes multiple samples at multiple stress levels to develop these lifetime distributions.

In view of these three downsides, this approach may lead to significant differences between the accelerated testing results and the field results, which can result in the unsuccessful prediction of product quality, reliability, durability, and maintainability during the product's usage time.

Given acceleration model approach

The "given acceleration model approach" utilizes one or several of the many previously developed models and equations to save time and cost of developing the testing model, and as a result, the reduction of time from design to the market. Peck's relationship, the Norris-Landzberg equation, and Booser's equation are all examples of common models used to describe the effects of stress on a unit's life for a particular failure mechanism [2.34]. Using one of these equations, one of the many others that have been developed, saves the time of creating a new model or finding parameters through extensive experimentation. In some cases, previous work may be available, and that previous work may provide the model and parameters for a specific new product.

At the least, these models may provide a framework for planning the ALT and a structure for the analysis.

While this approach uses fewer samples and resources than the previous approach [2.33], it does rely on more assumptions, and, as with the previous approach, it uses elevated stress conditions, which are unrealistic when compared to actual use conditions and, a critical aspect is that it does assume a suitable model exists.

As an example of this approach, assuming the failure mechanism is known and is properly described by the power-law model in the following equation

$$f_{\text{mean life}} = kv^{-n}$$

where k is a constant, v is a voltage, n is the power-law constant [2.35]. This model assumes that the power-law constant is known, and therefore, it permits the design of a single stress test that will meet the duration and precision requirements along with minimizing the sample size and cost considerations.

Step-stress approach

Step-stress ALT approach can be used, but only in situations when the damage leading to failure accumulates proportionally to the application of stress, i.e., higher stress results in greater damage. But the result of this approach often does not correspond to real-life failures or degradations.

The step-stress approach assumes the Markov property that the remaining life distribution depends only on the current stress and fraction failed, and that the damage accumulation only occurs due to the changing of the stress level. While a model based on this approach describing the failure mechanism's relationship to stress certainly aids in the design and analysis of the ALT, this approach has mostly theoretical interest, because it is unrelated to real-life product operation.

Shock testing approach

The shock testing approach was developed from the need to test the ruggedness of new products during the development process. The first shock pulses were generated by dropping products onto various impact mediums (e.g., sand) to reproduce shock inputs. These early testing methods were challenged by poor repeatability and consistency issues and were generally a pass/fail type of test.

Then the Damage Boundary theory was developed by Dr. Robert Newton in the late 1960s, which formalized a shock testing methodology that is currently the framework for the majority of shock testing. This theory independently identifies critical acceleration and critical velocity levels that cause damage. The result is a damage boundary test that characterizes a product's fragility, which can be used to either improve the product or to protect it by secondary means. Some professionals now believe that while these testing fundamentals remain much the same today, the shock testing approach and application methodology must continue to evolve to better meet the needs of rapid product development. Also, one has to realize that this type of testing is mostly related to materials that are to be used in future products.

One benefit of using this approach is the ability to increase testing speed [2.36]. To accomplish this, shock machines were developed that increase cycle rates, thereby reducing the overall test times for repetitive fatigue shock applications. Another benefit is the ability to simulate complex events, such as blast or crash effects, especially to electronic systems and vehicle occupants. Some consumer electronics, such as handheld devices, must be able to withstand harsh impact environments. Cellphone and tablet users expect their products to continue functioning even after they drop them. A particularly extreme example is associated with military applications. Consider the case of the fuze attached to penetrating ordinance. In order to provide the intelligence for detonating the munitions, it must survive extreme impacts and continue to function. While dropping munitions from an aircraft versus dropping a cell phone are obviously two significantly different dynamic events, from a shock testing perspective, wrote Peter Brown [2.36], both of these scenarios require a high-speed, high-fidelity shock testing apparatus.

For customer electronics shock testing, Lansmont and other manufacturers have different models available with varying table sizes and offering different performance levels. For example, the model HSX20 simulates the higher energy impacts—shock pulses with a maximum velocity change of 180 feet per second. This result from the shock table is equivalent to going from zero to 122 mph in less than 40 feet.

Lansmont has also developed a one-of-a-kind blast-effect simulator for seat testing applications. Army Research Labs (ARL) and Johns Hopkins Applied Physics Labs are both using vertical impact test systems (VITS) for research in seat designs to mitigate the effects of the high energy resulting from a blast.

Degradation approach

The degradation approach is based on the feature of some failure mechanisms to exhibit on the measurable deterioration of performance.

While the above approaches to modeling may permit the use of a single accelerating stress scenario, all the same considerations still apply related to the model's need to describe the product's actual failure mechanism. When the model to be used is not known, using multiple stresses will generally produce sufficient information to create or select an appropriate model. As with the Given Acceleration Model Approach, this will take more samples and test resources than situations where the failure model is known. The statistical fit is also more complex, so for the same sample size, there is generally a modest loss of precision.

All of the above approaches to accelerated testing are primarily based on a statistical (mathematical) base and are only partly related, as a secondary factor to the physical or chemical essence of reliability, durability, maintainability, life cycle, and other components of the product's performance characteristics.

Engineers generally prefer to use the statistical essence of failures more than the physical essence of the failure mechanism. It must always be remembered that degradation and failures of technical components are only a small part of the product's quality, reliability, maintainability, etc., and of the economic aspects, such as life cycle cost, profit, and others of a product's efficiency. Too often these aspects that can lead to the success or failure of a product are overlooked or ignored. This is often because of parochial interests or improper cost-saving philosophies by some of the professionals who are involved in the automotive and aerospace engineering product design and manufacture. If there is any question of this, just consider how infrequently in any of the cited literature there is any discussion or mention of the role of human factors in solving quality and reliability problems. Even in the literature on autonomous vehicles, where human factors should be the leading factor in designing systems of control, you will find people designing these systems primarily from the viewpoint of system control and maintenance, cost reliability, etc., with minimal emphasis or consideration to the human-induced variations, especially when the human inputs are not the ones predicted by the designers.

Therefore, the basic trend in the development of accelerated testing in automotive and aerospace engineering (as well as other areas of engineering, and not just in engineering, but in many other areas of technology), remains the need for the development of testing technology (methods and equipment) needed to provide accurate successful prediction of product performance.

If we would only consider the role of accelerated testing to be an interacted component of product efficiency, we could not only successfully improve product testing, but we would also improve product efficiency and organizational economic efficiency. Too often the alternative traditional approach, wherein accelerated testing (and testing in general) is considered as a distinctly separate discipline, without connection to the other components of product creation, design, and manufacture efficiency. While this is not to suggest that we cannot solve the problem of accelerated testing as a discrete discipline, this development does need to take into account the above connections.

As has been shown in the preceding, most available information on accelerated testing is related to trends in the statistical area of accelerated testing. But, we know, the basic area of accelerated testing that also needs development is related to the physical essence of reliability and fatigue testing. Development of this is essential for obtaining results that will provide high accuracy and correlation to the fatigue results experienced in real life during the product's service life.

Moreover, many of the statistical approaches to accelerated testing are primarily theoretical aspects of estimation, which do not take into account the nonstationary random character of the processes that are experienced in real life.

There are other classification systems that are included in Ref. [2.37]. One such is the specific accelerated stress testing (AST), whose related capabilities/ test methods are related to accelerated reliability testing.

This system includes:

- Accelerated reliability testing;
- Accelerated reliability using a simple premise and some complex math to achieve a reliability estimate of a product for particular conditions. With this system, as a particular source of stress is increased, the time to failure exponentially decreases, and the knowledge of this effect is used to design the accelerated reliability test;
- A simple example of this effect is the fatigue curves of steel. The logarithmic rate of change is affected by the change in the physics of accelerated reliability;
- In an accelerated reliability test, several sets of parts are tested at a stress level higher than that expected in the operational service level.

Another is Failure Mode Verification Testing (FMVT).

FMVT is a patented process to employ highly accelerated test methods. FMVT reveals inherent design weaknesses which are first predicted by using the FMEA (Failure Mode and Effects Analysis) process. By exposing a design to a combined set of amplified environments/stresses, multiple failure modes (and their sequence and distribution) are produced in as little as 1 day.

FMVT brings together the following disciplines:

- Design engineering;
- Reliability engineering;
- Computer modeling;
- Failure Analysis.

Advantages of FMVT

- Identifies failure modes and their root causes;
- Reduces testing time and costs;
- Complements planned design iteration sequences.

FMVT Applications

- Design verification;
- Reliability growth;
- Continuous improvement;
- Warranty reduction.

One more is Full System Life Testing (FSLT) [2.37].

Full System Life Testing by simulating "real-life" conditions, and accelerated stress testing methods provides the data necessary to evaluate recent or proposed product changes. The information goal of FSLT is to pass the testing (it is a Pass/Fail test).

The accuracy of an FSLT is based on:

- Known strength distribution;
- Known service conditions and distribution;
- Having a large sample size;
- Having an accurate reproduction of service conditions.

FSLT Product Applications:

- Automotive cockpit;
- Automotive door system;
- HVAC systems.

Discussion on the subject of accelerated testing.

In Ref. [2.38] one can read: "Perusal of quality- and reliability-engineering literature indicates some confusion over the meaning of accelerated life testing (ALT), highly accelerated life testing (HALT), highly accelerated stress screening (HASS), and highly accelerated stress auditing (HASA). In addition, there is a significant conflict between testing as part of an iterative process of finding and removing defects and testing as a means of estimating product quality. Below is review the basics of these testing methods and describe how they relate to statistical methods for estimation and reliability growth. It also outline potential synergies to help reconcile statistical and engineering approaches to accelerated testing, resulting in better product quality at lower cost."

Reference [2.38] defines an ALT method for stress-testing of manufactured products that attempts to duplicate the normal wear and tear that would normally be experienced over the usable lifetime of the product but in a shorter time period. If the product has a large number of components, component-level testing is generally required when the individual components are not fully tested or have not demonstrated proven reliability.

ALT difficulties:

- It is difficult to simulate actual operating conditions;
- In real life, most products interact with other components;
- There is little value in determining absolute reliability.

Test of Time: Accelerated Test Methods [2.38].

The use of accelerated test methods is one approach to handling the age-old problem of cutting testing time while still providing meaningful critical product details.

Formally developed over hundreds of years ago, accelerated methods speed up the testing process by using more intense test variables, such as higher temperatures, pressures and humidity, and harsher vibration schedules, and others.

Accelerated Testing: The Bigger Picture [2.38].

As the automotive industry continues to struggle, we can see increased use of accelerated test methods, such as HALT, and perhaps a shift to where more time is devoted to fixing problems during the development phase. Tools such as Finite Element Analysis (FEA) and Failure Modes and Effects Analysis (FMEA) can help support this shift. The increasing use of computer simulation may also prove to be beneficial, enabling test engineers to perform testing on parts that are closer to the production version of the product.

This author wrote in Ref. [2.38] that six things can be done to improve the success of a testing program, specifically:

A. Understand the product's realistic use and how it relates to test parameters. Many test engineers find themselves in the situation where they are testing to a certain specification, but they do not understand why the testing is being performed in a certain way. An understanding of why will not only give meaning to the test results but also give test engineers and technicians an understanding of what is important and why in the testing.

B. Regular communication between the testing professionals and the program manager so that everyone knows early on of any changes in the program timing. By doing this, if early in the process, the testing processes take more time than allotted, then the program manager will know that they must reduce time in other processes if the launch time is to be accomplished.

C. Obtain frequent updates of the product's schedule, and restate and remind involved parties of the needed time for product testing, especially if there is slippage in the program's timeline.

D. Test the product that is as close to the production model as possible. This also provides benefits in the planning for testing fixtures that simulate real use.

E. Familiarize yourself with the test specification. This may be difficult, especially if there are many specifications.

F. Although there are some widely used test specifications such as JEDEC, IEC, ISO, MIL, and USCAR, to name but a few, many of these testing specifications are company-specific. By familiarizing yourself early with the requirements of these testing specifications, precious project time can be saved.

Highly accelerated life testing (HALT) is testing in which, first, stresses higher than that applied to the product in the real world, and, second, uses simulation for only some of the many types of influences experienced in the real world.

HALT has a statistical difference with ALT [2.38].

Disadvantages:

- Not applicable in every case, for example, chemical degradation;
- Corrosion of a refrigerator door may not happen in a shorter time;
- Exposes test subjects to higher stresses than would be encountered in normal conditions, such as higher temperature than the highest temperatures that will be encountered in normal conditions;
- Uses only high humidity;
- Higher vibration that will be the highest vibration encountered in normal field conditions;
- Highly accelerated chemical/physical degradation which may not accurately reflect operating conditions, such as the weakening of insulation, for example, in motor windings due to high temperatures and moisture;
- Weakening of lubricants in bearings and other components due to high moistures and higher temperatures.

See also bibliography 1−12 in Chapter 4 about the above.

2.4 Fatigue accelerated testing

2.4.1 Common consideration

The term "fatigue accelerated testing" is often used in both the theory and practice of accelerated testing.

- Literature about proving ground testing had been published 50 and more years ago, such as the Kyle J.T., Harrison H.P. [2.39] brochure from the Nevada Automotive Test Center [2.40]. This and many others demonstrate that similar proving ground stress testing was used for obtaining initial information on machinery strength and fatigue evaluation in proving ground conditions. But qualified professionals now understand that this type of testing cannot offer the information necessary for accurate evaluation or prediction of the durability and reliability of the test subject in the real world. This is because this testing does not take into account all the environmental factors, such as temperature, humidity, pollution, radiation, etc., which influence the product's durability and reliability during its warranty period or service life. These limitations include:

- Often it does not consider the random character of real input influences that act on the product in the field;
- Often it does not consider the character of the input influences of the simulation's system of control, which cannot be simulated during proving ground testing;
- Does not consider the variables of influence presented by drivers and management on the reliability and durability of the vehicles;
- Does not consider the test subject's interaction with other vehicle components, especially units and details, for evaluation/prediction of their durability and reliability;
- And many other real-life situations that cannot be accurately simulated on the proving ground or in the laboratory.

But too often these limitations are ignored, especially in the literature of companies that design, produce, and use the equipment or methodology for this accelerated testing. Therefore, you will often find this flawed reasoning in the literature related to reliability or durability testing. Fatigue accelerated testing is provided for materials, details, units, and entire machines. Fatigue accelerated testing of materials one provides. Fatigue accelerated testing is usually provided during new materials development or when working with different materials. Accelerated testing of materials mostly uses standard methods and corresponding equipment in the laboratory. For accelerated testing of details, one mostly uses ASTM standards.

This book considers accelerated fatigue testing of machines, their units, and details in the laboratory and proving ground. These testing methodologies require a higher knowledge of physics, chemistry, theories of reliability, mathematics, statistics, and other disciplines.

Let us consider one such approach [2.41] **below. It is called** cyclic loading and fatigue [2.41].

Fatigue and structural integrity of aerospace systems

Historically, the community of individuals and organizations responsible for the design, inspection, and maintenance of aeronautical and aerospace vehicles have recognized that the potential for fretting can be a driving factor in system failures (Farris et al., 2000). Yet, certain air and space travel events, including the well-publicized catastrophic in-flight disintegration of a passenger plane fuselage and high-cycle fatigue (HCF) failures of gas turbine engines, have refocused efforts on understanding the fundamentals of the fretting damage mechanisms in aerospace structures and materials. To begin understanding the link between fretting damage and the structural integrity in aerospace systems, it is important to first review the paradigm shifts in the design and operation of both civilian and military aircraft sparked initially by several unexpected crashes of both American and British aircraft in the late 1940s and early 1950s. Each of these incidents was attributed to fatigue cracks introduced by cyclic loading of key metallic structures, including wing spar caps and components of the pressurized fuselage section.

Prior to these incidents, aircraft structures were designed and manufactured primarily with static strength and stiffness in mind, a fact echoed by an official from NACA, the National Advisory Committee for Aeronautics (the predecessor of the American space agency, NASA), who stated that interest in fatigue from cyclic loads "… was low because there had been no service experience to demonstrate that fatigue of the airframe was a serious problem" (Kuhn, 1956).

Initial attempts to account for fatigue damage by cyclic loads were accounted for by an approach dubbed "safe life" design. The safe life of an aircraft was based on laboratory fatigue lives of components that were subjected to applied load waveforms selected to be representative of in-flight conditions. A factor of safety of four was then applied to the observed cycles of failure of the test articles to account for variability in both materials and manufacturing quality of in-service airframe components. This safe-life approach was the basis for the first Aircraft Structural Integrity Program (ASIP) adopted by the United States Air Force (USAF) in 1958. By providing design and test methods for the prediction and prevention of structural fatigue failures, ASIP was designed to preclude structural failure of in-service and future weapons systems.

The second series of catastrophic losses of F-111, F-5, B-52, and T-30 aircraft in the 1960s focused scrutiny on the inability of the safe-life approach to account for the use of relatively brittle material in components subjected to high stresses. The resulting intolerance to defects introduced by either manufacturing processes or in-service damage was amplified by the inability to inspect and detect small cracks in many critical structural components. This renewed attention initiated a second paradigm shift in the approach to dealing with the relationship between material, manufacturing, and service-introduced defects, and the fatigue performance of airframes.

A damage-tolerant design acknowledges the presence of variability in initial manufacturing quality or damage accumulated during service. Such designs must be able to maintain their structural integrity or be "tolerant" of these defects or damage between scheduled inspection intervals. The USAF embraced the damage-tolerant philosophy in 1975 and continues to rely on its tenets for the basis of its ASIP program. The methodology was also incorporated into the design of civilian aircraft, as evidenced by the fail-safe approach adopted in the design of the primary structure of the DC-10 pressurized fuselage shell (Swift, 1971). In the two decades following the official adoption of the damage-tolerant approach, advances in the understanding of basic fatigue mechanisms, damage-tolerant structural design, nondestructive inspection techniques, and comprehensive tracking of individual aircraft usage have provided the USAF with confidence in this philosophy of ensuring the structural integrity of its fleet.

One can read in Ref. [2.42] their version of the accelerated fatigue test methods:

A. Testing Method Used at PSA
B. Staircase Method

This method [2.43] is a fatigue test procedure used in calculating the fatigue resistance for a sampling of parts. The method is based on an iterative principle, where several parts are tested. The test of the current part depends on the result of the testing of the previous one.

In the article, the authors present an equation for the calculation of the mean and the standard deviation of the component's fatigue limit.

In the literature, there are many ways to estimate the fatigue limit using the Staircase results.

2.4.2 Locati method

The Locati method [2.44] is usually applied when there are few available specimens. When using this method, a single specimen is theoretically enough. The aim is the same as with the Staircase's, but the Locati procedure is based on additional physical hypotheses.

The principle of the test campaign is the first part is tested during L cycles at an F stress level. At the end of this step, the level is increased by a stress increment, and the same part keeps being stressed during L cycles. This scheme is used until the failure appears. The equation for the evaluation of the fatigue strength is presented in Ref. [2.44].

This method provides an approximation of the mean and the standard deviation of the failure limit.

2.4.3 Staircase Loreti method

Each part that has not failed before the target number of cycles (censored data) using the Staircase pattern can be carried on with a Loreti procedure, which does not alter the Staircase results. Fundamentally, it is a Staircase procedure, and all of the parameters of the Loreti are based on the Staircase procedure, including:

- The first higher stress level using the Loreti method would be that after the Staircase. The level of the Staircase testing is increased by an increment of d.

• The stress step used in the Loreti method is the same as that used in the Staircase.

When a part does not fail after N cycles, it is tested at an increased stress level during L cycles, and so on with increased levels until failure. The fatigue limit distribution is then calculated in the locate method.

2.4.4 Methods of analysis by numerical simulation

The methods of analysis by numerical simulation is used in order to assess the efficiency of the previous methods. The numerical simulations are built up using this approach. Using this method provides a means of comparing the testing alternatives with respect to the quality of the estimation and take into account the cost and the time needed for every test plan.

The authors [2.44] proposed to generate samples, i.e., tests results based on

- A theoretical fatigue limit distribution;
- Some hypotheses, such as the number of parts and then using a priori distribution law.

Samples are simulated for the three methods, with sample size ranging from seven up to 1000 specimens. The computer simulation is designed to analyze up to 1000 discrete events for each procedure, providing up to 1000 individual results in order to provide a distribution of calculated averages and standard deviations for each sample.

When using this method, the simulations are first made according to "ideal" test conditions to check the convergence, which means a starting stress level equal to the fatigue strength average, and a step stress equal to the fatigue strength standard deviation. For methods using the Locati procedure, "ideal" conditions mean that the Basquin parameter is equal to the right hypothetical value.

Then the second process is to test any bias for each procedure, wherein the results are checked with offset test conditions, such as:

- The starting stress level is different from the true average;
- The step stress value is different from the true standard deviation;
- Using an erroneous Basquin's parameter value.

In the sequel to this paper, each chapter relates a specific application of the numerical simulation analysis to a testing procedure.

2.4.5 The numerical simulation of the Staircase

As was written in Ref. [2.44], by plotting the mean of each parameter estimate (mean and standard deviation) as a function of the sample size, it is possible to assess the convergence of this method.

The Staircase method converges the error trends toward zero as the sample size increases for the mean and standard deviation of the fatigue limit. But for very small samples, this method gives acceptable results for mean estimates only.

For the estimation of the standard deviation, the 3D analysis shows that for large samples, there is no impact on the estimation so long as the step is lower than twice the theoretical standard deviation. Beyond this limit, the standard deviation is overestimated, regardless of the first stress level. For smaller steps, the estimation is good only for a step value approximately equal to twice the theoretical standard deviation. When the step moves away from the value, the error on the standard deviation estimate increases or decreases, respectively, regardless of the first stress level.

2.4.6 The numerical simulation of the Locati

With the Locati method, there is no impact of the first stress level on the mean and the standard deviation estimation, while in the Staircase procedure, there is a slight impact for small samples.

Moreover, we still notice the same underestimation on the standard deviation estimates, whatever the first stress level.

Step stress effect.

The step stress size has no impact, either on the mean or the standard deviation estimates, whereas with a Staircase procedure a step that is twice the theoretical standard deviation provides a better estimation When using the Locati method the underestimation of the standard deviation is the same regardless of the size of the step.

2.4.7 The numerical simulation of the Staircase Locati

As with the Locati simulation, the Staircase-Locati method requires a fixed Basquin parameter.

Convergence. The Staircase-Locati method was proven to be a convergent method for the mean and standard deviation providing estimates, that are as acceptable as the Locati ones. The authors observed an average error on the mean of less than 0.2% and an average error on the standard deviation of about 7% for the most critical case, which was with seven samples.

From the Locati results, we observe can observe that there is no impact of the first stress level, both on the mean and standard deviation estimation, while there is a slight impact with a Staircase procedure for small samples.

Conclusion derived concerning the above methods

The Staircase-Locati and the Locati method are found to be insensitive to both the first stress level and the step stress. Thus they appear to be more robust than the basic Staircase method.

The mean and standard deviation calculations provided by these methods, respectively, give better estimation results for small samples (e.g., Seven sample size).

To determine the quality of estimation provided with a Staircase-Locati and Locati testing procedures, a complimentary analysis to further evaluate the Basquin parameter is required.

Another approach to fatigue accelerated testing can be found in Ref. [2.45]. It also appears in the *SAE International Journal of Vehicle Dynamics, Stability, and NVH*.

The authors wrote that this approach could evaluate the life characteristics of the vehicle by testing fatigue failure at higher stress levels within a shorter time period. Traditional laboratory testing of this type uses a rigid fixture to mount the component onto the shaker table. But, this approach is not actually accurate for durability testing as most of the vehicles, especially for those parts directly connected to the tire and suspension system. In this work, the effects of the elastic support on modal parameters of the tested structure, such as natural frequencies, damping ratios, and mode shapes, as well as the estimated structural fatigue life were studied through experimental testing and numerical simulations. First, a specially designed subscaled experimental testing bed with both rigid and elastic supports was developed to study the effects of the additional elastic support and the mass on the change of structural modal parameters. The significant modal parameter variation due to the additional elastic support was illustrated in the experimental results. Moreover, the modal parameters with elastic support were then used to build and tune the finite element model (FEM).

Afterward, the results of the accelerated testing profiles of both sine sweep and random vibration were applied to the FEM to compare the deviation of the cumulative fatigue damage between the tested structures with elastic and rigid supports. This work related and expanded the inaccuracy of the current accelerated durability testing system using a rigid support foundation, which introduces a significant amount of variation in fatigue damage as compared to that experienced with the elastic foundation for both sine-sweep and random loading conditions. The dynamic properties of the tested structure with rigid support were different from the real situation.

2.4.8 Fatigue testing for aircraft and satellites

Reference [2.46] provides another discussion regarding the approach to fatigue testing in aircraft. Fatigue testing of aircraft, components, and elements should not, and will not, change in the foreseeable future; other than that, there may be more required with the introduction of 3D printed parts. There is no way that the computer models that we have today will be sufficiently accurate to account for secondary stresses, embedded flaws in manufacturing or 3D printing, or accelerating damage effects for special processes in product manufacture, such as ion vapor deposition acid ending with a part with poor aluminum for corrosion protection and erroneous cost estimates. Besides, all structural finite element modes require boundary conditions, and although they are useful for bounding a problem by running analysis with and without various constraints, they are no match for a real test in finding the vehicle link. If the emphasis continues to be on safety and protection of the public, then aircraft full-scale fatigue tests need to be undertaken to prove the required performance and life, identity faults and provide a proven method to verify repairs, dispositions, and define inspection intervals, as well as maintenance actions for operators.

Reference [2.47] discussed the testing of satellites. Satellites are fragile, expensive, and unique structures, so vibration testing is vital to prevent their dangerous delivery trips, reducing them to useless, multimillion-dollar pieces of space-junk. The NASA rover *Curiosity*, which is now exploring Mars, also endured expensive vibration testing to ensure it arrived safely and would be operational after surviving the rigors of launch, a supersonic descent, and a 9g shock when its parachute deployed.

The testing of satellites is fraught with hazards and must be conducted as safely as possible to ensure that no damage occurs to the satellite, the launch vehicle, or the involved terrain. What's more, repeating a test after failing to capture data is simply not an option due to time restrictions and the fragility of the test object, making test reliability essential so that all necessary data is recorded in one test.

For mechanical satellite qualification and acceptance testing, the quality of the data is also of paramount importance. In these tests, first, the engineering model and finally the actual flight model of the satellite are placed on large shaker systems and precisely vibrated to check that the satellite matches up to its CAE design in terms of structural and modal performance. The flight model is also tested for its response to shocks and the structural resonances that can cause damage during the stressful launch phase. Typically, many hundreds of channels of vibration test data are acquired at high sampling rates.

Vibration test applications for satellite qualifications include acoustic fatigue, transient, random, and swept-sine analysis. The acoustic fatigue tests are carried out in large reverberation chambers, where extremely high-level sound excites the satellite, and its response is measured. The random and swept-sine tests are made on shakers to accurately determine the structural properties of the test object. To simulate the short shock transients experienced during launch, and the long transients, such as when solar panels are deployed, the response to the pyroshocks of the satellite and its subsystems must also be determined.

Qualification and acceptance testing are often performed at special facilities that hire time to the satellite manufactures. Such facilities benefit from an integrated test data stream that fulfills different functions at different levels, including displaying real-time results to clients. The measurement data is digitized immediately and piped around the facility over a LAN, allowing many users to access the same data stream as required. As well as recording large amounts of data to disk, remote LAN-based workstations enable simultaneous level and time-level and time-signal monitoring on all channels.

The complete test workflow is organized from a dedicated control console, where the test engineers plan the test, set up the data acquisition and the required analyses, calibrate the system, monitor the recording, initiate postprocessing, visualize the results, create reports, and archive data. Dedicated qualification test software guides them logically through the workflow with a clear user interface that controls and coordinates the many applications.

2.5 Vibration testing

The real vibration of mobile products is in 6° of freedom, as was demonstrated by the MTS Corporation (see Chapter 5 in Ref. [2.13]). This is visually depicted, and when using vibration testing, one needs to understand that the real-world vibration of mobile products is a result of the action shown in Fig. 4.23 in Ref. [2.13]. This includes:

- Features of the road, including the type of road (concrete, asphalt, cobblestone, earth, road profile—surface, density, and other properties);
- Test subject speed;
- Wind speed and direction;
- Air density and humidity fluctuations;
- Design and quality of the wheels and their coupling with the road surface;
- Design and quality of the whole product;
- And others.

The above factors affect stiffness, damper, elastic, and inertia qualities of the product that finally leads to the actual vibration characteristics of the mobile product.

HORIBA Automotive Test Systems is a leading supplier in the fields of engine test systems, driveline test systems, brake test systems, wind tunnel balances, and emissions test systems.

Figs. 2.8—2.10 demonstrate some of their shakers for the automotive industry.

The mechanical vibration facility used by the NASA Glenn Research Center is described in Ref. [2.48].

The facility consists of vibration table of components that include a horizontal shaker, a vertical shaker with spherical couplings, and a 20' modal floor. MVF is capable of 480,000 lbf vertically and 170,000 lbf in each lateral direction. The MVF requirements make it a higher capacity facility than any in existence—providing 50% greater payload capacity, 25% greater vertical force capacity, and 50% higher frequency range than ESTEC, the current larger capacity (aerospace) vibration system.

FIGURE 2.9   Horiba's vibration equipment. *The author received this photo from Horiba.*

FIGURE 2.10   Horiba's shaker with a system of control. *The author received this photo from Horiba.*

The horizontal vibration subsystem—actuator assembly (HAA) consists of:

- Horizontal actuator drives table horizontally;
- Composed of Horiz actuator and pad bearing (Horiz actuator mounting pedestals not Shawn);
- Pad bearings guides table vertically.

The Vertical Actuator assembly (VAA) consists of:

- Vertical actuator drives vertical vibration;
- Composed of vertical actuator and spherical coupling;
- Spherical coupling permits horizontal motion;
- Spherical coupling prestrains overturning moments (with vertical actuators locked down).

Andre Beltempo, of the National Research Council of Canada (NRC), and Dr. Waruna Seneviratne, of the National Institute for Aviation Research (NIAR), discussed in Ref. [2.58] how the cross-coupling compensation utility within the AeroPro Control and Data Acquisition Software can be used to improve the speed, accuracy, and efficiency of the most complex aircraft structural tests:

"Full-scale structural test rigs that feature highly cross-coupled actuation schemes present unique challenges for aerospace test engineers. In these complex configurations, multiple actuators exert different forces on the same portion of a test article simultaneously, raising the potential for errors that can jeopardize test schedules and delay development programs.

One tool for reducing such errors—thus improving test speed and efficiency—is C^3 Performance, a utility of MTS AeroPro Control and Data Acquisition Software. Also known as C-cubed, C^3.

Performance enables test teams to manage highly cross-coupled actuation schemes effectively, without having to build extra time into test schedules to resolve numerous, recurring stops and interlocks."

The C^3 Performance cross-coupling compensation technique was code-veloped by experts from the National Research Council of Canada (NRC) and MTS Systems Corporation. Today, it is a standard part of the testing protocol at NRC—Canada's full-scale structural test lab, where Andre Beltempo works as a structural test engineer.

"Typically, we use C-cubed on all tests," said Beltempo. "With full-scale fatigue tests, you see very complex actuator interaction. That's exactly why we need it."

Beltempo's team recently used C^3 Performance during a technology demonstration for a major helicopter manufacturer. The fatigue and static test program focused on a composite tail boom, fabricated using advanced manufacturing techniques. The tests were identical to those that would be used to certify the part for safe life tolerance FAA requirements.

The test involved six actuators and four million end points, running overnight for approximately 125 days, with an average frequency of 0.5 Hz. This test speed would have been impossible without C^3 Performance, according to Beltempo.

"We could not have done it as quickly without cross-coupling compensation," he said. "For an overnight test, we would have had to run the end levels at half the speed to not have to worry about unattended shutdowns. Our standard approach for any fatigue test is to tune the test, apply C-cubed, and tune it again right away. I've never seen a speed increase of less than a factor of two for a rigid article."

"The fact that it's so hands-free gives me the ability to focus on more important tasks," Beltempo said. "If you had to sit there and figure out all the coefficients on your own, it would be different. But you just click a button and it happens. It's a real productivity improvement."

While it is understandable that the organization that helped develop C^3 Performance, would apply it to every fatigue test as a matter of protocol, other labs are also discovering the problem-solving potential of this powerful utility.

For Dr. Waruna Seneviratne, a technical director in the Composites and Advanced Materials Lab at Wichita State University's National Institute for Aviation Research (NIAR), the problem was, testing causing staff sleep deprivation.

"We were running a fatigue test with a very distinctive combination of custom fixturing and actuation, and we had a very aggressive test schedule," Seneviratne said. "The test ran overnight and we were experiencing a large number of nuisance error limit triggers. Our team had to come down to the lab in the middle of the night to see what was causing test shutdowns and to restart the test. We were falling behind."

The test focused on evaluating the fatigue life of composite structures of F/A-18 Hornet aircraft, many of which are approaching retirement. Because the aircraft's replacement will not be ready until 2019, the U.S. Navy needs to safely prolong the service life of the aircraft's aging composite structures [2.58].

Seneviratne's research successfully determined that the composite-to-titanium bonded joints at the wing root had a great deal of life left in them. The new test expanded the study to include the entire inner wing, which has a composite skin. The test article includes the inner wing, trailing-edge flap, and center barrel, as well as simulated leading-edge flap and outer wing. NIAR's research team had to construct an elaborate high-strength steel rig with custom-designed fixturing and apply significant loads to recreate aggressive maneuvers.

"We were running 20% faster by the end of the first day with no mechanical modifications to the test rig," Cravens said. "C^3 Performance was very easy to learn and set up; it all occurs within the software. You simply create a cross-coupling matrix by applying a unit load on each load channel, which represents just one extra step for each actuator."

To streamline the test setup, C^3 Performance eliminates the time-intensive task of manually inputting cross-coupling data by employing unit load cases to generate automated cross-coupling compensation coefficients. Seneviratne characterized this process with an apt analogy.

With some extra tuning, Seneviratne and Cravens were able to increase the test rate by 24% and cut the number of stops significantly.

"It has greatly improved the performance of our test," Cravens said. "With the improved load tracking, small perturbations in feedback are much less likely to trigger error limits, resulting in fewer test stoppages. The reduction in error has also allowed us to run more segments per hour."

Before using C^3 Performance, NIAR was only able to achieve a maximum of 375 segments per hour, with an average of 97 stops and 55 interlocks per test block. With C3 Performance, this has improved to 480 segments per hour, with an average of 51 stops and 15 interlocks.

"The number of stops was a third of what they were, which was a huge gain for us," Seneviratne said. "That allowed us to run the test overnight, some nights without a single interruption. That's up to 10 h of testing we didn't have before, which helped tremendously with our schedule."

C^3 Performance saved weeks, enabling Seneviratne's team to provide the test results on time. In addition, lab productivity was enhanced: the team was able to operate on a more predictable schedule and had more time to devote to other projects. In any lab with finite resources, that is an important advantage.

But, as the author of this book will describe later, and as it has been described in many of his other publication(s [2.10,2.11,2.13], and others), vibration testing and fatigue testing are often wrongly called durability testing. As can be seen from the definitions, published in Refs. [2.10,2.13], and many others, fatigue testing by itself is not durability testing of the product.

2.6 Crash testing

Crash testing is considered separately in this subchapter because it relates to all—second, third, and fourth general directions of accelerated testing development.

Crash testing is a form of destructive testing usually performed in order to ensure safe design standards in crashworthiness and crash compatibility for various modes of transportation or related systems and components.

Crash testing is used for vehicles to help reduce losses such as deaths, injuries, and property damage from vehicle crashes. In the USA, one organization responsible for ensuring the correct undertaking of this task is the Insurance Institute for Highway Safety (IIHS), a nonprofit, independent scientific body, and educational organization. Another organization is the Highway Loss Data Institute, which supports the mission of the IIHS. Auto insurers support both organizations.

According to the information contained in Ref. [2.49], airbags are one of the most important safety innovations of recent decades.

They provide crucial cushioning for people during a crash. The devices are normally hidden from view but inflate instantly when a crash begins. Frontal airbags have been required in all new passenger vehicles since the 1999 model year. Side airbags are not specifically mandated, but nearly all manufacturers include them as standard equipment in order to meet federal side protection requirements.

Frontal airbags reduce driver fatalities in frontal crashes by 29% and fatalities of front-seat passengers age 13 and older by 32%. Side airbags that protect the head reduce a car driver's risk of death in driver-side crashes by 37% and an SUV driver's risk by 52% [2.49].

Some vehicles now have rear-window curtain airbags to protect people in back seats or front-center airbags to keep drivers and front-seat passengers from hitting each other in a crash. There are also inflatable safety belts aimed at reducing rear-seat injuries.

A crash test for vehicle safety is a type of destructive testing undertaken to ensure that standards for safe design with regards to crash compatibility and crashworthiness are followed for different transportation modes. There are multiple kinds of crash tests for vehicle safety undertaken to provide the necessary information and guidance to vehicle owners. Examples of crash tests are frontal impact tests, an offset test, side-impact tests, a rollover test, and roadside hardware crash tests.

Frontal impacts are test impacts against a solid concrete wall at a precise speed. SUVs are singled out from frontal impact tests. An offset test requires only a portion of the front of the car to impact the barrier or a vehicle. An offset test is important since impact forces in this type of test remain the same as those with the frontal impact test, but it is necessary for a small portion of the car to absorb all the force of the impact.

Side-impact tests as a crash test for vehicle safety are also important since side-impact accidents in vehicles result in a high fatality rate. This happens because cars usually do not have a significant crumple zone to cushion all the impact forces before the occupant is injured.

A rollover test verifies the car's ability to support itself, especially from the pillars supporting the roof, during a dynamic impact. Roadside hardware crash tests ensure that crash barriers and crash cushions protect the passengers of the vehicle from roadside hazards. This kind of crash test also makes sure that some appurtenances, such as signposts, guard rails, and light poles, may not serve as a hazard to vehicle occupants.

Many different crash test programs are practiced around the world, and all are dedicated to providing vehicle owners and drivers with data regarding the safety performance of new and used vehicles. These test programs provide safety performance based on real-world crash data [2.49].

The main advantage of having car airbags for passengers is that they provide an additional level of protection in the event of a car accident. This added protection can be the difference in some circumstances between life and death.

Risks of Airbag Deployment versus Risk of Injury or Death

Although some question the overall safety of vehicle airbags because of the instances of chest injuries and other impact injuries due to the nature of an airbag's deployment, these risks are lower than the risk of serious injury or death that may occur if the passenger is unprotected.

Disadvantages of Having Passenger Car Airbags

Unfortunately, these safety tools can have some significant liabilities.

Potential Injury

The biggest negative to airbags is that, although they are designed to protect, deploying airbags can injure occupants in some situations. The impact of a deploying airbag can injure a passenger who is improperly positioned. Deployment injuries can be most harmful to children and infants. Types of injuries from airbags include chest injuries, concussions, and whiplash.

Resetting Airbags

After airbags have been deployed, they may be difficult to reposition for the next deployment. You may spend a substantial amount of money at a shop getting new airbags after a deployment if there is only one occupant in the car, and multiple airbag deployments can be very expensive.

Virtual Dummies

The earliest automotive crash tests were morbid, often messy affairs. In the 1930s, in order to simulate the effects of high-speed collisions on drivers, researchers took human cadavers and subjected them to head-on crashes and vehicle rollovers. Anthropomorphic test devices (ATDs) emerged, and next, those gleaming, faceless crash test mannequins, which were then followed by computer simulations.

While early ATDs were able to provide data on around 20 points of the body using a mesh of accelerometers, force sensors and strain gauges, today's crash test computer simulations monitor the effects of a huge varieties of crashes to a far greater degree of accuracy across a wide range of human body types, ages, and driving positions.

Their main benefit is legibility [2.50]. In traditional crash tests, researchers must anticipate where to best place the cameras inside the vehicles to monitor what happens during impact.

Toyota has been at the leading edge of innovation in the crash testing field since the late 1990s when, in cooperation with Toyota Central R&D Labs [2.50], it began the development of its inaugural virtual crash test dummy, which is called Total Human Model for Safety or THUMS. Over the years, THUMS has been involved in thousands of virtual crashes, while slowly gaining new abilities. In 2004, it gained a face and bone structure. Then the 2006s version, through evolution, added precise modeling of the brain in order to see how it might be affected in a variety of crash scenarios.

The fourth THUMS iteration added detailed modeling of internal organs and the current version launched in 2015, added the muscle model. The result is a digital model that contains no fewer than 1.8 millions of elements that combine to reproduce the human form, from precise bone strength to the structure of organs, which can be used to evaluate injuries to both soft and bone tissue.

The THUMS is currently used in safety research by other major carmakers, including Audi, Volvo, Renault and Daimler, and numerous components suppliers. NASA has used this approach in the design process for Orion, a spacecraft that could lead the way for taking humans to Mars [2.50]. Researchers at Toyota continue to upgrade the above software.

Virtual crash test dummies are being used in new and expanded roles in testing. But while these virtual crash dummies may seem to have supplanted their physical testing rivals, they cannot take into account all the variances associated with advanced physical accelerated testing, especially for successful prediction of how all of the components of the product perform in a crash test [2.10].

While crash test dummies have been the subject of public service announcements, cartoons, parodies, and even the name of a band, real crash test dummies are true life-savers as an integral part of automotive crash testing. And, although cars get safer each year, and automotive-related fatality rates are declining, car crashes are still one of the leading causes of death and injury in the United States.

Some of the crash test definitions and short description from Wikipedia appear below:

- **Frontal-impact tests**: which is what most people initially think of when asked about a crash test. Vehicles usually impact a solid concrete wall at a specified speed, but these can also be vehicle impacting vehicle tests. SUVs have been singled out in these tests for a while, due to the high ride-height that they often have.

- **Moderate Overlap tests**: in which only part of the front of the car impacts with a barrier (vehicle). These are important, as impact forces (approximately) remain the same as with a frontal impact test, but a smaller fraction of the car is required to absorb all of the force. These tests are often realized by cars turning into oncoming traffic. This type of testing is done by the U.S.A. Insurance Institute for Highway Safety (IIHS), EuroNCAP, Australasian New Car Assessment Program (ANCAP) and ASEAN NCAP.

- **Small Overlap tests**: this is where only a small portion of the car's structure strikes an object, such as a pole or a tree, or if a car were to clip another car. This is the most demanding test because it loads the most force onto the structure of the car at any given speed. These are usually conducted at 15%−20% of the front vehicle structure.

- **Side-impact tests**: these forms of accidents have a very significant likelihood of fatality, as cars do not have a significant crumple zone to absorb the impact forces before an occupant is injured.

- **Rollover tests**: which tests a car's ability (specifically the pillars holding the roof) to support itself in a dynamic impact. More recently, dynamic rollover tests have been proposed in lieu of static crash testing.

- **Roadside hardware crash tests**: are used to ensure crash barriers and crash cushions will protect vehicle occupants from roadside hazards, and also ensure that guard rails, signposts, light poles, and similar appurtenances do not pose an undue hazard to vehicle occupants.

- **Old versus new**: Often an old and big car against a small and new car, or two different generations of the same car model. These tests are performed to show the advancements in crashworthiness.

- **Computer model**: Because of the cost of full-scale crash tests, engineers often run many simulated crash tests using computer models to refine their vehicle or barrier designs before conducting live tests.

- **Sled testing**: A cost-effective way of testing components, such as airbags and seat belts, is conducting sled crash testing. The two most common types of sled systems are reverse-firing sleds, which are fired from a standstill, and decelerating sleds, which are accelerated from a starting point and stopped in the crash area with a hydraulic ram.

Another leader in the field of safety research for automotive applications is Mercedes-Benz. At the end of the 1950s, Mercedes Benz began practical testing for safety research purposes [2.51]. Initially, individual components were tested by means of impact tests, for example, but entire systems were tested, by means of impact. Later impacts were also tested for components, such as the seat belt system, which became available in 1958. In 1959, spectacular crash testing was started using Mercedes-Benz vehicles as the basis for safety research. For these systematic crash tests, the test vehicles were first accelerated by means of a towing system, such as those used to launch gliders. With the Mercedes-Benz towing unit, sedans fresh from the assembly line could be launched into the air. This was necessary because right from the start of crash testing at Mercedes- Benz, the engineers did not just simulate collisions by running vehicles into a fixed barrier, they also simulated rollovers. For achieving this, the test vehicles were run at a speed of 75–80 km/h on to a so-called *corkscrew ramp*, which gave the automobiles the necessary twist so that they lifted off into midair and landed on their roofs. These tests led to the installation of stabilizing structures in the bodywork. From 1973 onward, crash tests became possible at the new test center in Sindelfingen, Germany. On the 65- meter acceleration track, a linear motor producing a thrust force of 53,000 N, accurately pulled the automobiles into a 1000-ton barrier, which rested on a very sensitive force measuring platform. This crash test facility was thoroughly renovated in 1998 with the creation of the Mercedes-Benz Technology Center (MTC). At an expense of 2.3 million euros, the facility was refurbished with state-of-the-art technology. The length of the acceleration track was increased to 95 m; as a result, all types of crash test variants were now possible at the facility. In particular, this included offset crashes, in which only part of the frontal width of the vehicle impacts the obstacle, and which in reality occur much more frequently than head-on vehicle collisions. The test sequences are no longer recorded by a high-speed film camera, but by video technology. And, with video, the very high frame rate has been maintained so that the crash tests can be evaluated in extremely slow motion. The facility was also equipped with a roof during the renovation so that passenger car and commercial vehicle testing can now be performed regardless of the weather.

From the beginning of crash testing at the company, not only vehicles were used to assess the effect of crashes, but measuring instruments in the head and chest of dummies provided information on the loads acting on the driver during an accident. Sandbags and mannequins initially took the place of the front passenger; soon, dummies were also collecting crash test data for the front passenger seat and rear bench seat. Individual dummy designs were used to measure specific injuries and to represent persons of different build and age. Increasing computer capabilities then allowed dummies to be replaced by mathematical multibody systems. The first digital crash computations with overall vehicle models were performed for the E-Class of the 124 series.

One aim of Mercedes Benz crash testing is to accurately represent real-world crash scenarios. Consequently, the head-on collision is increasingly being replaced by the offset collision. In 1992, an offset frontal collision was performed against a deformable barrier for the first time, providing results that corresponded even more closely to the behavior of a vehicle in a real-world accident. One outcome of this testing was a deformable barrier that was developed for this type of experiment in Europe; its design decisively influenced by test results from the Mercedes-Benz Safety Center in Sindelfingen. After the introduction of the offset barrier test, this new European test procedure represented another big step toward crash testing. In 1993, the offset crash against a deformable barrier made from metal honeycomb material at 60 km/h and with a 50% overlap became the new Mercedes-Benz standard.

The test vehicles also play an important role in safety research. From 1971 to 1974, Mercedes-Benz took part in the international ESV (Experimental Safety Vehicle) program. The aim of this project was to improve passenger cars in line with the safety criteria of the U.S. National Highway Traffic Safety Administration. The specifications called for a lower risk of passenger injury in the following cases:

- A head-on impact;
- Against a fixed barrier at 80 km/h, a head-on impact against another vehicle at 120 km/h, a side impact by;
- Another vehicle at 50 km/h, a rear-end collision by another vehicle at 120 km/h, and a rollover.

The developments were verified by means of actual crash tests.

Following the ESV era, Mercedes-Benz continued to use concepts and test vehicles to improve safety technology.

One example was the Auto 2000 research car, unveiled to the public in 1981, which was used to test seats with integrated belt anchors, an integrated seat child module, and with pedestrian-friendly bumpers. The Mercedes-Benz Development unit also set standards in the field of commercial vehicles with its test and concept vehicles, such as the 2004 safety study, which was based on the Sprinter van. This concept vehicle was distinguished by low-reaching windows and roll stabilization. In 2006, the Actros Safety Truck followed, although not as a design study. The vehicle, fitted among other things with Active Brake Assist, an emergency braking assistant, was made available on the market.

A helicopter crash test program was conducted by NASA, and the U.S. Navy, the U.S. Army, and the FAA hope to shed new light on seat safety and lightweight composites, while also exploring the value of new testing methods [2.52]. Engineers at NASA Langley's Landing and Impact Research Facility (LangD1R), in Hampton, Virginia, sought to improve the crashworthiness of seats and seatbelts, as well as to gather relevant data on the odds of surviving a helicopter crash For this program, the U.S. Navy provided the CH-46E Sea knight helicopter fuselage, complete with seats, which was \then fitted out with 15 "occupants"—13 instrumented crash test dummies and two instrumented mannequins. The Navy also contributed five of these crash test dummies, one mannequin and other equipment, while the Army provided a mannequin and a crash test dummy that was placed in a position representative of a patient in a medical evacuation litter. The FAA provided a side-facing specialized crash test dummy and part of the data acquisition system. NASA Langley added six of its own dummies, as well as being the leader in technical expertise, and provided the use of its own specialized facility, known as the Gantry. Engineers then used cables to hoist the helicopter fuselage into the air and swinging it above the ground like a pendulum. It was traveling at a speed of 30 mph when pyrotechnic devices separated the cables, sending the fuselage smashing into the soil below. The test article was fitted with 350 sensors to capture data on airframe accelerations and crash and test dummy loads. Over 40 high-speed and high-definition cameras recorded onboard and external movements.

Researchers also made use of a new photographic method available to them to help analyze the data collected from the crash test. Called "full field photogrammetry" it saw the helicopter fuselage which had been stripped of its usual coat of naval gray point in favor of camera catching paint scheme. "We painted more than 8000 dots on the side of the test article to measure global and local deformation on the fuselage skin," explained Annett, a leading expert in test filming with high-speed cameras. At 500 images per second, this was used to track each dot, ensuring researchers were able to plot and "see" exactly how the fuselage behaved under crash loads. (Incidentally, a similar approach was used by the author of this book in the 1980s for automobiles accelerated reliability testing. This experience was published by Wiley in the author's book *Reliability Prediction and Testing Textbook*, as well as in Refs. [2.11,2.53], and his other publications.)

Finally, when asked how testing could be further improved, Annett confirms the "the test, test, and test again" mantra of all test engineers "Frequent testing is always recommended, as there are always lessons to be learned with every crash test," he says. Progress toward safer travel is fragile. It can be disrupted by an ill-drafted law or by the exploitation of a loophole in existing regulations, allowing safer vehicle testing than they could be on the public roads. It is not always easy for authorities to remedy these failings quickly, but stopgap actions can help to minimize the potentially fatal consequences. The crash testing of heavy quadricycles is an example of promoting safety in spite of flimsy rules or flawed legislation (quadricycle refers to vehicles with four wheels). The first quadricycle crash tests by EuroNCAP were performed in 2014 with three electric and one petrol model. These vehicles were type-approved and met the minimal safety requirements set by European legislation. Nevertheless, the results were not encouraging [2.54]. A statement released by Euro NCAP at the time of the tests reads, "All of the quadricycles tested showed critical safety problems, although some fared better than others in the front or side impact tests." This was followed by a clear warning. "Consumers, however, should note that quadricycles in general offer a significantly lower level of occupant protection than is offered by cars" [2.54].

Despite the cataloging of weaknesses revealed by those initial Euro NCAP crash tests, quadricycle sales and marketing have continued in the 2 years since, with two models being launched and at least one old model being renamed. This year another four quadricycles were put through the same special crash tests to see if there has been any improvement and whether the drivers and passengers in these vehicles are getting better protection.

The test protocol is not as involved as for normal full-sized cars, and the star ratings are based on completely different criteria. It was the frontal test, which is a full-width test, with two front seat dummies.

Another difference from the crash tests on full-sized cars is the choice of dummies for side impacts. For normal cars, the newer WorldSID dummies are used, but the first quadricycle tests used the EuroSID 11 for all sides impact tests, and so they have stayed with the EuroSID 11, to keep the same level for these latest tests.

The vast majority of the data from the quadricycle tests comes from the dummies, and as there are fewer of them than in a normal car crash test, fewer cameras are needed.

All internal combustion engine quadricycles weigh around 450 kg or less. The three labs conducting the quadricycle tests are designed for crash testing vehicles up to 10 tons so that for that weight, they did not need special or modified equipment.

Let us discuss one more approach to crash testing. The Center for Advanced Product Evaluation (CAPE) in Westfield, Indiana (USA), is a unit of the advanced vehicular safety systems manufacturer (MMI). The center designs and builds test rigs that help to determine whether there is survivable space inside the vehicle and whether the vehicle's body-to-frame monitoring system is sufficient to withstand a rollover incident [2.55]. The testing performed at CAPE is typically designed to prove that the manufactured vehicle complies with standards set by different organizations.

CAPE completed the development of a test rig that can provide vehicle OEMs with roof crush testing up to 100 tons. It can be also used to test off-road vehicle roll cages and race car chassis. The rig uses four hydraulic actuators (cylinders), mounted at the four corners of a heavy-gage pressure plate, and controlled as four separate motion axes. At the core of this system is an eight-axis RMC150 electrohydraulic motion controller manufactured by Delta Computer Systems. CAPE used a special function of the RMC151 motion controller called "Virtual Gearing" to cause all four axes to move in precise synchrony to ensure that the pressure plate is kept completely level during a compression operation. The four "slave" axes follow a virtual "master" axis, which is set up to control the position of the pressure plate and the cumulative force applied in the testing. The typical compression cycle works as follows [2.55]. The hydraulic pump is turned on, and the transducers are initialized to zero values. Then the four compression cylinders are set up to be geared together, and the system is given a command to move the steel pressure plate up and out of the way. The vehicle cap/body is placed in the rig, and the pressure plate is lowered until it reaches a position that is just above the cab but not touching it. The command is then given to preload the rig to 500 lb, followed by the command to apply the full load, a process that takes between one and 5 minutes. The system is allowed to rest under load for 30 s, and then it is unloaded to zero pounds on the load cells. Finally, the press plate is moved completely off the cab, and the test data is downloaded from the motion controller to the network drive over the RMC's Ethernet interface.

Programming the motion steps was done using PMCTools software, provided with Delta's motion controllers. It enables programming the controllers using high-level commands, such as the Virtual Gearing arrangement. During testing operations, the Delta motion controller in the CAPE test rig performs the data acquisition and maintains all the test data internally. The PMCTools software can handle test system operator interface functions and data transfer to an attached PC. Unfortunately, all of these current types of crash testing generally do not take into account the human factors that are one of the basic reasons for real-life crashes. This author's books [2.11,2.13] cover what kind of human factors, and how to take them into account during crash testing, to human factors that are not taken into account during crash testing related to lighting problems on the road, especially in big cities.

Bibliography

[2.1a] Frank Murray S, Heshmat Hooshang, Fusaro Robert. Accelerated testing of space mechanisms. MTI Report 95TR29. April 1995.

[2.1]   Standard IEC 62506 Ed. 1.0 B: 2013: methods for product accelerated testing.

[2.1A] Masterson P. Liberty mutual insurance. Car's.com. January 7, 2019.

[2.1B] Jibrell A. Auto recall bill grew 26% to $22 billion in 2016, study says. Automotive News January 30, 2018.

[2.1C] Ewing S. Automotive recalls cost $22 billion in 2016. That's a 26 percents increase over the previous year. January 31, 2018 [Road show. Car industry].

[2.1D] Vortabe R. After second recall, Toyota Prius electrical system still overheating. CRS; April 14, 2019. The Twiliynt Zone.

[2.2]   Statistics Portal. Statista 2018.

[2.3]   U.S. CPSC — total civil penalties issued from 2006 to 2018.

[2.4]   Ridella, SA. NHTSA. Office of Safety Traffic Research, September 20, 2012. Springfield, VA.

[2.5]   Andrews W, Aisch G. A record year for auto recalls. New York Times December 30, 2014.

[2.6]   Shane D. Exploding airbag crisis in Australia: 2.3 million vehicles recalled. CNNMoney February 28, 2018. 1:10 AM ET.

[2.7]   BMW is recalling 1.6 million vehicles worldwide over potential fire risk. Time October 23, 2018.

[2.7a]   Isidore C. Ford recalls 1.4 million cars because steering wheel can come off. CNNMoney March 14, 2018.

[2.8]   For Hyundai and Kia, Risk Goes Beyond Fire. Hard-won quality reputations at stake. Automotive News October 20, 2018.

[2.8a]   Auto Recalls. NHTSA should take steps to further. Report to Congressional Committees. December 2017. https://www.gao.gov/assets/690/688714.pdf.

[2.8a-1] Panait M. Subaru extends JDM recall over new cases of inspection cheating. Autoevolution November 5, 2018. https://www.autoevolution.com/news/recalls/.

[2.8a-2] Patrascu D. Ford recalls 1.5 million focus models in North America. Autoevolution October 25, 2018. https://www.autoevolution.com/news/recalls/.

[2.8a-3] Patrascu D. BMW EGR fire recall Grows to 1.6 million cars globally. October 23, 2018. https://www.autoevolution.com/news/recalls/.

[2.8a-4] Penait M. Toyota recalls millions of hybrid vehicles. October 7, 2018. https://www. autoevolution.com/news/recalls/.

[2.8a-5] Panait M. Shanghai GM recalls more than 3.3 million vehicles in China. September 30, 2018. https://www.autoevolution.com/news/recalls/.

[2.8b]   Lion Air jet had same airspeed problem on last 4 flights. http://a.msn.com/01/en-us/ BBPm3kX?ocid=se.

[2.8c]   Beech H, Bradsher K. At doomed flight's helm, pilots may have been overwhelmed in seconds. New York Times November 8, 2018.

[2.8d]   Vlasic B, Stout H. Auto industry galvanized after record recall year. New York Times December 30, 2014.

[2.8e]   Koenig D. Boeing's troubled jet is costing $ 1 billion to fix so far. The Associated Press; April 24, 2019.

[2.8f]   Gameron D, Tangel A. Boeing sees more 737 costs. Wall Street Journal April 18, 2019.

[2.9]   Shepardson D. Editing by David Gregorio. Reutors. Senators to press automakers, regulators on Takata air bag recall. New York Times March 20, 2018.

[2.10]   Klyatis LM, Anderson EL. Reliability prediction and testing textbook. Wiley; 2018.

[2.11]   Klyatis LM, Klyatis EL. Accelerated quality and reliability solutions. Elsevier; 2006.

[2.12]   Horiba, MIRA. Vehicle durability testing.

[2.13]   Klyatis LM. Accelerated reliability and durability testing technology. Wiley; 2012.

[2.14]   Mezger S, Deng M. Efficient functional testing of flight attendant panels. Aerospace Testing International; ShowCase 2018.

[2.15]   Honda civic fleet and accelerated reliability testing — July 2005. INL/EXT 06-01262. Energy efficiency and renewable energy. U.S. Department of Energy.

[2.16]   Arguets FJ, Wehrey SJ. Field operations program Toyota RAV4 (NiMH) accelerated reliability testing — final report. ENEL/EXT 2000 — 00100. Idaho National Engineering and Environmental Laboratory Automotive Systems and Technology Department; March 2000.

[2.17]   Hybrid electric vehicle end-of-life testing on Honda insight, Honda gen 1 civics and Toyota gen 1 Priuses. INL/EXT-06-1262.

[2.18]   Birch S. 24 million KM of testing for Mercedes C-class. Automotive Engineering April 2007.

[2.19]   LMS Supports Ford Otosan in Developing Accelerated Durability Testing Cycles. http://www.lmsintl.com/LMS -Ford-Otosan-developing accelerated-durability-testing-cycles.

[2.20]   Mezger S, Deng M. Efficient functional testing of flight attendant panels. Aerospace Testing International; ShowCase 2018.

[2.21]   Ralph D. Kimberlin: flight testing of fixed-wing aircraft. AIAA Education Series 2003.

[2.22]   Gold I. Waiting for the 'bus'. Aerospace Testing International September 2014.

[2.23]   Mezger S, Deng M. Efficient functional testing of flight attendant panels. Aerospace Testing International; ShowCase 2018.

[2.24]   Dynamic research, Inc. (DRI).

[2.25]   The promise of virtual testing.MTS System, Inc.

[2.26]   7 layers Co: test engines.

[2.27]   Reilly T. Satellite and spacecraft vibration testing control. Aerospace Testing International SnowCase 2018.

[2.28]   Fatemi SZ, Guerin F, Saintis L. Development of optimal accelerated test plan. RAMS Proceedings 2012.

[2.29]   Flight test programme - EASA − Europa EU. Example document for LSA applicants − v1 of Feb. 17, 2016 - [3] ABCD-GD-00. General Description Document. [4] ABCD-WB-08-00. https://www.easa.europa.eu/.../ABCD-FTP-01-00%20-%20Flight%20test%20program.

[2.30]   Pradeep L, Pecht MG, Hakim E. Influence of temperature on microelectronics. CRC Press; 1997.

[2.31]   Hobbs GK. Accelerated reliability engineering: HALT and HASS. Wiley; 2000.

[2.32]   Introduction to accelerated testing types. ReliaSoft.

[2.33]   Schenkelberg, F. Determine and design the best ALT. RAMS 2012 proceedings.

[2.34]   Condra LW. Reliability improvement with design of experiments. New York: Marcel Dekker; 2001.

[2.35]   Yuan T, Liu X. Bayesian planning of optimal step stress accelerated life test. In: 57th annual reliability and maintainability symposium (RAMS) proceedings; 2011.

[2.36]   Brown P. The industry view from Lansmont corporation. TEST Engineering and Management June/July 2013.

[2.37]   Intertek. www.intertek.com/AST.

[2.38]   Rogers, R. Accelerated life testing (ALT). NTS Detroit Laboratory.

[2.39]   Kyle JT, Harrison HP. The use of accelerometer is simulating field conditions for accelerated testing of farm machinery. ASAE Paper #60-631. Memphis; 1960.

[2.40]   Nevada automotive test center (NATC).

[2.41]   Farris TN, Matlik JF. Comprehensive structural integrity. 2003.

[2.42]   Beaumont, P. Gudrin, F. Lantieri, P. Matteo L. Facchinetti. Borret, GM. Accelerated fatigue test for automotive chassis parts design: an overview. RAMS 2012 proceedings.

[2.43]   Dixon WJ, Mood AM. A method for obtaining and analyzing sensitively data. Journal of the American Statistical Association 1948;43:109.

[2.44]   Boitsov BY, Obolenskii EP. Accelerated tests of determining the endurance limit as an efficient method of evaluating the accepted design and technological solutions. Strength of Materials 1983;15.

[2.45]   Rahman, E. Wu, N. Wu, C. Automotive components fatigue and durability testing with flexible vibration testing table 10-02-01-0004.

[2.46]   Aerospace Vehicle. An overview. https://www.sciencedirect.com/topics/materials-science/aerospace-vehicle.

[2.47]   Bruel & Kjaer Sound & Vibration. Good closed-loop satellite vibrations. Aerospace Testing International December 2012.

[2.48]   Otten KD, Suarez VJ, Le DK. Status of design features of the new NASA GRC mechanical VibrationFacility (MVF). IEST; 2010.

[2.49]   Insurance institute for highway safety. Highway loss data institute. Overview.

[2.50]   Parkin S. Crash test geniuses. Crash Test Technology International September 2016.

[2.51]   https://media.daimler.com/marsMediaSite/.../Crash-testing-for-safety-research.xhtml.

[2.52]   Klyatis L. Successful Prediction of Product Performance. quality, reliability, durability, safety, maintainability, life-cycle cost, profit, and other components. SAE International 2016.

[2.53]   James A. Hit the dust. Aerospace Testing International September 2013.

[2.54]   Edmonts S. Quarcycle testing. Crash Test Technology International September 2016.

[2.55]   Coons B. A secret weapon for roof-crush testing. SAE Automotive Engineering May 2018.

[2.56]   HORIBA instruments − HORIBA. www.horiba.com/us/en/scientific/horiba-instruments/.

[2.57]   Novi, Michigan Automotive testing expo. 2018. https://testing-expo.com/usa/.

[2.58]   MTS. Force & Motion. Aerospace testing. NO. 50. October 2018.

Exercises

1. Describe the basic content of the international standard in accelerated testing.
2. Describe the standard IEC 62,506 in accelerated testing methodology.
3. Why are recalls reliable metrics about product effectiveness?
4. Demonstrate the dynamics of the changes in recall numbers during the last several years in the U.S. market.
5. Describe the basic general directions in accelerated testing usage and development.
6. Describe the first general direction in field and flight accelerated testing and its specific.
7. Describe the second (computer/software simulation) general direction in accelerated testing development.
8. Describe the third (laboratory and proving grounds) general direction in accelerated testing development.
9. Describe the fourth (accelerated reliability and durability testing) general direction in accelerated testing development.
10. Discuss accelerated fatigue testing.
11. Describe trends in accelerated crash testing.

Chapter 3

Developments in studying real world conditions for accurate simulation and successful accelerated testing

Abstract

This chapter considers the groups of input influences that need to be studied for accurate simulation. It considers in detail the multi-environmental (climatic) influences,
why real world simulations are usually not accurate, and why testing results from the
laboratory or proving ground do not correspond with real world results. It will
demonstrate typical multi-environmental coverage, checklist various environmental
pairing scenarios, and considers the connections of environmental factors with machinery and people. It also describes climatic characteristics as external conditions of
machinery use, including the classification and characteristics of world climate for
engineering analysis, the characteristics of the radiation regime, the thermal regime,
daily variations of air temperature, air humidity and rains, wind speed, atmospheric
phenomena, biological factors; the influence of climatic factors and atmospheric phenomena on the properties of the materials and the system "operator-machine-subject of
the machine influence". It also provides consideration of the influence of daily and
yearly fluctuations of air temperatures and rapid changes of climatic factors, including
the influence of water (moisture), air humidity, fog, and dew, as well as the characteristics of the combined influences of basic climatic factors. This chapter also considers climatic models that act as mathematical representations of the interactions
between atmosphere, oceans, and land surface, such as ice—and the sun.

3.1 Introduction

A basic need for successful accelerated testing of automotive, aerospace, and
other products is obtaining accurate information of the real field/flight conditions. Without this information it is impossible to accurately simulate these
conditions in the laboratory. The early availability of test subject(s) during the
design and manufacturing stages is especially important for accelerated reliability and durability testing technology. These are important components of
the successful prediction of a product's quality, reliability, durability, safety,

Trends in Development of Accelerated Testing for Automotive and Aerospace Engineering.
https://doi.org/10.1016/B978-0-12-818841-5.00003-9

maintainability, life cycle cost, and other performance characteristics. Coupled with some other advanced solutions these basic requirements are leading to positive trends in the development of accelerated testing in automotive and aerospace engineering.

We are seeing an increased awareness that the success of accelerated testing depends largely on how careful and successful the study of the real world conditions is in providing the correct information necessary for simulating them in testing. Unless you have the correct information, the simulation of ground or flight conditions will not be accurate, and accelerated testing results will not correspond to field or flight results.

The ground or flight conditions necessary for accurate testing consist of

- Input influences
- Human factors
- Safety assurance
- Others

All of these are necessary in order to accurately simulate their real interactions and the resulting product's performance. This chapter will consider in detail, how to perform these critical data acquisition procedures. For example, critical ground or flight conditions, that often are called environmental (multi-environmental) conditions are but one of the basic interacted groups of ground or flight input influences (Fig. 3.1). As can be seen in this Figure, a more appropriate term for these conditions is multi-environmental, because there are many conditions in the real world that need to be simulated in the laboratory for meaningful reliability, durability, quality, etc. accelerated testing.

In the real world, other groups of conditions act in conjunction with the multi-environmental (climate) group of influences. It is a cumulative reaction to all these multi-environmental influences that create the output variables (Fig. 3.2) that lead to product degradation and failures. And these failures then lead to reliability, durability, safety, life cycle cost, and other negative impacts (see Chapter 4). This can be seen in detail in Refs. [3.1, 3.7, 3.16], and others.

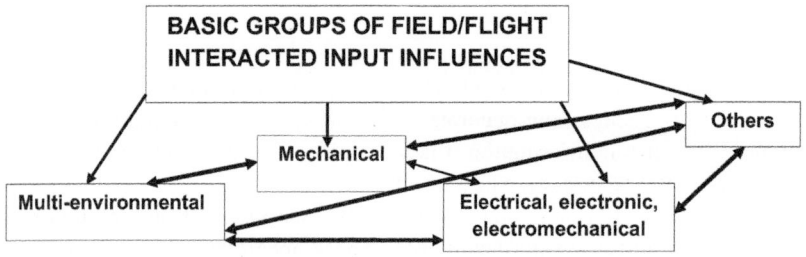

FIGURE 3.1 Scheme of basic interacted groups of ground or flight input influences.

1. MULTI-ENVIRONMENTAL GROUP OF INPUT INFLUENCES:

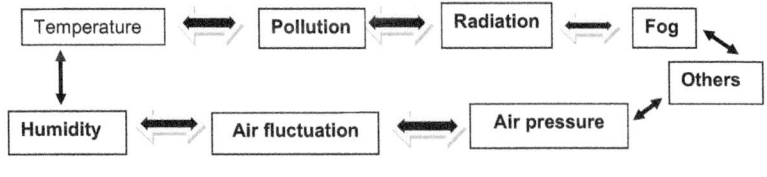

2. OUTPUT VARIABLES:

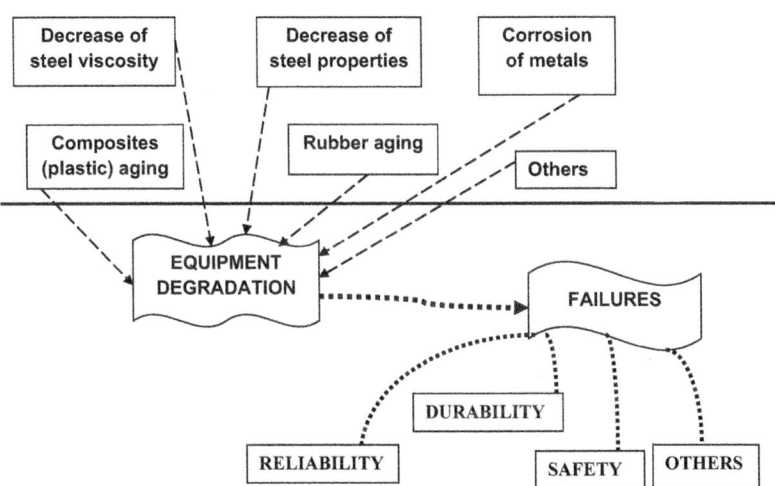

FIGURE 3.2 Scheme showing the complex interaction of some basic ground and flight climatic (multi-environmental) groups of input influences on equipment degradation, failures, reliability, durability safety, and others.

Therefore, the accurate study of the field conditions, including multi-environmental conditions, is critically important for the development of the accelerated testing in automotive and aerospace engineering.

Each of the above groups consist of numerous sub-components, which can be seen in greater detail in this author's other books (Refs. [3.1, 3.16]).

3.2 Multi-environmental factors

While there are many publications on environmental and multi-environmental testing, unfortunately, many of them do not adequately consider the real operating climatic conditions of the product. That is, there is not authentic initial information to enable accurate simulation for testing the basic parts in real world conditions. This lack of accuracy in real world data will influence the product's reliability, as well as other aspects of its real world performance.

As a result, too often simulations are not accurate and the testing results in the laboratory or proving ground do not correspond to the real world results. This Chapter demonstrates how to improve this through carefully studying the multi-environmental (climate) conditions for accuracy their simulation and accelerated testing.

Given that the dependence of equipment reliability is associated with the operating conditions encountered during the life cycle, it is important that such conditions be identified accurately at the beginning of the design process. The multi-environmental factors that influence machinery reliability, including automotive and aerospace products, are included in Table 3.1, which provides a checklist for typical environmental conditions. Combined environmental factors may be more detrimental to reliability than the effects of any single environmental factor.

TABLE 3.1 Typical multi-environmental coverage checklist [3.1].

Natural	Induced
Albedo, planetary IR	Acceleration
Clouds	Chemicals
Electromagnetic radiation	Corona
Electrostatic discharge	Electromagnetic, laser
Fog	Electromagnetic, radiation
Freezing rain	Electrostatic discharge
Frost	Explosion
Fungus/mold/mildew	Icing
Gravity, low, normal, high	Magnetics
Hail	Moisture
Humidity, high	Nuclear radiation
Humidity, low	Shock, pyro, thermal
Ice	Space debris
Ionized gases	Temperature, high, aero, heating, fire
Lighting	Temperature, low, aero, cooling
Magnetics, geo	Turbulence
Meteoroids	Vapor trails
Pollution, air	Vibration, mechanical, microphonics
Pressure, high	Vibration, acoustic
Pressure, low, vacuum	
Radiation, cosmic, solar	
Rain	
Salt spray	
Sand and dust	
Sleet	
Snow	
Temperature, high	
Temperature, low	
Wind	

It is fundamental to the design process that the testing criteria consider both single and/or combined environmental factors to anticipate the operating conditions and hazards which must be properly included in the system design profile. One often overlooked such example is an item that may be exposed to a combination of unusual conditions such as abnormal temperatures, humidity, altitude, shock, and others, not while in operation, but during transportation. The design conditions need to account for and make acceptable these unusual but real world effects [3.1].

Tables 3.2 and 3.3 provide reliability consideration for pairs of environmental factors.

Each of these environmental factors that may be present requires a determination of its impact on the operational and reliability characteristics of the materials, details, units, and their effect on the entire machine's design.

Packaging techniques should be identified that afford the necessary protection against degrading factors.

The environmental stress identification process that precedes the actual selection and identification of environmental stress simulation techniques, requires considering stresses associated with all life intervals of the product. This includes pre-operational and post-operational environments. Maintenance environments, when stresses imposed on the test subject during manufacturing assembly, inspection, testing, shipping, and installation, may have significant impacts on the products performance (reliability and other components).

Stresses imposed during the pre-operational phase are often overlooked even though they may present a particularly harsh environment that the equipment must withstand. The environments to which a system or product is exposed to during shipping and installation may be more severe than those encountered during normal operating conditions, but are easily overlooked. It is also probable that some of the environmental stress factors experienced in a system design pertain to conditions that will be encountered in the design and manufacturing phase rather than during actual operation.

Implementation method: To ensure a reliability-oriented design, it is necessary to accurately determine the needed environmental resistance of the product in all of these conditions.

3.3 Environmental factors and machinery

Most mobile products in the automotive and aerospace industry as well as many products in other areas of industry are primarily used outdoors and therefore are exposed to the naturally occurring elements.

The outdoor use of these products often subjects them to unfavorable influences from naturally occurring environmental factors. Most of these influences are atmospheric related, such as: high and low air temperatures which vary by location, daily and yearly, fluctuations of these temperatures, solar radiation, humidity, pollution (mechanical and chemical), rain, wind, etc.

TABLE 3.2 Various environmental pairs.

Low temperature and humidity	High temperature and ozone	
Relative humidity increases as temperature decreases, and at a lower temperature may induce moisture to become frost or ice.	Starting at about 300°F (150°C), the temperature starts to reduce ozone. Above about 520°F (270°C), the ozone cannot exist at normally encountered pressures.	
Low temperature and solar radiation	**Low temperature and low pressure**	**Low temperature and salt spray**
Low temperature tends to reduce the effects of solar radiation and vice versa.	This combination can accelerate leakage through seats, etc.	Low temperature reduces the corrosion rate of salt spray.
	Low temperature and sand and dust	**Low temperature and fungus**
	Low temperature increases dust penetration.	Low temperature reduces fungus growth. At subzero temperatures, fungi remain in suspended animation.
Low temperature and shock and vibration	**Low temperature and acceleration**	**Low temperature and explosive atmosphere**
Low temperature tends to intensify the effects of shock and vibration. However, it is generally a consideration only at very low temperatures.	This combination produces a similar effect as with low temperature and shock and vibration.	The temperature has minimal effect on the ignition of an explosive atmosphere but does affect the air-vapor ratio, which is an important consideration
Low temperature and ozone	**Humidity and low pressure**	**Humidity and salt spray**
Ozone effects are reduced at lower temperatures, but ozone concentration increases with lower temperatures	Humidity increases the effects of low pressure, particularly with electronic or electrical equipment.	High humidity may dilute the salt concentration, which could affect the corrosive action of the salt by increasing the salinity. It may increase the coverage of the spray, thereby increasing the conductivity.

TABLE 3.2 Various environmental pairs.—cont'd

	However, the actual effectiveness of this combination is determined primarily by the temperature.	
Humidity and fungus	**Humidity and sand and dust**	**Humidity and solar radiation**
Humidity helps the growth of fungus and microorganisms but adds nothing to their effects	Sand and dust have a natural affinity for water, and the combination increases deterioration.	Humidity increases the deteriorating effects of solar radiation on organic materials.

Other unfavorable influences on the products can be atmospheric phenomena such as fog, snowstorms, frost, ice on the ground, dust-storms, water-storms, and others. Because of the damaging influences of these environmental factors, the quality of the materials, the design details, and the operational considerations must be taken into account, or they may cause deterioration of the machinery's reliability and effectiveness. The above factors are influenced not just by engineering factors, but also by human factors that are introduced by the people who operate and maintain the machines, and by the external influences such as the conditions of the roads, the operating atmosphere, contaminants, etc. The inputs created by anticipated changes in human conditions inputs and the deterioration of the land, roads, air, and space also influence reliability. Therefore, the reliability of machinery must be modeled as a complex system and must account for all of the "operator-machinery-object of machinery actions" influences. The successful functionality of the system depends also on the operator's actions. The recognition of the importance of the operator's actions, intended and unintended, must be an important factor.

The operational climatic factors are characterized as wide ranging from arctic to subtropical for ground vehicles and even more extreme for air and space vehicles and their components. The nature and character of the operating environment will have a major influence on the selection of internal material and the overall reliability of the machinery. But predicting these factors is very complicated. The reliability and effectiveness of machinery that is used outdoors or in air and space applications will depend heavily on a level of corresponding, and suitable design for use under those climatic conditions.

TABLE 3.3 Various environmental pairs.

High temperature and humidity	High temperature and low pressure	High temperature and salt spray
High temperature tends to increase the rate of moisture penetration. The general deterioration effects of humidity are increased by high temperatures.	Each of these environments depends on the other. For example, as pressure decreases, outgassing of constituents of materials increases; as temperature increases, outgassing increases. Hence, each tends to intensify the effects of the other.	High temperature tends to increase the rate of corrosion caused by salt spray.
High temperature and solar radiation	**High temperature and fungus**	**High temperature and sand and dust**
This is a naturally occurring combination that causes increasing deterioration effects on organic materials	A certain degree of high temperature is necessary to permit fungus and microorganisms to grow. However, fungus and microorganisms cannot develop above 160°C (320°F).	The erosion rate of sand may be accelerated by high temperatures. However, the high temperature reduces sand and dust penetration.
High temperature and shock and vibration	**High temperature and acceleration**	**High temperature and explosive atmosphere**
Since both environments affect common material properties, they will intensify each other's effects. The degree to which the effects are intensified depends on the magnitude of each environment factor in the combination. Plastics and polymers are more susceptible to this combination than metals unless extremely high temperatures are involved.	This combination produces the same effect as high temperature and shock and vibration.	The temperature has minimal effect on the ignition of an explosive atmosphere but does affect the air-vapor ratio, which is an important consideration.

This means that design must guarantee optimal reliability in different environmental conditions that in turn requires the development of environmental accelerated testing along with a generalization of the accumulated experience-proved design and testing of machinery, which has been designed for such specific environmental conditions [3.18]. Typically, electronic, plastic, or elastomeric products are the most sensitive to environmental actions.

3.4 Determining climate characteristics as external conditions for machinery use

3.4.1 Establishing a classification system with characteristics for world climate as an engineering tool

Global climate has a large range of variance. Territorial climate depends on regimes of solar radiation, circulation of atmospheric components, moisture-rotation, physical-geographical specifics (relief, surface, etc.), human influences on the climatic conditions (development of water tanks, hydroponics, and others.). The characteristics of these factors are largely determined by the geographical location (its geographical width, the distance from the ocean, lakes, etc.).

World climatic characteristics can be characterized into six basic micro-climatic regions (Table 3.4) [3.4].

In using these tables for the determination of the character and intensity of the influence of climatic factors on the interior and exterior areas of a product's materials and the machinery's reliability, consideration must also be given to the specific characteristics and the distribution of these factors within any given region.

3.5 Characteristics of the radiation regime

Solar radiation is the electromagnetic radiation of the sun (radiant energy) which is called solar radiation. The solar radiation that reaches the earth's surface consists of wavelengths between 295 and 3000 nm (a nanometer is one-billionth (1 multiplication 10^{-9}) of a meter).

Ozone in the stratosphere absorbs and essentially eliminates all radiant energy below 295 nm. While extremely sensitive instruments may detect radiation below 295 nm, the amount is considered negligible by most experts.

This terrestrial radiation is commonly separated into three main wavelength ranges [3.4]:

- wavelengths between 295 and 400 nm (6.8% of the total radiation) are known as the ultraviolet (UV) portion of the solar spectrum.

  Ultraviolet (UV) according to ASTM G 113-94. *Terminology Relating to Natural and Artificial Weathering Tests of Non-metallic materials*, is

TABLE 3.4 Classification and characteristics of world climatic regions for technical applications [3.4].

Region	Characteristics
Air temperature	A territory where the median of annual absolute maximums air temperature is equal to or less than 40°C (104F), the median of annual absolute minimum temperature is equal or less than −45°C.
Cold	A territory where the median of the annual absolute minimums of the air temperature is less than −45°C (−49F).
Tropical, humid	A territory where the combination of air temperature is 20°C (68F) or higher, and relative humidity 80% or higher, exists for 12 h or more during the day for a continuous period of more than 2 months a year.
Tropical dry	A territory where the median of the annual absolute maximums of the air temperature is higher than 40°C (104F) and which is not related to the microclimatic region with a humid tropical climate.
Temperate-cold marine	Waters of seas and oceans whose location is more than 30° north o/and less than 30° south.
Tropical marine	Waters of seas and oceans whose location is between 30° north and 30° south.

radiation for which the wavelengths of the components are shorter than those for visible radiation

- wavelengths between 400 and 800 nm (55.4% of the total energy of radiation) are the visible (VIS) portion of the solar spectrum
- wavelengths between 800 and 2450 nm (37.8% of the total energy of radiation) are known as the infrared (IR) portion of the solar spectrum.

The spectral range for the UV portion and its sub-components is not well defined. However, the CIE (Commission Internationale de I'Eslairage) E−2.1.2 committee makes the following distinction: UV-A = 315−400 nm; UV-B = 280−315 nm; UV-C < 280 nm.

Visible light (the radiation the human eye can detect) is between 400 and 800 nm, making up just over half of the solar spectrum. About 40% of the radiation from the sun is contained in the infrared portion of the solar spectrum beyond 800 nm.

The defined break between the UV and VIS portion of the spectrum may be different depending on the source of information. Some consider the break to be at 400 nm, some at 385 nm and others at 380 nm. While this might be considered a minor variation, it must be understood when calculating radiant dosages for exposure, whether in outdoor or artificial conditions. The variance between a break at 385 nm and a break at 400 nm could be more than 25%, which could be extremely significant in attempting to estimate the service life of a material (see Table 3.5).

Irradiance can be defined as the radiant flux incident on a surface per unit area, commonly expressed in W/m^2. For this parameter it is necessary to indicate the spectral range in which the measurements were taken or for which the values were calculated, such as 295−3000 nm (total solar) or 295−400 nm (total UV). If we turn our attention to narrow wavelength intervals, we obtain the spectral irradiance, measured in $W/m^2/nm$. Most radiant exposures are measured in either kJ/m^2 or MJ/m^2 to convert this energy into numbers to which we can more easily relate (see Table 3.5).

The terms used for measuring solar radiation can be thought of as analogous to a bathtub being filled with water. Irradiance would be the rate the water is coming out of the faucet, while radiant exposure would be how much water

TABLE 3.5 Global solar spectral irradiance at sea level[a] (in accordance with CIE Pub. 85, Tab. 4).

Spectral range	nm	Irradiance, W/m^2
UV-B	280−315	2.19
	280−320	4.06
UV-A	315−380	49.43
	315−385	54.25
	315−400	72.37
	320−400	70.50
Total UV	≤380	51.62
	≤385	56.44
	≤400	74.56
Total UV + VIS	≤780	658.53
	≤800	678.78
IR	780−2450[a]	431.87
	800−2450[a]	411.62
Total	≤2450[a]	1090.40

[a]*Limit of CIE Pub. 85, Tab. 2.4.*

is in the tub at any specific time. Spectral irradiance, which defines the wavelength range, would be the quality of the temperature of the water used to fill the tub [3.3].

The ratio between direct and diffuse radiation reaching the earth's surface is strongly influenced by atmospheric conditions Water vapor (humidity) and pollution will increase the amount of radiant energy found in the diffuse component. A desert climate has a much higher percentage of radiant energy than that found in a subtropical climate. This occurs because there is much less water vapor in the desert than in subtropical climate. By contrast, the locations with high levels of pollution have dramatically reduced amounts of direct radiant energy.

Based on Rayleigh's law, shorter wavelengths of radiation are more likely to be scattered than long wavelengths. Therefore, the percentage of UV will always be less than that of total solar radiation. This difference can be seen in graphs comparing the percentage of direct irradiance between total solar radiation (including all regions of the solar spectrum) and UV only (Fig. 3.3).

The effects of direct and diffuse radiation are an important consideration when considering radiant energy received at different orientation to the sun [3.7]. Because of the high level of vapor in a subtropical climate such as south Florida, about 50% of the UV radiation is diffused even on clear days. Many days in Florida are not clear which results in an even greater percentage of radiation in the diffuse component. A desert climate such as central Arizona would have a greater percentage of UV radiation in the direct component (as much 75%).

Most of the active parts of solar radiation strike the earth's surface as parallel rays, which we call direct solar radiation (S). Most of this radiation

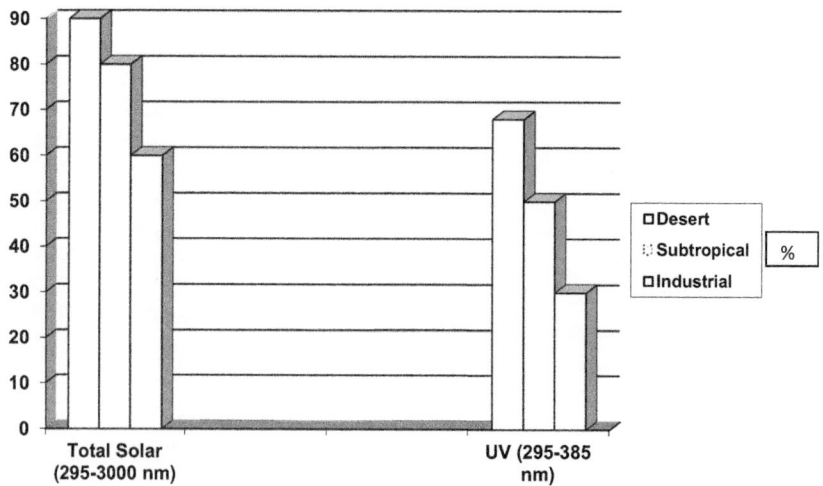

FIGURE 3.3   Direct solar area [3.1].

strikes in the southern regions. The part of solar radiation, which is dispersed by the air molecules and aerosols and then strikes the earth's surface, is called dispersed radiation (D). The direct and dispersed solar radiation are related to short-wave radiation (the length of the wave is 0.17−4 μm). The sum of direct and dispersed radiation is evaluated as total solar radiation (Q). The distribution of the annual sum of solar radiation on the earth is analogous to the distribution of direct solar radiation.

The solar radiation is redistributed in the atmosphere and on the earth's surface. The part of solar radiation that reflects from the atmosphere and the Earth's surface is called reflected high-frequency radiation (R). The other part of radiation which is absorbed by the earth's surface is called absorbed short-wave radiation. The quantity of absorbed and reflected radiation depends on color, structure, moisture, and other attributes of the surface where the solar rays fall. The characteristic of the reflection capacity of the surface is called albedo (A) (%). Albedo is the ratio of the amount of radiation reflected by a surface to the amount incident upon it. It is the ratio of reflected radiation from the surface (R) to the total of radiation which comes to this surface (Q):

$$A = \frac{R}{Q} \times 100$$

If the surface reflects more of the radiation, then A will increase.

The value of albedo (%) of some materials and surfaces is shown in Table 3.6.

The albedo surface of the territory changes during the year and depends on the appearance of the snow, paint, and influences of other factors. The albedo of desert areas has an insignificant change.

In addition to the short-wave radiation reaching the earth surface, there is also a long-wave radiation to the earth's atmosphere, which is called oncoming

TABLE 3.6 The value of some albedo materials [3.4].

Type of material	%
Snow: fresh dry clear moisture	80−95 50−55
White new paint	75
Limestone	50−65
Light sandstone	18−40
Light dry sand	30−35
Steel painted in red color	34

radiation E_a. The self-radiation E_s emanates from the ambient heat which warms the solar radiation at the earth's surface. The difference between E_a and E_s is the effective radiation E_{ef}.

The basic characteristic of the action of solar radiation is its radiation regime. The radiation regime B of a region is evaluated by measuring the amount of change during the year and the geographical distribution of radiation balance [3.4]:

$$B = (S + D)(1 - A) - E_{ef} = S^1 + D - E_a - R - E_s$$

where: S, S^1 are direct solar radiation corresponding to perpendicular rays and horizontal surface; D is dispersed radiation; A is albedo of the surface; E_{ef} is effective radiation; E_a is the long-wave radiation of the earth's atmosphere; R is short-wave radiation reflected from the earth to the atmosphere; E_s is long-wave radiation from the earth's surface.

The intensity of solar radiation depends on geographical latitude, height of the sun, and clarity of the atmosphere.

3.6 Characteristics of the air thermal regime

The thermal regime of outdoor air is characterized by its distribution and changes in temperature. The main source of heat for the lower layers of the atmosphere is caused by the warmth of surface influences (land, water, plant, etc.). On the other hand, active surfaces obtain their warmth from the sun.

Temperature changes during the year in each of the earth's regions depend on the quantity of solar energy which reaches the surface. They also depend on other factors such as atmospheric circulation, sea current, surface of the locality, the fundamental composition of the surface, and other factors [3.6].

While the median values of air temperature are its basic characteristics for categories, maximum low and maximum high air temperatures are important influences on a machine's reliability. This may be characterized by median values of minimal $t_{min.m}$ and maximal $t_{max.m}$ temperatures; absolute (external) values of minimal $t_{min.abs}$ and maximal $t_{max.abs}$ temperatures.

The highest air temperature on the earth ($+58°C$ [136F] in the shade) has been registered in Libya, the lowest ($-88.3°C$) in Antarctica.

3.7 Daily variations of air temperature

Machinery's reliability is influenced not only by low and high air temperatures, but also by changes in the speed of the temperature change over a length of time. It can be evaluated by the days temperature variations (amplitudes).

There is a difference between the average values for the warmest month and the coldest hour of the day (periodical variations) or the difference between the mid-maximal and mid-minimal daily air temperature over the period of a month (non-periodical variation).

The greatest daily air variations of temperature are specific to regions with a high elevation of continental climate. The lowest is for regions with a low continental climate (the regions of ocean and sea influences).

The greatest impacts on the machinery's materials and tensions in the machine's components will be affected by the maximal daily temperature amplitudes.

3.8 Air humidity and rain

Air humidity depends on many factors, including the distance from the ocean and sea, air temperature, time of year and day, quantity of rain, etc. Therefore air humidity depends on the region, time of day, etc.

Variations in the value of the humidity primarily depend on the geographical location of the region.

Most of the world's rain (12,660 mm/year) is in East India, while the lowest amount falls in the Sudan.

One example of the effect of such phenomena is the damage to the copper windings of electric arc furnace (EAF) transformers that has been observed in climatic regions with high relative humidity [3.7]. Such damage is caused by the build-up of heat inside the transformer. As a result of this known degradation process, a special program—for simulating electricity and heat flow in an EAF transformer has been developed. The model has been validated by testing the operation of an EAF transformer in the Ahvaz Steel Making Plant. The results of the analysis indicate that the simulation model can be applied for controlling the hot spot temperature of the transformer. This provides an appropriate mechanism for increasing the reliability of EAF transformer and for preventing damage to its cooper windings in regions with heavy rainfall.

3.9 Characteristics of wind speed

Wind speed and direction depend on the character of the air mass circulation near the earth, as well as differences in air pressure, time of year, time of day, surface relief, and other factors.

Atmospheric circulation influences climate and weather, and depends on the transfer of air masses. Weather changes depend on the motion of cyclones and anticyclones. Wind speed and the variability that influence air pressure are the important characteristics that affect machines that work outdoors as well as exterior structures such as buildings, bridges, etc. Changes in air direction have often influenced weather changes or the appearance of a storm. As a result, air speed influences machinery reliability and structural integrity.

3.10 Atmospheric phenomena

Atmospheric phenomena such as fog, dew, frost, ice or snow on the ground, air fluctuation, storms, and dust storms exert significant influence on machinery's reliability. Unusual atmospheric conditions such as gases, pollutants, and acid rain may cause entirely new reactions to the product. In highly industrial areas, acid rain may be the primary element driving the weathering process and one that affects a wide range of materials.

Blowing dirt and dust may have effects on the weathering process without reacting with the actual molecular structure of the materials. These effects include the screening of ultraviolet radiation from materials by dirt, which absorbs the ultraviolet portion of the spectrum. Semi-permanent varnishes can form on the surface of materials in certain climates. Mold, mildew, and other microbiological agents may play a significant role in material degradation or a machine's operational reliability (e.g., microbial contamination in diesel fuel, particularly in tropical and subtropical climates), although they may not be generally thought of as weathering factors.

3.11 Biological factors

Some biological factors that influence technical products are mold, insects, and rodents. These factors also have an influence on reliability especially in tropical regions (frequently during machinery storage) and for ships in the water.

Mold is related to the lowest plant forms that lack the property of photosynthesis, and forms as a result of interconnection with materials in which it secretes the products of metabolism that consist of different acids, which in turn decompose insulating materials and plastics. The most favorable condition for mold development is relatively humid air (50%−85%) and temperature [20−30°C (68−86F)]. If the humidity is lower and the hygroscopic nutrition is absent, the mold is unable to develop. The production of mold can also be accelerated by its fast speed of development and huge variety (about 40,000 species).

Some types of insects, especially termites, feed on electrical conductor insulation, which then causes machinery failures. The same is true from the actions of rodents.

3.12 The influence of climatic factors and atmospheric phenomena on the materials and on the system

Not all climatic factors and atmospheric phenomena have a significant influence on a product's reliability, durability, safety, maintainability, and other operational or performance aspects.

The form and compositions of clouds, the time of the first frost, the temperature of upper layers of the soil, etc. are climatic factors that generally have little or no influence on product reliability and durability.

Typically the most important climatic influences on vehicles are solar radiation, high and low air temperatures and temperature variation, humidity, changes in wind speed, fog, air pollution (chemical, dust storms, etc.), etc.

The effects of these phenomena on a material's properties also depend on the intensity and the duration of the influence of the above factors and their cumulative adverse combination.

Climatic factors are often major reasons for the failures of products that are used outside. The physical and chemical properties of the materials which are selected in the design (metals, plastics, electronic, etc.) must be compatible with the climatic conditions encountered, or the choice will result in a decrease in the reliability of the product. This is the reason that consideration of the climatic factors which influence the product is an important element in the selection of the materials used in the product.

3.12.1 Influence of solar radiation

The primary impact of solar radiations action on the elements of metallic machinery is an increase in the temperature of these elements and of the air entering these elements (car body, the speed controller, etc.). More complex processes may occur in plastics which can result in more rapid aging.

Furthermore, as solar radiation is the basic factor of the thermal regime's interface between the atmosphere and the earth's surface, the influence of low and high air temperatures on the properties of materials is largely a result of the effect of solar radiation on the thermal regime of the air.

Photochemical reactions are usually accelerated at elevated temperatures. In addition, temperature determines the rate of subsequent reaction steps. These secondary reactions can be sometimes approximately qualified (but not exactly) using the Arrhenius equation.

A general rule of thumb assumes that reaction rates double with each 10°C (50F) rise in material temperature. However, this may not be seen in physical measurement or changed appearance.

Also, thermo chemical reactions that may be initiated at higher temperatures may not occur at all or at a very low rate at lower temperatures.

The temperature of metallic parts of equipment [that are not subject to internal or external thermal events (e.g., fluid flow through pipes or combustion chamber heat)] is a function of the ambient temperature, metal solar absorptivity, solar irradiance, and surface conductance. This is the reason that in the presence of sunlight, the surface temperature of an object usually becomes considerably higher than the temperature of the air.

Solar absorptivity in both the visible and infrared regions is closely related to color, varying from about 20% for white surfaces to over 90% for black surfaces; thus, material of different colors will reach different temperatures on exposure. This surface temperature dependency on color can have secondary (non-thermo chemical) effects on materials as well. For example, as a result of different surface temperatures, mildew, and other biological growth will form and accumulate at different rates on materials of different colors. White or lighter colored materials tend to "grow" more mildew than darker colored materials.

Much higher temperatures are obtained on painted or coated metal surfaces than in the bulk of a plastic material because the thermal conductivity and heat capacity of metals are generally higher than plastic substrates. Ambient air temperature, evaporation rates, and the convective cooling from the surrounding air during exposure all play a role in the temperature of a material, and therefore influence its degradation rates.

3.12.1.1 Thermal influence of solar radiation

The intensity of solar radiation is evaluated by the quantity of heat in joules that falls on 1 cm^2 (J/cm^2) on an absolutely black surface which is exposed for 1 min to perpendicular rays. An absolute black surface is one whose surface absorbs all solar radiation.

The quantity of energy E that irradiates an absolute black surface of an object may be evaluated by the law of Stephan-Boltzman [3.5]:

$$E = \sigma FT^4 \tag{3.1}$$

where: σ is the coefficient of proportionality or a constant of radiation; F is the surface of the objects; T is the absolute temperature of a radiated surface.

The application of the Stephan-Boltzman's law can be extended to natural "gray" surfaces. By more accurate calculation of a surfaces radiation capacity in Formula (3.1) one can introduce the relative coefficient of radiation δ. Formula (3.1) then becomes

$$E = \delta \sigma FT^4$$

The warming of a body by solar radiation depends on the intensity of solar radiation, the outdoor temperature, and the reflective capacity of the body's surface. Reflective capacity depends on the color and roughness of a surface: more radiation is reflected by a body with a smoother surface.

If the body becomes warm, it also becomes a source of radiation. One can track the regularity of surface heat exchange in the example of the metallic housing heat exchange. In matte black housing, that has no inner source of warmth (for example, the body of an excavator which is not working) the radiation energy can be shown schematically in Fig. 3.4. The housing walls are assumed to be thin, and therefore the temperatures of the external and the internal wall surfaces are equal.

The upper cover of the housing absorbs the warmth of the solar radiation as well as the inside of the housing (σT^4_s). The lower wall of the housing (bottom) absorbs warmth from the upper cover and radiates it both inside and outside (σT^4_D). By placement of the housing on the ground, the bottom of the wall radiates its warmth and can receive warmth from the ground (σT^4_s).

Accordingly, the temperature equilibrium for the system is shown by the following mathematical dependences [3.4]:

$$\sigma T^4_B = \sigma/2 \cdot (T^4_D - T^4_S); \quad \sigma T^4_D = 1/2 \cdot (1.6 + \sigma T^4_B)$$

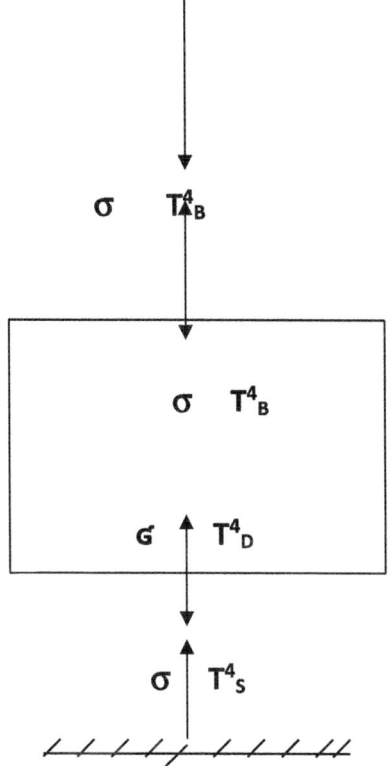

FIGURE 3.4 Schematic diagram for the definition of the heat balance of housing wall radiation [3.4].

where:

TB is the temperature of housing cover;

TD is the temperature of the housing bottom;

TS is the ground temperature; and σ is the constant of radiation.

The temperature of the body's surface is determined by the difference between the absorbed heat and the radiation inside and outside the body. This is the warmth balance of the body. It can also be the heat balance of the surface and the warmth balance of the body.

3.12.1.2 Thermal balance of the surface

In general, the body's surface absorbs heat or radiates it to the environment. Both processes may occur simultaneously. The quantity of heat Q which is absorbed by the surface consists of:

1) the heat from all types of radiation (short-wave, long-wave, and reflected) Q_E;

2) the heat from environmental heat exchange; self-heating of the surface radiation Q_A;

3) the loss of heat as a result of evaporation Q_v and condensation Q_K;

4) and also as a result of the heat of conductivity inside of the body Q_L.

The quantity of heat which is necessary for evaporation is deducted from the body, but the quantity of heat which is necessary for condensation must be added to it. In general, the quantity of heat Q which passes through the body surface can be estimated from the formula:

$$Q = Q_E + Q_S - Q_A - Q_v + Q_K \pm Q_L$$

Still air is a heat insulator; therefore an insignificant quantity of heat can be diverted from the surface or brought to it when the air is motionless. This situation can be changed when the air is in motion.

There is a heat exchange between the body's surface and the outside air. The character of this heat change depends on separate types of heat transfer. Corresponding to this are surfaces that absorb solar radiation or radiate warmth, or intermediate surfaces (Fig. 3.5). By absorption of solar radiation without evaporation, the resulting heat is transferred directly to the heating surface. Some part of this heat is then radiated to the outside air, another part is transferred inside the body. The heat absorption is usually greater than its radiation and is absorbed by the body.

Therefore, the surface temperature is increased (curve a).

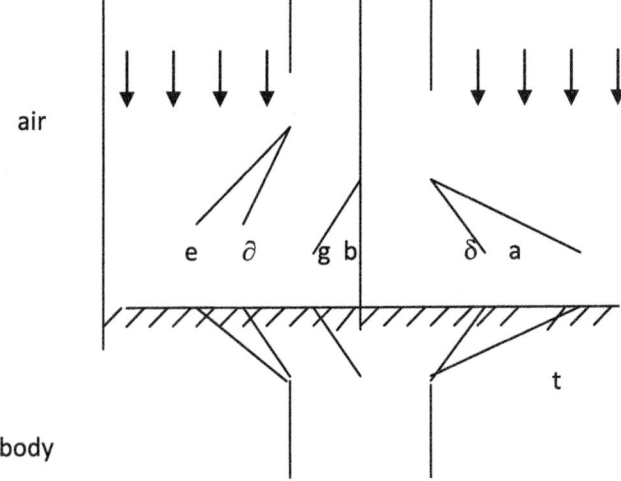

FIGURE 3.5   Changing the surface body temperature by different types of radiation [3.4]: **a** is the absorption of rays; **t** is the absorption of rays with simultaneous evaporation of moisture from the surface; **b** is radiation; **g** is radiation with the formation of dew or hoar-frost on the surface; **∂** is clear evaporation without influence of radiation; e is under normal temperature.

The following equation can be used to determine thermal equilibrium:

$$(Q_E - Q_A) - Q_S - Q_L = 0$$

By absorption of rays with simultaneous evaporation, part of the heat is transformed into evaporation, and therefore the temperature of the surface and, as a result, its radiation decreases compared to its clear absorption (curve δ). The condition of the thermal equilibrium for this surface is:

$$(Q_E - Q_A - Q_v) - Q_S - Q_L = 0$$

A surface with normal temperature that equals its environment occurs when the heat conductivity of a body is not sufficient to eliminate heat from the surface or when the body is cooling to a degree that the surface temperature decreases as a result of evaporation. When the heat from evaporation comes from the air, as well as the condition of warmth equilibrium:

$$-Q_v + Q_S = 0$$

3.12.1.3 Action of solar radiation on plastics

More and more, plastics are included in the design of modern machines and equipment. The trends in the development of these products show that in the future there will be a wider use of plastics in different areas of industry.

Complicated photochemical processes react on plastics, rubber, and their combinations under the action of solar radiation. These processes decompose chemical structures. As a result, there is a change in the quality of both materials and products.

Solar radiation, especially its ultraviolet part, often destroys numerous, very strong bonds in the molecules of plastics. Therefore, the aging of plastics is accelerated and is followed by product failures.

The aging process of plastics is accelerated by warmth, moisture, air oxygen, radiation of high energies, and other factors. The rate of the aging of plastics due to solar radiation depends on its intensity, the percentage of ultraviolet radiation in the solar spectrum, and the absorption properties of the plastics.

Research has shown that the breakdown of molecular connections and the aging processes of plastics are activated when the intensity of the radiation is more than 16.8 kilo joules (kJ) [3.6].

It is accepted that there are two simultaneous processes in the aging and destruction of plastics. There is a breaking of the bonds in the molecules resulting in the formation of molecular fragments which can then form new bonds between the atoms and molecular fragments. As a result of the aging process, the plastic is changed in its mechanical and electrical properties, color, etc.

The influence of solar radiation on the properties of plastics can be evaluated by placing plastic specimens in radiation chambers or under specialized outdoor conditions. For example, for the second type of specimen testing it would be advisable to use stations provided by ATLAS [3.6, 3.17], which is a division of the Weathering Services Group in the different climatic areas of the USA and Europe.

In most cases the complex of climatic factors acts on plastics under natural conditions. These influences can be simulated in the laboratory for the use of accelerated evaluation of these factors on material up to destruction.

Changing of frictional and dielectric properties of materials may also exert an influence on the reliability of the machinery. For example, the braking time [3.11] for one brake design (KSP-1) by friction braking without radiation was 4.0 s, after radiation for 15 h it was 4.6 s, and after radiation for 30 h it was 5.5 s. As a result of the aging of the material the friction coefficient of the plastic material decreased and the braking time increased.

In some cases the aging process of plastics can be decreased by changing the ray absorption capability (increase its degree of stability in light as well as making it heat-proof) and injecting these materials with special stabilizers that decrease the processes of destruction.

3.12.2 Influence of high temperatures

The temperature of materials is raised as a result of the direct influence of solar radiation, and The heat exchange between air and liquids or gases that heat the material to higher temperatures. This heat must be dissipated when the equipment is working. The sources of heat in working equipment (machinery) are the engine, the units of friction where heat arises as a result of the action of friction and is transferred from mechanical energy to thermal, and electrical conductors which dissipate heat from the resistance to be passing current.

High temperatures have the greatest effect on the properties of plastics and some other materials. High air temperature influences the elasticity of rubber. For example, when the temperature increases from 0 to 50°C (122F) [3.4].

The heat dissipation from semiconductor devices is another factor that elevates the operating temperature of these devices [3.12]. Unfortunately, most electronic devices are susceptible to failure at elevated temperatures, so the reliability of the device is affected by its operating temperature.

Virtually all the failures of mechanisms will increase at higher temperature.

Common failures that are due to increased operating temperature result from:

Thermal coefficient of expansion (TCE)
Creep in the bonding materials
Corrosion
Electromigration
Diffusion in the devices

Publication [3.12] details the effect of the operating temperature on the failure rates of some typical electronic devices. These curves clearly show a very strong dependence of material reliability on the operating temperature.

In order to account for this, operating temperatures (also called junction temperatures) should be limited to below 100°C (212F) as a worst case operating condition [3.13]. However, in systems requiring very high reliability, the use of junction temperatures not exceeding 85°C may be desirable.

This selection of a maximum allowable junction temperature is generally suggested by the manufacturer of the component and is based on its power dissipation and reliability requirement. In designing an electronic system, the cooling technique and the system cooling configuration are based on the maximum allowable component temperatures, heat dissipation rates, and environmental specifications [3.12].

In many products the greatest quantity of heat is generated from friction that is given off by braking devices.

In these braking devices plastic brackets with high frictional properties are widely used when normal and only slightly increased temperatures are encountered. But the frictional properties of these materials are decreased by increased temperatures of these materials, such as in the heat caused by braking, and the heat of the sun's radiation. The result is plastic materials may bind or soften under the increased temperatures, and the liquid fractions come to the surface all of which result in the destruction or degradation of the plastic [3.1].

The insulation material used in electrical conductors (cables, wires, bindings of electrical machinery and apparatus, etc.) absorb the heat from the environment (sun radiation and hot air) and the heat that is emitted from the conductors. Many different types of plastic, rubber, and paint are used as insulating materials. The aging properties of these materials depend on the actions of high temperatures, sun radiation, humidity and oxygen in the air. Heating and aging effects on insulation plastics rapidly decrease their dielectric strength and reduces their longevity. Therefore, in the design of insulation one can often see the use of inorganic fillers, thermosetting plastics, and other devices such as varnishes for binders, impregnates, and coated compositions, etc. specified, but even so the life of these insulators decreases if subjected to high temperatures (100−180°C).

The viscosity of combustible liquids, grease, solvents, etc. also decreases at higher temperatures. This decrease in viscosity diminishes the quality of the lubricating properties of grease because it decreases the thickness of the film between the lubricated surfaces. This accelerates abrasion, which wears out the surface.

Increasing temperature also decreases the viscosity of industrial liquids that are used in engines, hydraulic systems, and braking fluids. It increases the wear on hydraulic motors, cylinders, and apparatus, ultimately resulting in the leakage of liquids from spaces with high pressure to spaces with lower pressure.

The oxidation and aging of liquid oils in industrial liquids are also accelerated under high temperatures. This aging may also be accelerated by the evaporation of less dense fractions from oils and other liquids. As a result their structure or performance characteristics may be changed.

3.13 The influence of daily and yearly fluctuations of air temperatures and of rapid changes of climatic factors

Low and high temperatures exert opposing influences on materials. And rapid changes of temperatures (during one day or several hours) increase the negative effects of the temperature fluctuation on the machinery.

Additional tensions can occur on the product's metallic components through rapid changes in air temperature, which can induce different rates of thermal expansion of these elements [3.14].

These thermal tensions arise more frequently in thin elements, which have flexible contours, because of the changed in the length of the elements which increases more quickly as compared to thick elements.

And, there is the factor of irregular cooling or warming of the machines more massive components as a result of rapid changes in air temperature. This leads to additional tensions in the materials. Most tensions arise through rapid cooling of the components. The relative elongation or compression of discrete layers of the material may be evaluated by the following equation [3.4]:

$$\varepsilon_t = \alpha_t(t_2 - t_1)$$

where: α_t is the coefficient of linear widening; t_1 is the temperature in the first layer; t_2 is the temperature in the second layer; $t_2 = t_1 + (\partial t/\partial l)\Delta l$ (Δl is the distance between the layers).

The dependence of the materials specific conductivity as influenced by its temperature can be evaluated from the equation [3.15]:

$$\sigma_e = \sigma_{eo}e^{at} \approx \sigma_{eo}[1 - \alpha t],$$

where: σ_{eo} is the specific electrical conductivity by $t = 0°C$; and α is the thermal coefficient.

Rapid changes in the above temperatures decrease the service life of electrical machines, especially electric motors.

The dielectric permeability is affected by air pressure, humidity, and temperature. Low and high temperatures together with the corresponding changes in air humidity influence breakdown of an air clearance's dielectric under the same air pressure (Table 3.7).

TABLE 3.7 Correction coefficients for tensions of the breakdown of air clearance [3.4].

Pressure, MmHg	Temperature, °C					
	−40	−20	0	20	40	60
845	1.07	0.99	0.93	0.87	0.82	0.77
1013	1.25	1.17	1.10	1.03	0.97	0.91
1182	1.43	1.34	1.26	1.19	1.12	1.05

When air is an insulator, such as with electrical cranes, excavators, etc., rapid changes in temperature, humidity, and air pressure are adverse factors affecting the workings of electrical devices. Another effect resulting from rapid changes in temperature is the cracking of protective paint.

The varying thermal expansion causes the paint and steel to delaminate. As a result, there is scouring and the removal of the paint layers from metallic surfaces.

High pressure in air chambers decreases the work of carburetors in the engines and oil transformers.

3.14 Influence of water (moisture), air humidity, fog, and dew

Water is everywhere in our environment, whether in the form of humidity, rain, dew, snow, or hail.

Virtually all materials used outdoors are exposed to these influences.

There are two ways in which water affects materials. First, water may be absorbed by synthetic materials. And, second, by being coated from humidity and direct wetness. In the first method, as the surface layers absorb moisture, there is a volume expansion that places stress on the dry subsurface layers. Following the drying period, or the desorption of the water, the surface layers will experience volume constriction. As the hydrated inner layers resist this constriction, it leads to surface stress cracking. This fluctuation between hydrated and dehydrated states may result in stress fractures.

The freeze-thaw cycle is another physical effect. Because water expands when it freezes, absorbed moisture in a material causes expansion stresses that cause peeling, cracking, and flaking in coatings. Rain, which periodically washes dirt and pollutants from the surface, has an effect on the long-term rate of deterioration that is determined more by its frequency than its amount. When rain strikes surfaces, evaporation processes cool the surface rapidly, which may cause the physical degradation of a material. Frozen rain, or hail, may also cause physical degradation of materials because of the strong kinetic energy associated with its impact.

Water also can be directly involved in degradation involving chemical reactions. The chalking of titanium dioxide (TiO_2) in pigmented coatings and polymers is one such example.

While the structure of a polymer may be changed by radiant energy, the actual release of material on the surface can be enhanced, if not caused, by the cyclic action of chemically absorbed moisture.

Contact with water in any of the above-mentioned methods can also accelerate the rate of oxidation.

Moisture may also act as a pH adjuster, especially when the moisture is combined with other environmental influences, such as when considering the effects of acid rain, which may cause an etching of many paints and coatings.

TABLE 3.8 Characteristics of the influence of air humidity on the internal aspects of materials and the working conditions of equipment [3.6].

High humidity	Low humidity
Corrosion of metals	Cracks appear and microdestruction of the insulation materials
Saturates the mineral oils, technical liquids, and fuel oil	–
Changes the consistency of grease with the formation of an emulsion	Grease becomes more solid
Decreases the volumetric resistance of insulation materials	Sealing materials dry out
Decreases the surface resistance of insulation increases the dielectric	Plastic details deform
Permeability of the air	–
Growth of mold	–

The amount of air humidity which exerts a negative influence on materials depends on the percentage of moisture (Table 3.8). If there is more moisture in the air (more than 90%), it either decreases the quality of materials, penetrates inside these materials, or constitutes the film of moisture on the materials surface. If the content of moisture in the air is less than 50%, the moisture from the materials evaporates in the air and results in internal changes to the materials: they become fragile and develop cracks.

Hygroscopic materials absorb moisture from the air, for example, insulating materials, which are produced from cotton or paper. Moisture can penetrate materials in three ways: through capillary condensation, through penetration into a polymer's structure (intermolecular interval), and by entry through cracks and large pores in the material.

As air temperature increases, the rate of moisture penetration into the material increases.

Moisture, which penetrates the material, decreases its solid resistance.

Moisture may settle on the material's surface forming a film. As a result, the surface resistance of the materials decreases enormously (Fig. 3.6). The most serious decreases are in the surface resistance of insulators, which are greatly affected by the pollution by water films contaminated with gases and dust.

Moisture also creates favorable conditions that accelerate atmospheric corrosion of metals from the moisture settling on the metallic surface. This type of corrosion is responsible for 50% of the common loss of metals. Details on atmospheric corrosion can be seen in Ref. [3.4].

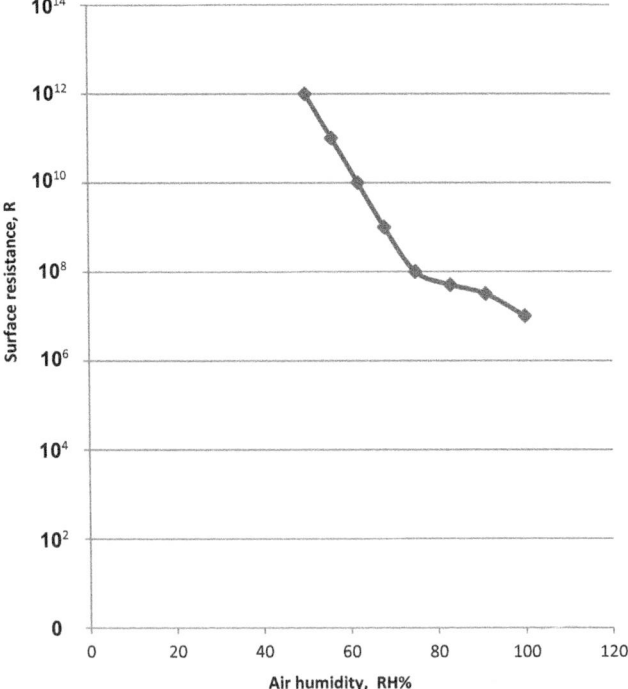

FIGURE 3.6 Dependence of surface resistance (R) on the insulation of ceramic details from air humidity RH [3.1].

Air moisture also reacts with liquid mineral oil. Air moisture in oils results in a decrease in the lubricating and anticorrosive qualities of oils. The interconnection of the water moisture with lubricating oils and greases forms an aqueous emulsion with diminished lubricity.

While these problems relate to high humidity, low air humidity presents its own issues. Low air humidity can cause a drying of materials, which may result in their drying out and buckling. When moisture is decreased in insulation such as in electrical windings, the resistance of the insulation increases, but the drying out of this insulation may result in cracks and, as a result, flaking and destruction of the insulation.

3.15 The characteristics of combined influences of basic climatic (environmental) factors

As previously discussed, in real life situations these different climatic factors act simultaneously and in combination thereby effecting the reliability of the product (Fig. 3.7). Moreover, the effect of their action depends heavily on the interconnections between these factors. Unfavorable combinations often have an adverse influence on product reliability.

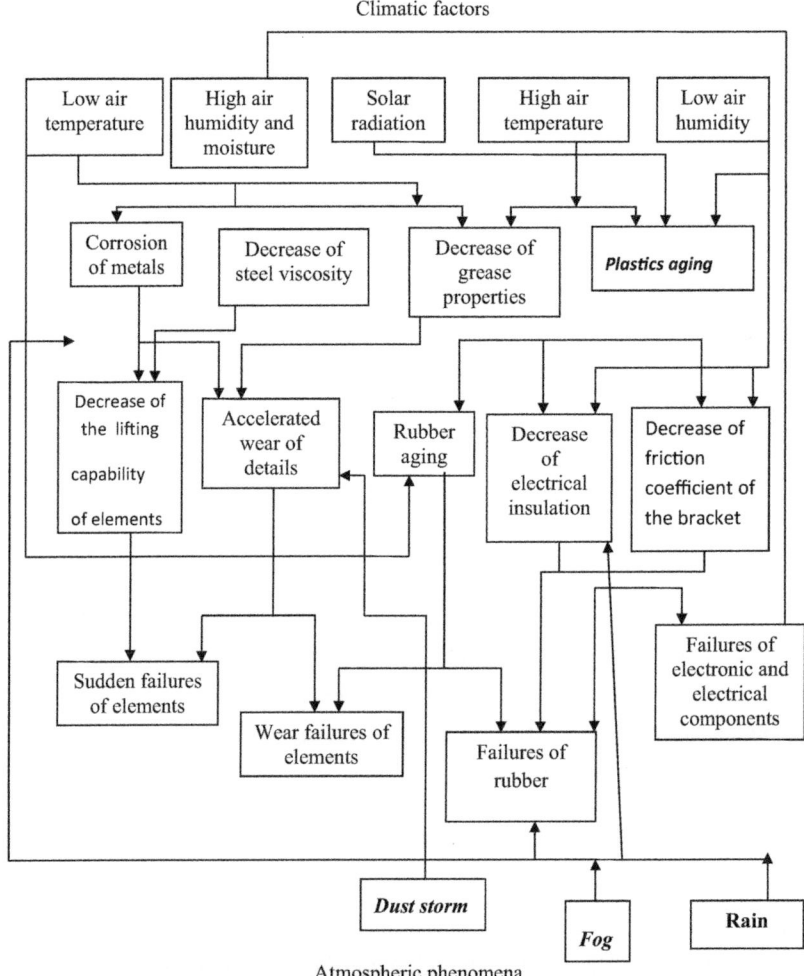

FIGURE 3.7 Scheme of complex influence of basic climatic factors and atmospheric phenomena on the properties of materials and machinery reliability.

Examples of unfavorable combinations of influences are factors such as low temperature of air combined with wind, moisture, low or high air humidity, etc.

The effects of the combination of different factors of influence on the interiors of materials and equipment are best evaluated by using methods of passive experiments under field conditions and actual experiments of accelerated actions in test chambers or under field conditions.

In considering the roles that solar radiation, temperature, moisture, and their secondary effects play on products, we must recognize that these factors work together in degrading materials. If one simulates these factors independently, it is very unlikely that the resulting degradation will resemble that of the material that is exposed to outdoor conditions, where all of these factors play a role in the degradation process [3.19].

The synergistic effects of these main climatic factors vary, depending on the materials being exposed.

Even small changes to a product's formulation, such as the addition of stabilizers, flame retardants, fillers, etc., can drastically change the degradation characteristics of that material. The use of recycled material, impurities in the polymer matrix, and the variations of product processing are additional variables to weathering performance. While there are literally thousands of publications that examine the durability characteristics of pure polymers, stabilizers, and specific aftermarket products, the knowledge of any material's durability to weathering is not an exact science. It is safe to say that a perfect understanding of the effect of weathering factors on every material will never be achieved.

Professionals have used the first method for many years, and there are special test facilities to simulate many different climatic conditions. For example, ATLAS [3.17] has a division (Weathering Services Group) which has three primary facilities in the United States and internationally (ISO/IEC Guide). There are also dozens of sites around the world, providing the widest range of climatic and environmental conditions for materials and product testing.

Static weather testing capabilities include direct exposure using fixed or variable angles, and backed or unbacked racks; under-glass exposure for interior materials; and black box exposure for paints and coating materials. As an example, Table 3.9 shows the average monthly UV and the total radiant exposure for Phoenix, Arizona and for Southern Florida [3.10], Table 3.10 shows annual climatological data of ATLAS stations for different sites around the world.

However it is well to remember that one cannot duplicate climatic factors simply with passive experiments. The aging of polymeric materials is influenced by various atmospheric phenomena, such as solar radiation, high and low air temperatures, humidity, ozone, and tension of the material. As a result, there may be a decrease in mechanical strength, electrical resistance, friction coefficient, and other properties. Fig. 3.8 is an illustration of the changing moment of friction for the braking of a steel couple against the press-mass as a result of the atmospheric aging of the press-mass.

It is impossible to evaluate the influence of each climatic factor of the changes to a material's properties due to atmospheric aging if the specimen is exposed to only one climatic area for atmospheric aging [3.11].

While it is possible to establish common characteristics and intensity of the complex influence of climatic factors on the property of the same materials by their disposition in different climatic areas, it is evident that different climatic conditions influence the above indexes. An analogous influence on these conditions is the relative prolongation of polychloride plasticity (Fig. 3. 9) and

TABLE 3.9 Average monthly UV and total radiant exposure for Phoenix, Arizona, and Southern Florida (MJ/m^2).

Month	Phoenix (AZ) 34° South		Southern (FL) 26° South	
	UV[a]	Total	UV[a]	Total
January	20.1	490	20.0	505
February	19.8	546	22.5	545
March	24.7	633	26.5	618
April	33.3	755	28.0	612
May	38.6	786	28.0	609
June	36.8	770	25.7	543
July	35.1	745	24.7	532
August	32.5	756	24.0	543
September	29.3	711	22.3	540
October	25.8	705	21.7	555
November	19.2	582	18.0	490
December	18.3	525	18.6	496
Annual total	333.5	8004	280.0	6588

[a]Below 385 nm wavelength.

how long it takes before cracks appear caused by deformation of the rubber (Fig. 3.10).

Another consideration is that some plastics are chemically changed through the actions of climatic factors on their volatile substances resulting in the escape of these volatile elements into the environment. This also changes the quality of the original plastics.

Figs. 3.9 and 3.10 demonstrate how the aging of materials over time decreases their coefficient of lengthening property. It should also be noted the character of the above decrease also depends on the original quality and the intensity of the climatic factors. And, the aging of materials is almost always an irreversible process.

By selecting differing approaches of input influences and different intensities for each influence, the testing method for active experiments offers the possibility of examining a material's long-term properties under various climatic conditions. Accelerated testing of this type in the test chambers allows researchers to control the input influences on the studied product. However, the

TABLE 3.10 Annual climatological data, domestic and international remote sites.

Location	Latitude	Longitude	Elevation (m)	Average ambient temperature (°C)	Average ambient RH (%)	Rainfall (mm)	Total radiant energy (MJ/m²)
Louisville, KY	38° 11′ N	85° 44′ W	149	13	67	1092	5100
Jacksonville, FL	30° 29′ N	81° 42′ W	8	20	76	1303	5800
Prescott, AZ	34° 39′ N	112° 26′ W	1531	12	65	1093	7000
Lochem, the Netherlands	52° 30′ N	6° 30′ E	35	9	83	715	3700
Hoek van Holland, the Netherlands	51° 57′ N	4° 10′ E	6	10	87	800	3800
Sanary, France (Bandol)	43° 08′ N	5° 49′ E	110	13	64	1200	5500
Singapore (Changi Airport)	1° 22′ N	103° 59′ E	15	27	84	2300	6030
Melbourne, Australia	37° 49′ S	144° 58′ E	35	16	62	650	5385
Townsville, Australia	19° 15′ S	146° 46′ E	15	25	70	937	7236
Ottawa, Canada	45° 20′ N	75° 41′ W	103	6	73	1910	4050
Sochi, Russia	43° 27′ N	39° 57′ E	30	14	77	1390	4980
Dhahran, Saudi Arabia	26° 32′ N	50° 13′ E	92	26	60	80	6946

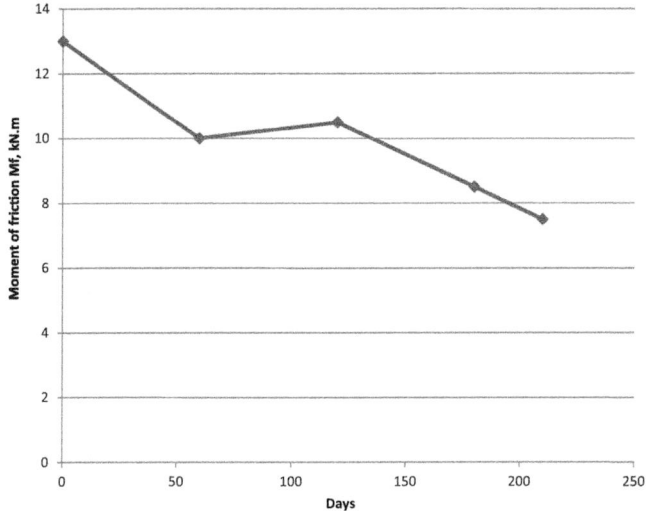

FIGURE 3.8 Changes in the moment of friction in the frictional braking 35,567 press-mass brackets from the duration of atmospheric aging [3.1, 3.11].

test chambers needed for such testing must be complicated to achieve this goal.

The complex task of simulating the Earth's climate is carried out by computer programs designed to detect long-term climate trends based on large-scale forces. Unlike weather prediction models, climate models are not intended to predict individual storm systems.

And, while climate models are based on what we know happened in the past, they may not accurately forecast the future. Some climate models are also

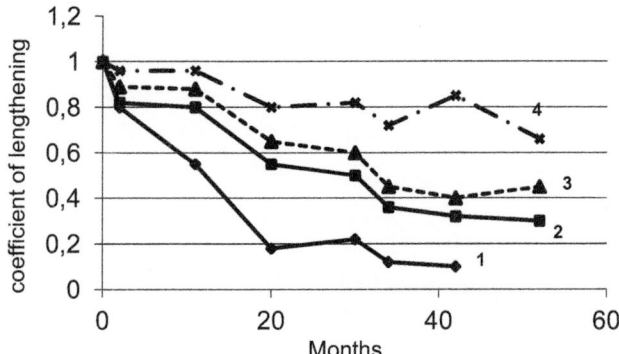

FIGURE 3.9 Changes in the material characteristics in different climatic areas [3.1]: coefficient of the relative prolongation K_2 of a plastic specimen with a thickness of 1 mm; 1 is the initial value; 2 is the mean latitude; 3 is the subtropical humid; 4 is dry subtropical.

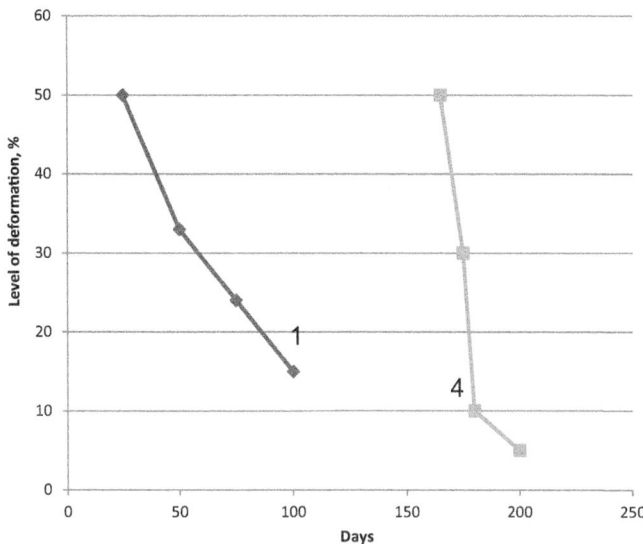

FIGURE 3.10 (continuation of 3.9). Changes in the materials characteristics in different climatic areas [3.1]: 1 is the initial value; 4 is dry subtropical.

used for predictions, for example, the models that helped to predict the climatic response for, a cooling influence that lasted for several years (Fig. 3.10).

Models need to be developed to address the question of how the climate system will react to additional greenhouse gases. It is believed that these models have correctly predicted effects subsequently confirmed by observation, including greater warming in the Arctic and over land, greater warming at night, and stratospheric cooling. Far from overestimating future climate change, climate models are more likely to be conservative in their predictions.

While the above depicts mostly strategic common situations, it is necessary to use specific real world input influences and to conduct specific types of testing of the subject using corresponding climatic conditions of this subject's use.

3.16 How reliable are climatic models for use in accelerated testing?

Some involved in climate models produce predictions differing from those presented in the text of this chapter.

Climate models are attempts at mathematical representations of the interactions between the various aspects of the climate system, including the atmosphere, oceans, land surface, ice, and the sun.

Clearly, this is a very complex task, so models are built to estimate *trends* rather than specific events [3.21]. For example, climate model can predict if it will be cold in winter, but it cannot tell you what the temperature will be on any specific day—that's weather forecasting. Climate *trends* are weather, averaged over time—usually 30 years and more. These trends are important, because they eliminate or "smooth out" single events that may be extreme, but quite rare.

Climate models have to be tested to find out if they work. Because we cannot wait 30 years to see if a model is good or not; models are tested against the past, against what is known to have happened. If a model can correctly predict trends from a starting point to somewhere in the past, the theory is that it could be expected to predict with reasonable certainty what might happen in the future.

So all of these models are first tested through a process called *Hindcasting*. If the models used to predict future global warming can accurately map past climate changes, the assumption is that if they get the past right, there is no reason to think their future predictions would be wrong. And, because the models could not simulate what had already happened unless extra CO_2 was added to the model the testing suggested CO_2 must be the cause for global warming. All other known forcing factors are adequate in explaining temperature variations prior to the rise in temperature over the last 30 years, while none of them are capable of explaining the rise in the past 30 years. CO_2 does explain that rise, and explains it completely without any need for additional, or as yet unknown forcing factor.

Some models which have been running for a sufficient time have also been proven to produce accurate prediction.

The models successfully predicted the climatic response after the eruption of volcanoes realizing large amounts of CO into the atmosphere.

All models have limits and uncertainties as they are modeling complex systems, however, all models can improve over time, and with increasing sources of real world information such as satellite obtained data. The output of climate models can be constantly refined to increase their power and usefulness.

As climate models have already predicted many phenomena for which we now have empirical evidence. Climate models can provide a reliable guide to potential climate change.

Mainstream climate models have also accurately projected global surface temperature changes [3.21].

We can read another description of this problem in Ref. [3.22].The complex task of simulating Earth's climate carried out by computer programs designed to detect long-term climate trends based on large-scale forces is seen in Ref. [3.22].

Bibliography

[3.1] Klyatis LM, Klyatis EL. Accelerated quality and reliability solutions. Elsevier; 2006.
[3.2] MIL-HDBK-217E. Reliability prediction of electronic equipment. 1990.
[3.3] SAE G-11. Reliability, maintainability, and supportability guidebook. 1990.
[3.4] Koh PI. Climate and reliability of machinery. Moscow: Mashinostroenie; 1981.

[3.5]   Klyatis LM. Accelerated evaluation of farm machinery. Moscow: Agropromisdat; 1985.

[3.6]   ATLAS, Materials Testing Solutions. Weathering testing guidebook. 2001.

[3.7]   Klyatis L. Successful Prediction of Product Performance. quality, reliability, durability, safety, maintainability, life-cycle cost, profit, and other components. SAE International; 2016.

[3.8]   Klyatis, LM. Use of simulation in solving biological engineering problems —evaluating Agricultural product quality. Paper No. 94-3612. Written for presentation at the 1994 ASAE International Winter Meeting. Atlanta, GA, pp. 1—7.

[3.9]   Klyatis LM. Environment and reliability of agricultural machinery. In: 10th annual agricultural conference. reliability evaluation & engineering session. Cedar Rapids, Iowa; 1995.

[3.10] Weathering services. ATLAS weathering services group, Bulletin AWSG, vol. 10; 1997.

[3.11] Kragelsky IV. Friction and wear. Moscow: Mashgiz; 1981.

[3.12] Protection of electronic apparatus from influence of climatic conditions. 1970. By edition of G Ubish, Energy, Moscow.

[3.13] Eruchimovich SV. Research on plastics in process of aging. Moscow: VNEEAM; 1988.

[3.13a] Holman JP. Heat transfer. New York: McGraw-Hill; 1991.

[3.14] Klyatis LM. Climate and reliability. In: Proceedings 56th ASQ annual quality congress, Denver; 2002. p. 131—40.

[3.15] Ireson W, Coombs G, Clude F, Moss RY. Handbook on reliability engineering and management. New York: McGraw-Hill; 1995.

[3.16] Klyatis LM. Accelerated reliability and durability testing technology. WILEY; 2012.

[3.17] ATLAS. Materials testing solutions, weathering, Lightfastness corrosion, ATLA materials testing Technology (USA).

[3.18] Klyatis LM. Establishment of accelerated corrosion testing conditions. In: Reliability and maintainability symposium (RAMS) proceedings. Seattle, WA; January 28—31, 2002. p. 636—41.

[3.19] Scott G., Wong I., Chen, W.-Z., Walters, E., Lucas, C., WasynczukO., Distributed simulation of an uninhabited aerial vehicle power system, Proceedings SAE 2004 Power Systems Conference, November 2—4, 2004, Reno, Nevada, P-391, SAE International, pp. 235—241.

[3.20] Sunningham K, Foster JV, Shan GH, Stewart EC, Rivers RA, Wilborn JE, Gato W. Simulation study of a commercial transport airplane during stall and post-stall flight. SAE International 2010.

[3.21] How reliable are climate model? Sceptical Science. https://www.scepticalscience.com/climate-models.htm.

[3.22] Climate communication. How reliable are climate models? https://www.climatecommunication.org/questions/reliable-climate-models/.

Exercises

1. What are the ground and flight conditions necessary for accurate testing?
2. Show the scheme of the basic interacted groups of ground and flight input influences.
3. Describe the path from the multi-environmental group of input influences to failures and reliability/durability, safety, and others.
4. Describe a typical multi-environmental checklist.

5. Describe some environmental testing pairs.
6. Describe the classification and characteristics of world climatic regions for technical applications.
7. Discuss some of the characteristics of the solar radiation regime.
8. Describe the characteristics of the air thermal regime.
9. Describe daily variations of air temperature.
10. Describe some of the characteristics of wind.
11. Describe some of the atmospheric phenomena such as fog, dew, frost, ice, or snow on the ground?
12. How do biological factors influence technical products?
13. How do climatic factors influence the materials and the system "operator-product-subject of the product influence?
14. Show and describe the equation for the Stephan-Boltzman law.
15. Describe elements involved in the thermal balance of a surface.
16. What affects can solar radiation have on plastics?
17. Describe the effects of high temperature resulting from solar radiation on materials.
18. Describe some of the influence of daily and yearly fluctuations of air temperatures and of rapid changes of climatic factors on products.
19. Describe some influence of water (moisture), air humidity, fog, and dew.
20. Describe the characteristics of some of the combined influences of climatic factors.
21. Show the complex influence of basic climatic factors and atmospheric phenomena on the properties of the materials and machinery reliability.
22. Discuss the reliability of climatic models for use in accelerated testing.

Chapter 4

Basic negative and positive trends in the development of accelerated testing

Abstract

This chapter considers how the trends in the development of accelerated testing may be positive or negative. This chapter includes the review of the negative trends impacting the employment of accelerated testing, the basic negative factors that lead to the slow improvement of accelerated testing, the common negative trends in accelerated testing development that relate to automotive and aerospace engineering, the specific tactical negative trends in accelerated testing development, the trends in using virtual (computer) simulation testing to replace product and physical simulation, erroneous use the exponential law of distribution for accelerated testing. It analyzed the positive trends in the development of accelerated testing that lead to improvements of the product's effectiveness, including common positive trends and specific positive trends in accelerated testing that relate to any specific types of testing in automotive and aerospace engineering.

4.1 Introduction

Usually, there are both positive and negative aspects of each new technology or product. If the positive aspects of the new technology or product are sufficient (quantitatively and qualitatively), this new technology or product is adopted and considered as commonly positive. If not, and it is a net negative, its success is doubtful.

As was demonstrated earlier, product recalls have been increasing over the years, and especially so in recent years. This trend can be seen not just in recalls but also in decreased reliability, maintainability, safety, the number of deaths and injuries as a result of road/flight incidents, and increased life cycle cost.

The number of deaths as a result of incidents is a very important metric. The basic cause for all of these is the inaccurate prediction of the complicated product's life performance. And, poor life performance prediction is often a result of failing to adopt the new accelerated testing development, and especially the very slow implementation of accelerated reliability and durability testing technology.

Trends in Development of Accelerated Testing for Automotive and Aerospace Engineering.
https://doi.org/10.1016/B978-0-12-818841-5.00004-0
121

There are several actions that can be employed to eliminate many of the reliability, durability, and safety problems found in the automotive and aerospace engineering, including different types of testing whose objectives should be:

1. Finding the real reason, not the result of the problem;
2. Studying what the reason for the problem is, and if your finding is correct or incorrect;
3. Deciding on how you can eliminate, or as a minimum, mitigate to an acceptable level the reasons that are creating the problem; and
4. Implementing the new procedures and using the more effective knowledge about the role of the testing level needed for assuring product effectiveness.

Examples of these actions using can be found in Refs. [4.1,4.2].

The referenced publications also provide different ideas for the improvement of the current status of accelerated testing. For example, in Ref. [4.3] the authors briefly review their statistical and other concepts for accelerated testing. They also give an overview of some of the current and planned research to improve accelerated test planning and methods.

Recently there are also many worldwide conferences and symposiums, especially in the area of Autonomous Vehicle accelerated testing development, which are available resources. Unfortunately, in such programs and presentations, we seldom find relevant information about developing testing trends. For example, in the Novi, USA, October 2018 Symposium "Autonomous Vehicle Test & Development" (North America's only conference dedicated to test & validation of autonomous vehicles & self-driving technology) [4.4], the agenda of this specific symposium/conference, which was nominally devoted to testing, included 45 presentations, but not a single one of the titles was in the trend in the development of testing. This situation reflects the common lack of attention to the importance of the trends in the development of testing.

4.2 Negative trends in the development of accelerated testing

4.2.1 Basic negative trends in the development of accelerated testing

Computer simulation of field conditions variables and testing the product using this simulation is simpler and cheaper than physical simulation and testing, but it cannot take into account all of the interactions of either the product's components or the complete product's operation in the real world.

There are two basic groups of causes responsible for the negative aspects occurring in accelerated testing development (Fig. 4.1).

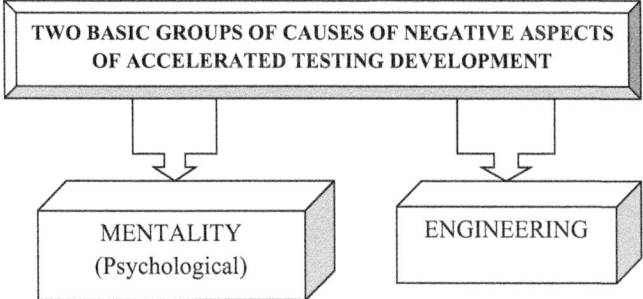

FIGURE 4.1 The basic groups of causes of the negative aspects of accelerated testing development.

Their brief description can be seen in Fig. 4.2. These negative aspects lead to the two basic causes of recalls, which are fundamentally a result of employing the negative aspects of accelerated testing, and therefore, resulting in unsuccessful prediction, which is depicted in Fig. 4.3, while the path from unsuccessful prediction to recalls can be seen in Fig. 4.4.

4.2.1.1 Mentality causes

- Most professionals, especially senior managers who make the decision about an investment, erroneously think that using simpler less expensive accelerated testing, which traditionally has been called accelerated life

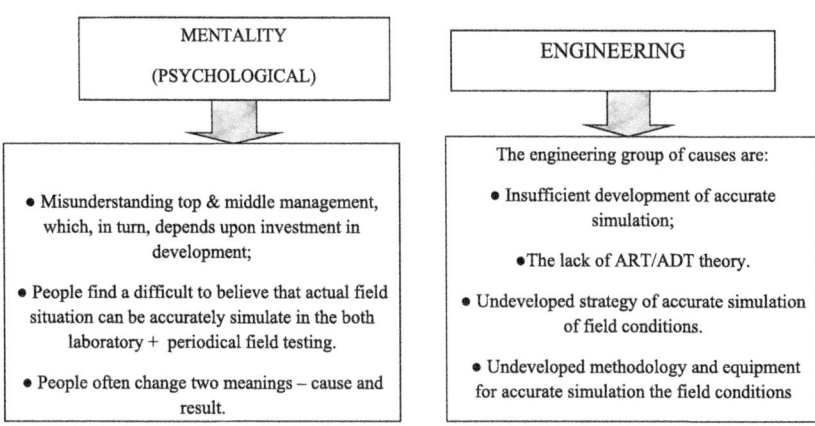

FIGURE 4.2 Brief depiction of two basic groups of causes of negative aspects of accelerated testing development.

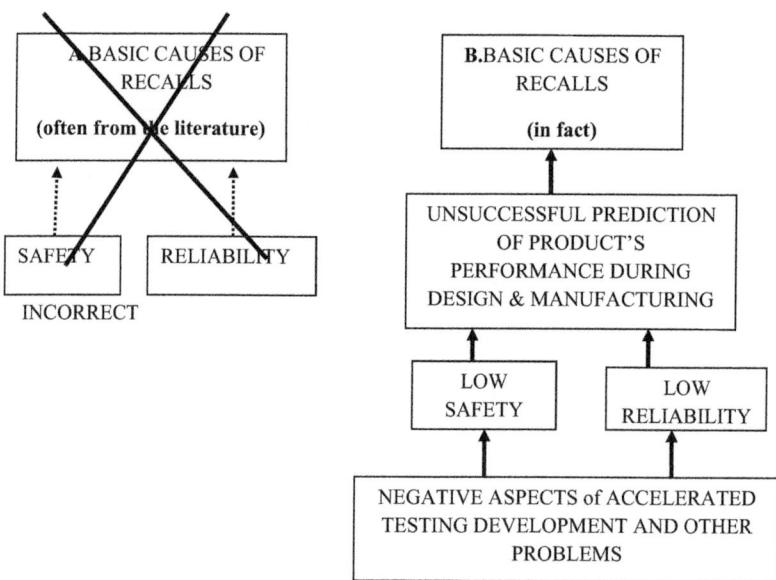

FIGURE 4.3 Depiction of the basic causes of recalls (A—as is often seen from the literature; B—the actual facts).

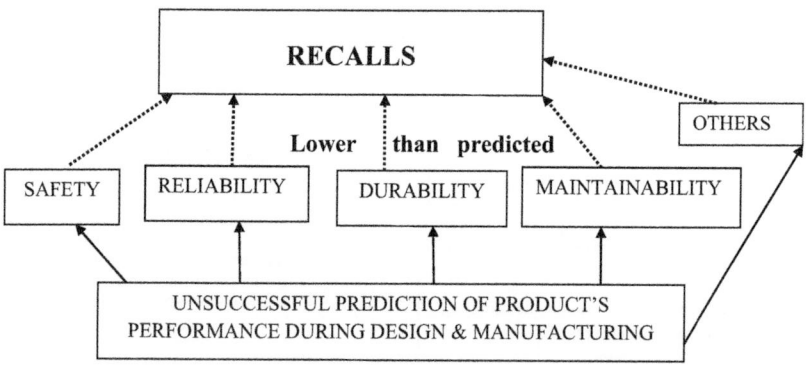

FIGURE 4.4 The path from unsuccessful prediction to recalls through lower than predicted safety, reliability, durability, maintainability, and other performance components.

testing, is adequate, and therefore they are reluctant to invest in more complex testing even though the test subject is far more complex. They think this because they consider testing to be a separate and expensive process. They do not take into account the subsequent incurred costs and procedures during design and manufacturing that are a direct result from failed predictions. For example, simple vibration testing is far less expensive than accelerated reliability and durability testing (ART/ADT) with full simulation of the field conditions in the laboratory. But vibration or vibration/temperature testing is based on only simulating one or two elements of the field conditions and ignores other elements of these conditions, as well as their interactions. Therefore, the simple vibration (or vibration + temperature) testing cannot provide the accurate results needed for studying the product's full real field/flight performance. As a result, the testing cannot provide the initial information necessary for accurate prediction of reliability, safety, durability, and other performance components.

- Often it is difficult for managers to believe that actual field situations can be accurately simulated in the laboratory, even when it is coupled with periodical field testing. But as it turns out, when properly performed, there is only a minor difference between effective accelerated reliability and durability testing (ART/ADT) and field testing. Through accurate multi-influence simulation, one can study most of the components of nature that act on the product in the laboratory instead of field/flight testing, but much more rapidly. Normally, there should be a low level of difference between the laboratory (plus periodical field) results and the real-world results. This assumes accurate simulation by transferring the field/flight influences to the laboratory with a high degree of accuracy.

    (Of course, even with this some field/flight-specific influences may need to be investigated as an additional testing component). Another factor that is rarely accounted for is the savings in development time for the product when accelerated reliability and durability testing (ART/ADT) has been utilized in comparison with that obtained using traditional methods. Using ART/ADT provides a shorter timeline and cost for the development of the full complexity of the life cycle, from design through manufacturing and usage.

- People often confuse causes and results. Frequently, you will read in the literature that safety and reliability problems are the causes for recalls. But, in fact, the safety and reliability problems are not the causes of the recalls but are the result of failures in the prediction of the product's performance.

Therefore, the failures to properly predict real-world operational failures during the design and manufacturing phases are the real causes of the recalls (Fig. 4.3).

4.2.1.2 Engineering causes

- Poor understanding of, and development of the theories of accurate simulation of field conditions.

- Failing to develop a strategy of accurate simulation of the field conditions.
- Poorly developed methodology for the accurate simulation of field conditions by failing to include all significant input influences, safety, and human factors in the laboratory plus field/flight product testing.
- Equipment that is not suitable for accurate simulation of the field conditions in the laboratory.

Shown below are the basic reasons why recalls are increasing from year to year (the number of recalls from 1980 can be seen in Ref. [2.2]). Fig. 2.5 demonstrates that vehicle recalls increased significantly in 2017 [2.8a].

A **basic negative influence that slows the implementation of advanced accelerated testing** is the mindset by all involved in the testing process managers, engineers, and academia, such that minimizing the cost and simplicity of testing are the organization's primary goal. Too frequently, their objective is to implement accelerated testing in a less expensive and simpler way. Often this is a test for a test, but not for the product's checking and development. This approach does not take into account important real-world realities, such as:

- More complex products necessitate more complex test methods and test equipment for the accurate simulation of real-world conditions necessary for successful accelerated testing.
- More complex products need more complex and accurate simulations for successful accelerated testing.
- The accuracy of, or the failure to identify the true degradation mechanism from testing influences all subsequent steps of design, manufacturing, and usage. The cost associated with failures resulting from inaccurate prediction is rarely accounted for. Only direct testing costs and time are considered. The myriad other design, manufacturing, and usage problems that are directly related to failed testing are rarely considered or quantified.
- While it appears that money has been saved in testing, it ultimately leads to increased expenses in the improvements to the design and manufacturing processes during use and in remediating customer complaints or necessitating product recalls.
- Reducing the time from design to market, but doing this by relying on inaccurate simulation and testing processes.
- Using the expression "real world" to describe testing protocols that, in many cases, are very far from real-world conditions or operations.
- Using old approaches, such as the Monte Carlo, exponential distribution, and others that are used for approximations, but do not accurately reflect the nonstationary random processes of real life.

The author has demonstrated examples of these in his books [4.1,4.2,4.6].
There are both strategic and tactical negative trends that are slowing the development of accelerated testing.

Strategic negative trends in the development of accelerated testing have a global character and are not dependent on the type of testing machines or their units. Strategic negative aspects for accelerated testing include a wide range of the following factors commonly found in engineering practice:

- Excessively rapid implementation of virtual accelerated testing, which leads to using more computer simulation for physical products. As has been shown, computer simulation is not yet ready, and therefore useful for complex products, such as complete cars, trucks, aircraft, satellites, and space research devices for interplanetary exploration. While virtual (computer) simulation appears to lead to savings in time to market and reduced costs of testing, the quality (accuracy) of real-world simulation is less. Compared to advanced types of physical simulation and testing, the accuracy of computer simulation for complex products is often less.
- Frequently, management looks upon the testing processes as nonrevenue generating costs without taking into account the financial impacts of inaccurate simulation, which results in increases to the final costs of design, manufacturing, and usage. This parochial approach too often re-sults in increased expenses for design, manufacturing, and product support, but this is never related back to it being the result of ineffective testing.
- Real-world conditions are complicated and more expensive to duplicate than virtual simulation, so there is a mindset toward decreasing the development of accurate physical simulation.
- A wide-ranging practice of using the wrong definitions associated with accelerated testing. As a result, there are misunderstandings when discussing and using accelerated testing, fatigue testing, accelerated reliability testing, reliability testing, accelerated durability testing, durability testing, proving ground testing, vibration testing, corrosion testing, and other types of testing.
- The increasing gap between the technical progress in the design and manufacture of new products and that of accelerated testing, because there is less attention focused on the development of the physical testing requirements.
- Narrow thinking about the need for and the value of, accelerated testing as an important component of a new design's effectiveness. This was described in detail in this author's previous publications [4.1,4.2,4.6], and others.
- Status Quo thinking in not wanting to initiate the institutional challenges and work involved in developing new physical testing approaches. Why should you be the one to rock the boat? Anyway, it worked before, so, why change it?

Moreover, the speed and complexity of the technical progress in the design and manufacturing of new products are increasing at a rapid rate. We see this especially in the development of new products in the automotive and aero-space areas.

But, during this same time, the speed of technological progress in the area of accelerated testing (testing) has been moving forward very slowly. In fact, much more slowly than the technological progress in the design and manufacturing areas. This is easily demonstrated by looking at the developments in both design and testing in automotive and aerospace engineering for the last dozen years. When comparing the development of accelerated types and equipment for testing for the same time period, in all types of accelerated testing (testing), the products have been moving very slowly (Fig. 4.5). Further Fig. 2.6 considers "modern" types of testing, such as HALT, HASS, AA, and other testing techniques are only a combination of some of the separate influences encountered in real life which involves many influences. And, often in using these methods, the loads and the influences used are greater than the maximums of those experienced by the actual field loads, and these assumptions change the physics-of-degradation process.

This means that the development of testing needs to keep pace with and reflect the advances in product design and development [4.1]. Simply stated, the more complicated design brings about the need for more careful and complicated testing.

As a result of the above, the gap between the speed of development of the design and testing is increasing (Fig. 4.5), resulting in an increasing difference between design complexity and the accuracy of the testing.

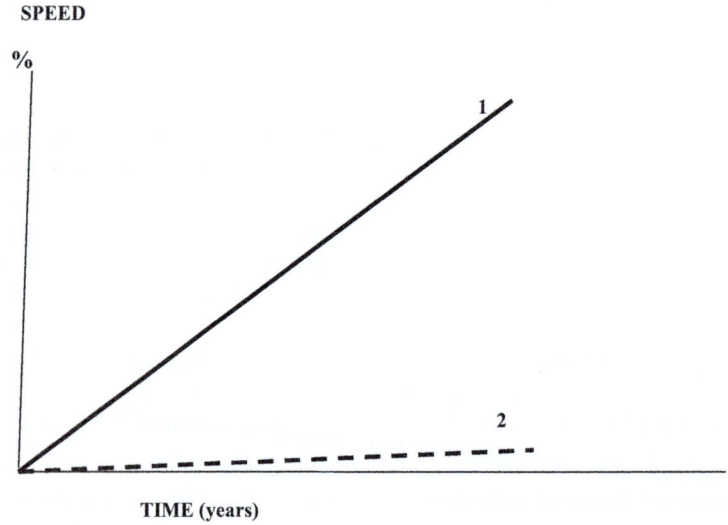

FIGURE 4.5 Common trends in the development of design process (1) and accelerated testing (2) during the last dozen years.

Often a primary negative influence on effective testing is the goal of saving money for the testing process. Management seeks to reduce the costs associated with the testing, without taking into account, the costs of poor quality of the product that are the direct results of the cheapened testing processes.

An analysis of the curve 2 demonstrated that:

- The increase in the advances in testing during the last dozen years has been very small.
- The increases that do exist in curve 2 during the last dozen years is primarily a result of increasing the implementation of modern systems of controls in testing.
- This increase in the systems of control is primarily related to the widespread implementation of electronics in these systems.
- The real technology of accelerated testing is not improving; in many cases it is actually decreasing.
- The progress in accelerated testing is deteriorating because the emphasis on field/flight simulation is decreasing.
- The effectiveness of simulation is decreasing, because there are fewer investments in accurately simulating the field/flight real conditions, especially with real and full-size components.
- Analysis of the contents of presentations at the World Congresses and Symposiums in engineering, such as SAE World Congresses, Reliability and Maintainability Symposiums, ASQ World Congresses (and World Conferences on Quality and Improvement), International Conferences on Engineering and Natural Science (ICENS), SDPC International Conferences on Sensing, Diagnostics, Prognostics, and Control, and many others will demonstrate these trends. While the above meetings are at a global level, there is scant evidence that lower level meetings are any different.
- The final basic reason is the desire to save money in the development of testing technology, which hinders the increase of the quality of the testing technology.

Too often, the primary goal of testing development is saving money for the project by reducing the cost of the testing, but without taking into account the questionable quality of product resulting from the cheapened testing processes. This is a serious and major process.

An example of this philosophy is that while aerospace projects are very expensive long-term projects, companies in this industry are constantly seeking to save money by reducing the cost of testing. When designing and manufacturing the Boeing 787, Boeing had contracts with companies in over 40 countries around the globe. For example, while companies in Israel designed and produced some of the components, they did not have contractual provisions or Boeing's permission concerning the testing of these components. Situations like this result in the lengthening of the time from design to market. So, in an attempt to save money in testing, the end result is a higher expense and a lengthened time for the entire project.

4.2.2 Common negative trends in accelerated testing development related to automotive and aerospace engineering

One common negative trend is the rapidly increasing use of virtual testing and accompanying this with a decrease in physical testing volume. This situation can be seen in Fig. 4.6.

This author has written about this in many of his previous publications, including [4.1, 4.2, 4.6], and others. The basic substance contained in these publications are:

- Virtual testing is easier and less expensive.
- The algorithms presently employed in the software programming of the test subject cannot still consider the varied interactions present in complicated machinery. Generally, they separate their simulations to discrete components or units that are not accurate representations of real-world interactions. Some researchers sometimes try using mathematical analysis, such as the Monte Carlo distribution model, but even these cannot accurately simulate the real interactions of a product's components (details and units), especially in complicated machinery.

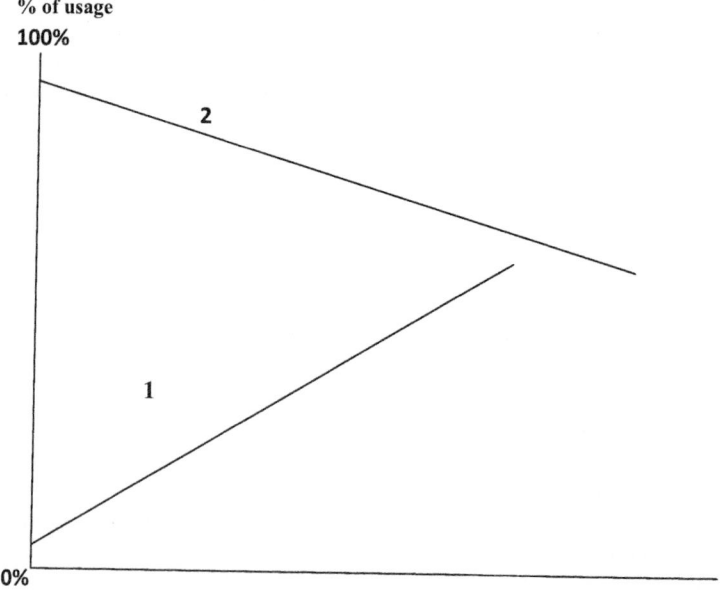

TIME (last 20-30 years)

FIGURE 4.6   Common trends of increasing virtual testing (1) and decreasing physical testing (2).

- Algorithms need to be developed for software programming that better simulate real-world conditions, such as the combined effect of solar radiation, real components, surface variations, fluctuations, and many others.
- While the virtual simulation of real-world conditions can be a very useful tool in the predesign process, companies are increasingly relying on virtual simulation for the design and the manufacturing phases.
- This increase in the use of less expensive virtual accelerated testing also encourages decreased investment and development of the more difficult and expensive physical simulation of field conditions and the investments necessary for the different types of accelerated testing. This can be seen by comparing test Expo's accelerated testing methods and equipment of 20—30 years ago to that presently available.

These observations and other problems can also be seen in the publications of other authors.

Although many professionals maintain the belief that virtual testing provides major benefits, primarily in economy and time to market savings, this assumption neglects to account for the disadvantages inherent in virtual testing as compared with physical testing. Specifically, this includes the use of inaccurate simulation of real-world conditions, their interaction(s) with the real-world environment, and interactions with other components of the product.

Physical testing has evolved with a long history of developed processes aimed primarily in the quantity area and much less on the quality area. Only some of this testing takes into account the complexity of both the product's development and the many interacted components and conditions encountered in real-world field conditions (see Refs. [4.1,4.2,4.6], and others). But seldom do these approaches raise to the level appropriate for true accelerated reliability and durability testing technology (ART/ADT), which is based on accurate physical simulation of real-world conditions. To do so, it must include the interactions of associated components and other products, taking into account the complexity of both the product's development and the many interacted components and conditions encountered in real-world conditions. Virtual testing rarely takes this into account. As a result, the virtual testing methodology is not presently ready to provide accurate simulation, and as a result, does not provide successful prediction, which then leads to increased recalls and other negative impacts, as was detailed in the preface of this book.

Consider this example. John Wilson wrote in Ref. [4.7] that the government web site that provides recalls information, https://www.recalls.gov, lists more than 60 automobile recalls, plus approximately 24 for tires in 2017. The recalls involved a wide range of manufacturers and vehicle types. In addition to the automotive-related recalls, approximately 300 consumer product recalls were issued in 2017. He wrote that military and aerospace products, which generally are not included in consumer recall data, also continue to have problems as new high-tech designs are put into operation. New, improved launch vehicles, missiles, aircraft, weapons systems, and ground vehicles are still experiencing excessive field failures.

Then John Wilson asked: "…what has happened to quality design and manufacturing? Why do we see so many recalls?"

He then provides the answer [4.7]: "…several factors came to mind. There seem to be more new products introduced than in the past. Many products are more complex than in the good old days," and more complexity means more things to go wrong. Just look under the hood of the late model car.

Also, today's products include more "high-tech" features that may not be as tried and true as they should be before being put on the market. Everything from dolls to light bulbs and appliances plus all of the "Internet of Things" gadgets have problems. Another factor is more diligent monitoring and reporting.

Twenty or 30 years ago, there were only a few product safety monitoring and reporting agencies, so consumers were just stuck with faulty (and sometimes unsafe) products. Nobody knew and nobody kept track and reported.

In this reference, he further writes that while these are all contributing factors, it is also true that we also have a more sophisticated design and quality assurance tools. The same technology that leads to the introduction of modern products can also be used to improve these products and to reduce costs. For example, thanks to integrated circuits and digital electronics products use far fewer discrete components to accomplish the same functions as 20 or 30 years ago. Theoretically, this should produce greater reliability due to fewer components, fewer and sturdier connectors, and a greater understanding of the failure mechanisms.

He further wrote: "… if technology is such a benefit, why do we still have so many inadequate or unsafe products? As an old greybeard engineer in the testing business, I have a tendency to blame all of the fancy, high-tech electronics. However, I think a thorough statistical analysis will prove that false. Many, if not most, problems are, in fact, mechanical. Some component breaks, binds, slips, wobbles, or wears prematurely. So, what is the cause of most of these problems? Why do these components not function as intended, as the computer models suggested?"

He postulates further that too much physical product testing has been eliminated from the product development process. This is largely because it is so much faster, easier, and less expensive to model and analyze on computer. Unfortunately, virtual testing is only as thorough as their programs, algorithms, and assumptions; and they are not able to foresee or account for all of the nearly infinite variety of conditions misuses, abuses, and manufacturing variables that the products experience in real-world operations.

Finally, John Wilson did not propose doing away with computerized design and analysis, or even reducing the increased rate of their use. However, he did propose that the extent of hands-on testing of physical products should be reinstated and even improved and that the feedback from such testing can further improve the computer models. Unfortunately, it is also true that regardless of the progress made in testing, where the aspect will always remain that until the products are actually used and misused and abused, we will continue to see a large number of recalls.

Another example comes from Joachim Linday, the Mercedez-Benz' Senior Manager, of Overall Testing for E-Class cars who said in Ref. [4.8]: "We can computer simulate each under-hood component and the temperature in the area in which it operates and how cooling air can be introduced. For example, we can do that for the wiring harness. But all that is design: to be certain that everything works, it still has to be physically tested because **in the computer you do not have a precise picture of the real world.**"

Then he said [4.8]: "I believe that **we always need to do final testing physically, because the customer doesn't buy an electronic program, he buys a car!**"

Fig. 4.7 demonstrates one more common negative aspect in accelerated testing—the separate simulation of field conditions inputs for testing during design, manufacturing, and usage. This does not take into account the real-world situation, where different influences (temperature, humidity, pollution, radiation, air fluctuations, field surface, human factors, and others) act simultaneously and in interaction on the product.

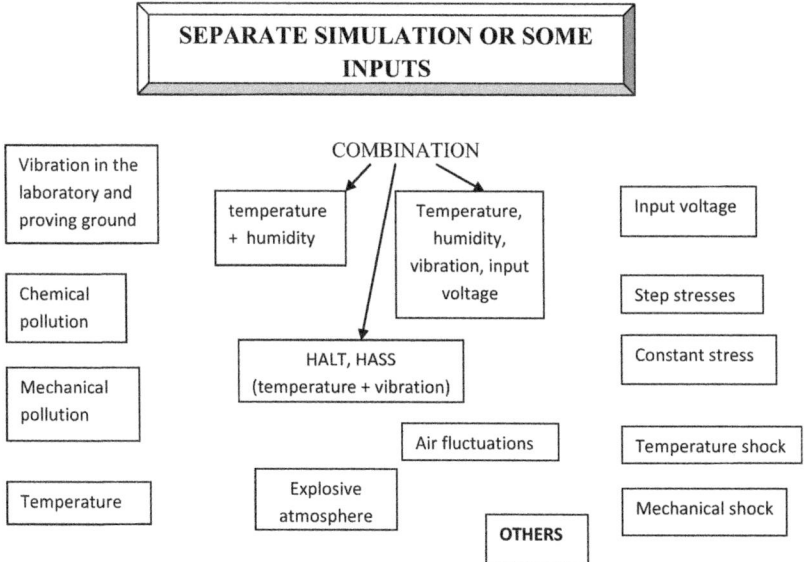

FIGURE 4.7 Illustration of some of the separate inputs (or some inputs) appropriate for the simulation of field inputs for testing during design, manufacturing, and usage.

IN REAL WORLD INTERACTED

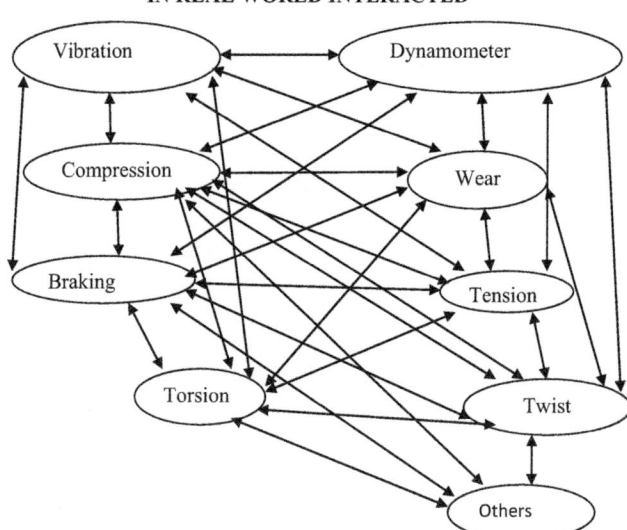

But SIMULATED AND TESTED SEPARATELY IN THE LABORATORY

FIGURE 4.8 Depicts the complex influences that should be included in mechanical accelerated testing, but are not commonly practiced for testing materials and vehicle components.

Fig. 4.8 depicts the various elements necessary for consideration in just one group, specifically the mechanical group, in the simulation, and testing of field conditions for materials, details, and units in the real world. As is shown, the various and different mechanical types of influences and testing act simultaneously in their interactions, and not separately as is usually done in the laboratory testing.

Similar situations exist for other groups (multienvironmental, electrical [electronics], etc.), of conditions for accelerated testing.

One more negative aspect of typically practiced accelerated testing is the inaccurate simulation of field conditions not just for the entire vehicle or complicated components of machines, but also in the inaccurate simulation of these conditions in accelerated testing of simple units, details, and materials.

In a similar fashion, an example of the many influences concerning corrosion testing shown in Fig. 4.9. As depicted, in the real world, corrosion is a result of the actions of the following influences:

- Chemical pollution;
- Mechanical pollution;
- Moisture;
- Temperature;
- Vibration;
- Deformation;
- Friction;
- And others.

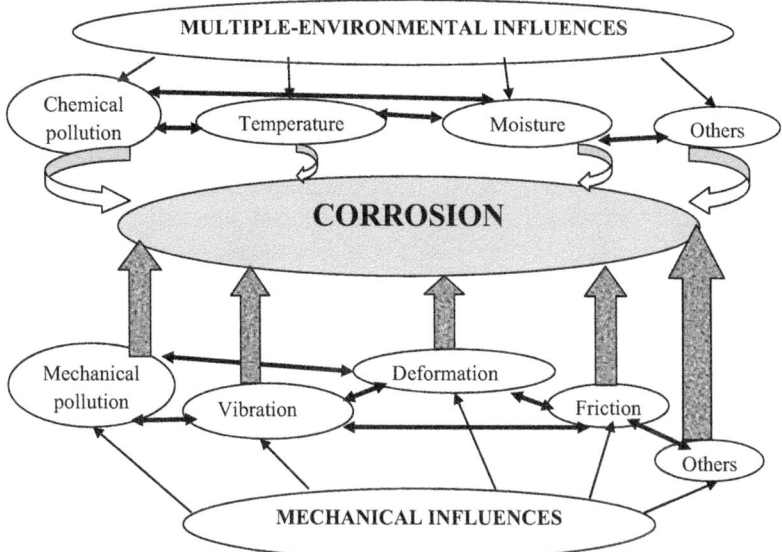

FIGURE 4.9 Diagram showing how corrosion is a result of interacted multienvironmental and mechanical input influences.

But in practice, corrosion testing as performed by many companies and organizations usually involves only one influence, chemical pollution, or sometimes combined with humidity and/or temperature only. So, it is apparent that this laboratory corrosion testing does not include real-world influences, and the simulation is not accurate. The corrosion testing results do not correspond to real-world corrosion. The final result is: the corrosion protection system designed on the testing results may not be effective. This is a continuing negative trend in development corrosion testing.

A similar negative situation is with the discrete testing of units, such as engines, transmissions, and other units, which does not simulate the vibration and environmental factors of the whole unit, which is necessary for accurate accelerated testing.

4.2.3 Specific tactical negative trends in accelerated testing development

Some of the specific tactical negative trends in the development of accelerated testing include the lack of or limited availability of:

- Specific test subjects—components or complete machines (units or details);
- Specific test subjects—complete machines or equipment;
- Specific simulation of full-field conditions;
- Specific simulation of some, or only partial influences of field conditions;

- Specific methods of field simulation;
- Specific approaches to simulation (physical, virtual, and others);
- And other specific considerations.

A major reason for each of these specific negative trends that are frequently found in academia and organizations is they do not want (or do not have the resources) to invest in the expensive and complicated equipment needed for physical simulation, and especially for the accurate simulation, of the real-world conditions. One example of this would be the comparative ease of the use of theoretical methods as compared to physical testing. This practice may be utilized by many professionals involved in testing and is not just related to units that are parts of complete equipment but also to complete equipment.

This is also true when accelerated testing is used for present production equipment or units. This type of testing is also typical for further testing of equipment at the end of the design or manufacturing phases for the product. Of course, when this tactic is employed, some components could be subjected to more modern processes for the future development of these test subjects.

Another basic negative aspect of the trends in the development and use of modern systems of accelerated testing in many industries is the continued reliance on very old and narrowly established (for example, old chemistry research data) and inaccurate statistical approaches, such as the Arrhenius distribution and the generally used exponential distribution, which are not reflective of real-world situations (see Section 4.2.5). This negative aspect can be found in many industrial companies and organizations.

Fig. 4.10 is an illustration of these negative aspects.

According to Capes Jones as stated in Ref. [4.9]: "Poor software quality costs $150 + billion per year in the United States and over $500 billion worldwide." Many of these software quality issues, come from poor test data quality. And, "According to NIST the average testing team spends between 30% and 50% of their time setting up test environments rather than on actual testing and the estimated number of projects with significant delays or quality issues is 74%."

Fortunately, one solution to the test data quality problem is through the use of a technology called virtual data. Similar to the way that virtual machines create virtual copies of physical computing resources, virtual data creates multiple lightweight virtual data copies from a single, full-size copy.

The value of virtual data can be seen in Ref. [4.9]: "The problem with test data is that fully testing if code is ready for production requires a parity copy of production data, yet creating full parity copies of production data is often too onerous for most QA teams to manage. In order to more easily manage test data in development and QA, teams often use subsets of production data. Then before releasing final code to production the code is run on a full size copy of production data. This final testing might be done in the last weeks of a multi-month project. What typically happens is that this final production parity

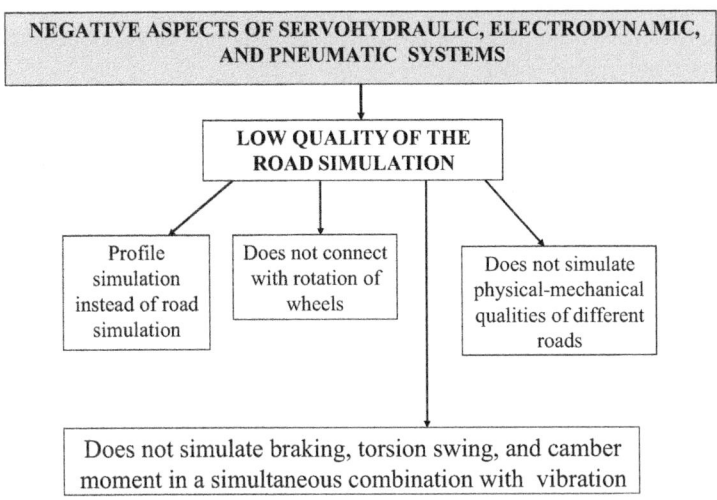

FIGURE 4.10   Some negative aspects of current vehicle laboratory vibration testing.

testing flushes out more bugs than can be fixed before the release date. Thus the release has to either be delayed or released with bugs."

Another example can be seen in the book [4.10]. This book, which contains 24 Chapters and 560 pages, includes 177 subchapters. In this book, text relative to testing is found in only seven subchapters. Specifically:

- The role of testing (3 pages in Chapter 5);
- Product testing (2 pages in Chapter 8 "Reliability Growth and Testing");
- Accelerated tests (2 pages in Chapter 8);
- Tools and systems that support testing (5 pages in Chapter 8);
- Equivalence testing (4 pages in Chapter 15);
- Some concepts and significance tests (1 page in Chapter 15);
- Accelerated life testing (4 pages in Chapter 22).

It is interesting that with this limited discussion of testing, the authors provided this description for their publication: "The authors begin by presenting broad insights and high-level strategies for improving product quality. Next, they demonstrate how to implement robustness and reliability strategies that complement existing governance and decision processes. A section on tools and methods shows how to institutionalize best practices and apply them consistently. Finally, they tie strategies, decisions, and methods together through a case study project."

And "Readers are introduced to many thought leaders whose writings can be sources of further learning."

From instances like these, it is easy to see why the trends of accelerated testing are moving so slowly.

Another example of these negative aspects of accelerated testing development can be seen in Ref. [4.11]. In this publication, the authors formulate the definition of accelerated life testing as a method for stress testing manufactured products that attempt to duplicate the wear and tear that would normally be experienced over the useable lifetime of the product, but in a shorter period of time.

This definition includes the common formulation which relates to any testing, where one obtains test results in a time frame that is shorter than that experienced in normal usage. But this definition of accelerated testing is too general and does not provide any specific requirements. Formulation, such as this do little to advance the art and science of testing.

The authors actually confirmed this when they wrote:

Problems

- It is difficult to simulate with operating conditions;
- In real life, they interact with other components;
- Little value in determining the absolute reliability.

The authors then formulate their definition for highly accelerated life testing (HALT) as:

● A test in which stresses are applied to the product well beyond those experienced in normal shipping, storage, and in-use levels.

And they contend that:

● HALT is scientific;
● HALT has statistical differences with ALT.

But they offer little evidence as to what is meant by scientific or what specific statistical differences exist between HALT and HASS.

The authors further write:

"Accelerated Factor AF > 1:

● Exposing tests units to more than normal stresses.
● Higher temperature;
● Higher humidity;
● Higher vibration.

Accelerated Chemical/Physical Degradation"

But they remain mute on other real-life factors that influence product degradation and failures.

These are a few of the examples of how some current publications are not embracing the advances needed in accelerated testing.

Excerpt from the seminar titled Autonomous Vehicle Test and Development Symposium.

Engineers and managers involved with product development are constantly challenged to reduce time to market, minimize warranty costs, and increase product quality. With less and less time for testing, the need for effective accelerated test procedures has never been greater. This course covers the benefits, limitations, processes, and applications of several proven accelerated test methods, including accelerated reliability, step stress, FSLT (full system life test), FMVT (failure mode verification testing), HALT (highly accelerated life testing), and HASS (highly accelerated stress screening).

A combination of hands-on exercises, team activities, discussion, and lecture are used throughout the course. Participants will also receive a copy of the instructor's book, Accelerated Testing and Validation Management, *which includes numerous hands-on exercises and a CD with analytical spreadsheets.*

The program for this seminar provided another example of the poor attention given to the role and importance of testing. While this symposium title is "Autonomous Vehicle Test and Development Symposium" (which was offered in conjunction with Autonomous Vehicle International magazine [4.12]), but among the many presentations only three were related to testing. And, all three were related to proving ground testing that, as will be discussed later, can evaluate the product in proving ground conditions, but cannot provide the information for real-world reliability, durability, safety, and maintainability evaluation, and prediction. This is because proving ground testing cannot account for real-life environmental influences, including customer usage changes during the product's service life.

Proving ground data also does not take into account the real-life combinations of actions from input influences, human factors, and safety factors.

This symposium was an integral part of the Autonomous Vehicle Technology World Expo.

As written in the Symposium program, "Advanced driver assistance systems giving rise to fully automated driving vehicle technology is nothing new. Since the final meeting of the Eureka PROMETHEUS Project in Paris in 1994, it's been clear that fully autonomous self-driving vehicles are set to become a reality. Yet 20 years later, the final stages of testing, validation, and failsafing pose a huge challenge to the automotive industry. The rigorousness and thoroughness of the testing processes need to be at an altogether higher level than anything that has gone before, if the final reality is to be achieved with complete safety and integrity guaranteed."

The Autonomous Vehicle Test and Development Symposium is part of Europe's largest autonomous vehicle event.

Topics under discussion at this symposium included:

- Public road testing
- Virtual testing
- Simulation
- Traffic scenario testing
- Embedded software testing
- Reliability testing of software and hardware systems
- Safety and crash testing
- Fail-safe testing
- Cyber-threat testing
- Validation and verification
- Autonomy software
- VeHIL
- V2V and V2X testing
- Robotics
- Testing legislation
- Safety standards and legislation
- Human factors and HMI testing
- Case studies
- Possibilities
- Best practices

This also appears in Ref. [4.13], where it was written "Accelerated durability testing of automotive components has become a major interest for the ground vehicle industries." The authors think that this approach can predict the life characteristics of the vehicle by testing fatigue failure at higher stress levels within a shorter period of time. Currently, the tradition of laboratory testing includes a rigid fixture to mount the component to the shaker table. The authors are correct when they write that this approach is not accurate for the durability testing of most vehicle components, especially for those parts that are connected directly to the tire and suspension system. In their work, the effects of the elastic support on model parameters of the tested structure, such as natural frequencies, damping ratios, and mode shapes, as well as the estimated structural fatigue life in the durability testing, were studied through experimental testing and numerical simulations. First, a specially designed subscaled experimental testing bed with both rigid and elastic supports was developed to study the effects of the additional elastic support and the mass on the change of structural model parameters. The significant model parameters showed variation due to the additional elastic support, which was illustrated by the experimental results. The model parameters with elastic support were then used [4.13] to build and tune the finite element model (FEM).

Another example can be seen in Ref. [4.14], which focused on the approach to ground vehicle durability testing development. The analysis presented in this publication is important because it demonstrated some of the basic negative trends in accelerated testing development, especially as applied to ground vehicle testing. In fact, their methodology uses very old (over 100 years) ground vehicle testing approaches for the evaluation of modern product durability in the real world. The authors wrote in Ref. [4.14] that a test vehicle should be subjected to rigorous testing in the field and on proving ground tracks and that the vehicle's response to each is captured. The acquired road load data is analyzed, and pseudo damage values are estimated for both field and proving ground data. Using this methodology, which is in current practice, proving ground sequences are derived based on the correlation of the damage produced in customer usage conditions and from accelerated proving ground tracks data. This correlation required that the structural frequency responses gathered from the field should also be incorporated in the proving ground testing, and the combination from the two provide product durability validation.

In this paper, improvement of the quality of accelerated proving ground (PG) durability testing, was derived by a frequency-based relative damage spectrum (RDS) method. This methodology was executed and validated in a passenger vehicle model in which the process was as follows:

- Test vehicle preparation;
- Customer usage profiling;
- Road load data measurement in both field and PG testing;
- RDS correlation.

The authors provided customer usage profiling through the combination of simulation of test vehicle use based on the real-world customer usage and terrain conditions while undertaking data acquisition trials in the field. The information about payload, road conditions, the period of operation, and turnaround time were collected by marketing surveys with customers (Table 4.1).

TABLE 4.1 Customer usage profile for passenger vehicle application as per Ref. [4.5].

Road type	Rated load	Unladen	Total
Highway (smooth terrain)	53%	23%	76%
Bad road (rough terrain)	13%	6%	19%
Off road (nonasphalt)	4%	1%	5%
Total	70%	30%	100%

A finalized usage profile and intended performance profile was developed based on the surveys and was finalized, as presented in Ref. [4.15]. The actual customer routes that were used in the field trials were selected subjectively. It is represented in the usage profile [4.16]. These publications wrote that the next step would require road load data development in both the field and proving ground conditions.

For proving data measurements, trials were to be undertaken at different speeds in each track to capture a wide range of frequency responses.

The next important negative aspect of this approach is the development of correlation methodology for comparing proving ground testing results and field results. The following phases were utilized to establish the "optimized?" proving ground durability test sequence:

- Terrain classification;
- Mixed calculation and target setting;
- PSD analysis;
- Frequency-based damage correlation.

For the last bulleted item—frequency-based damage correlation—they use a calculation of the potential damage incurred from both proving ground and field road load data by using a common artificial SN curve and slope, as was published in Ref. [4.17].

The authors then concluded in Ref. [4.17] that "this optimized proving ground durability test sequence can be developed for each commercial vehicle model by developing a frequency based correlation of road load inputs of customer usage. These test cycles give a competitive edge to vehicle manufacturers over others for providing the customers with a more durable product and for incurring less developmental and warranty costs."

But, as was stated earlier by this author, proving ground test does not take into account the multienvironmental and other factors that cannot be simulated on the proving ground, because it ignores these factors, proving that ground testing is not an adequate approach for durability prediction or degradation estimates for field operations. Any proposed methods of recalculating proving ground test results as a substitute for actual field test results cannot represent the physical essence of real-world actions. But, as proving ground testing is less expensive and simpler than serious durability (using reliability and other test methods) testing, this negative trend in the development of accelerated testing continues.

The below example relates to erroneous use vibration testing and call this as reliability testing [4.17a].

The 14th Institute of the China Electronics Technology Group Corporation is the birthplace of China's radar industry. It has developed many pieces of high-end radar equipment and its products have been exported to dozens of countries and regions.

The 14th Institute of the China Electronics Technology has been using "m + p" international's industry accepted VibControl vibration control systems since 1998 to conduct sophisticated radar reliability tests.

In this article was written:

"The vibration tests carried on shakers and controlled by "m + p" VibControl systems, therefore, play a key role in reliability verification. The 14th Institute has more than 10 sets of "m + p" VibControl Software with m + p VibPilot and m + p VibRunner acquisition hardware, m + p VibPilot is a compact, rugged 4/8 −channel hardware platform. For higher channel counts, m + p VibRunner hardware is the first choice. It can be used as a desktop instrument or mounted into a 19 in. rack and supports distributed measurements. Equipped with 24-bit sigma-delta A/D converters and a sampling rate up to 204.8 KHz, the m + p acquisition hardware allows for alias-protected measurement in a frequency range up to 80 kHz and with more than 120 dB spurious-free dynamic range.

Following the strategy of always striving to be a leader in the aerospace and defense testing sector, m + p international has integrated advanced control capabilities, such as nothing/force limiting into its VibControl software. Many safety features ensure reliable closed-loop vibration control - from pretest checks to abort checking, nothing, and controlled shutdown.

The radar's reliability level is constantly increased using these tests. Finally, the expected value of mean time between failures is reached and its operating cost is reduced.

In many radar reliability tests, m + p international's products played an important role. The responsible technical engineers of the 14th Institute expressed their satisfaction with the test results and gave a high evaluation of m + p international's products [4.17a]."

As we can see, this Institute uses vibration testing, which is one from many components of reliability testing, but call them "reliability testing." This is a continuation to wrongly use separate influences simulation instead of complex simulation, as was described earlier in this book, as well as in other author's books, real-life reliability testing.

4.2.4 Trends in using virtual (computer) simulation and testing as a replacement for field/flight conditions

There are several driving forces that are reducing field or physical product testing and replacing them with computer simulation. These include:

1. Publications regularly emphasize the trends in automobile engineering development that are based on a product's development that is accomplished through computer simulation and accelerated testing.

For example, James Truchard, President and CEO, National Instrument wrote [4.18] that for most of the last 100 years, a special focus has been placed on quality and testing that was traditionally physical testing. However, this has been changing over the past decade. This is partially due to significant emphasis in reducing automotive product cycles from the previous norm of 4–5 years to present requirements for 12–18 months. This change, called "zero prototyping," was only recently introduced.

As a result of this trend, simulation needed to be added formally into the design cycle, resulting in a "simulate-build-test-ship" design cycle. The dependence on simulation early in the design cycle intensified the need to institute simulation by mathematical models with test results much earlier in the product development cycle.

2. Another recent trend is the increasing complexity of vehicles. This is frequently coupled with the replacement of mechanical systems or structures with electronics. As the vehicles' electronic content increases, most of the testing, which was previously primarily mechanical in nature, now incorporates the added complexity of electronics testing. This presents the requirement for a common test platform that addresses seamlessly both the mechanical and electronic test environments.

This increased complexity also means product designers and manufacturers have had to rise to a multitude of challenges, as product specifications have become more complex. One example of this is the addition of tire pressure monitoring systems to vehicles. Also, due to globalization, a typical vehicle platform, which was originally designed only for the U.S. market, must be reengineered for European and Far-Eastern markets. It is shared across different brands worldwide and ultimately becomes the basis for vehicles in several classes manufactured in different plants around the world and complying with a multitude of regulatory requirements.

3. The integration of physical testing with simulation. In the product-development process, testing appears in two areas: first upstream as a means for establishing product designs and performance requirements, and later downstream as a pass/fail criteria on the product prior to its release for production.

Although the types of measurements in both instances are similar, the tests' purposes and who and how an organization uses the results of the testing are substantially different.

A bidirectional flow of information between the simulation used in the design and the information gained from product performance testing is critical for success. Traditionally engineers compared testing data from previous models or components against simulation results and used the information to calibrate them, thereby increasing the confidence in their simulation predictions for the current designs. Test data from previous models and components can also be used as inputs for new simulations to improve the fidelity of their results. Simulation can also provide insights to permit minimizing and optimizing testing. Determining the optimum location of sensors, actuators, and exciters is one such example.

Prior to physical testing, simulations can help engineers identify the optimal designs or challenges that exist with the prototype.

Another important aspect of this integration is the need for an early test platform to deliver connectivity to the design and simulation tools. Simulation software vendors need to build better connectivity with test platforms moving forward as has happened over the past decade with the Computer Aided Design (CAD). This should also include addressing the growing need to provide integrated methods that allow better visualization of, and comparison of test and simulation data, especially through the increased use of video. But it must always be fundamental that this virtual simulation and integration testing needs to include the correct simulation of the real-world processes, as has been detailed earlier.

4. The multidomain and multidiscipline nature of modern automotive system design and testing is another growing trend that greatly affects testing methodologies. Consider, for example, the automotive entertainment system.

This once simple discrete component, a radio, has been completely transformed into an automotive information and media center. Today's "radio" design and testing must now include assuring the performance of TV display(s), MP Player, DVD and CD players, FM and satellite radio, GPS navigation systems, cell phone and e-mail access, electronic games, remote diagnostics, satellite-based car alarm and control, and other features. This is a major change from the original AM radio function.

Similar challenges exist with other systems, such as power train management, climate control, intelligent braking and handling systems, and other features. The quality assurance and fail-safe design of such complex systems require multidomain measurements and excitation capabilities in a modular and expandable test platform that must be executed in a coordinated and time-critical manner.

5. The need for next-generation physical test platforms. With the significant changes occurring in the automotive accelerated testing methodologies, it is unlikely that the past fragmented approaches to testing will be able to scale appropriately. There is a growing need for an easy-to-use robust modular, customizable commercial off the shelf (COTS) test hardware and software testing platforms with plug-and-play architecture, akin to CAD. It must also be robust enough to alleviate hardware connectivity issues, making the hardware intuitive or transparent. It should also provide seamless connectivity to the environment surrounding the test, using

accepted standards wherever possible through the use of communication buses, such as CAN (controller area network), sensors and actuators with Transducer Electronic Data Sheet (TED), design and simulation Product Life Cycle Management (PLM) software, Enterprise Resource Planning (EKP), manufacturing execution systems, and others. The National Instruments (NI) LabVIEW [4.18] graphical test software platform is one such example. LabVIEW integrates seamlessly with hardware and provides build-in connectivity to third-party devices and the test environment.

One ongoing need is for standardized terminology and the homogeneous application of this standardized nomenclature by everyone involved in the testing profession. An example of the present state of this issue can be found in Ref. [4.19]. The article began by using misleading terms for vibration testing and environmental testing. For example, the article states "Environmental dynamic testing is a technical discipline that includes all vibration tests conducted on most engineering structures. The goal is to simulate the effects of the operational environment on a given object. A car clutch, a dishwasher pump, or an airplane altimeter are just a few objects required to pass a dynamic environmental test prior to use."

Then it continued, "In general, there are three main stages in vibration tests: first is test setup. This phase is critical to test success and the actual component lifespan relies on a good test setup. Two tasks need to be accomplished: define a test profile that represents the operational vibratory environment; and fix the test item to the shaker in a way which represents its real operational mounting. In many cases test profiles are taken from standards."

This is problematic from several aspects. First, a "vibratory environment" is a combination of two different influences—vibration and environment that must be applied in a combined fashion. Second and especially troubling, to duplicate real-life conditions, the test profile used must be based on the real field operations, not from those contained in a standard.

It continues: "The second stage is the test itself. During the test, the vibration controller is the main player. And there are lots of questions. Single or multiple inputs? Response limiting? How many statistical degrees-of-freedom (random test) or which compression factor (sine test)?"

But qualified professionals know that vibration in the real world has a random character. Importantly, the real-world inputs are multiple, not singular; and that for a 21st Century testing protocol, the real vibration of mobile equipment must have six degrees of freedom.

As can be seen in this example of a publication in a respected international magazine, the lack of common terminology and practices is actually in conflict with the title of the article, which is "Back to Environmental Basics" and is more aligned to meaning "Go back to the 20th Century."

4.2.5 Erroneous use of the exponential law of distribution in accelerated testing

The major question is, why are so many using some of the mathematical laws of distributions that are not appropriate for use in accelerated testing?

Introduction

Analysis of the exponential growth-based models

Exponential growth is exhibited when the rate of change—the change per instant or unit of time—of the value of a mathematical function is proportional to the function's current value, thereby resulting in its value at any time being an exponential function of time, i.e., a function in which the time value is the exponent. Exponential decay occurs when the growth rate is negative. In the case of a discrete domain of definition with equal intervals, it is also called **geometric growth** or **geometric decay**. With either exponential growth or exponential decay, the ratio of the rate of change of the quantity to its current size remains constant over time.

The formula for the exponential growth of a variable x at the growth rate r, as time t goes on in discrete intervals (that is, at integer times 0, 1, 2, 3, ...), is

$$X_t = X_0 \, (1 + r)^t$$

where x_0 is the value of x at time 0. This formula is transparent when the exponents are converted to multiplication.

It should be obvious that the exponential growth or decay models are rough and very far removed from real-world processes. The use of the exponential law of distribution often results in errors of 1, 2, and three multiple degrees. Also, this law does not account for the interdependence between real failures.

A basic justification that researchers often invoke for using this law for real-world analysis is that they are starting with minimal available information—typically from studied processes with no more than a dozen or so experiments. But in such cases, it does not matter which law is used, because there will be a need to adjust and trim the methodology more as learned about the product's real-world performance.

Events in which there is an exponential law of distribution correlation

Several hundred data points are needed to accurately confirm the validity of applying any law of distribution.

Fig. 4.11 depicts the defects distribution and causation during a typical product's life cycle. While the early and late phases of the product's life cycle have higher defects, the working period is characterized by a longer period of time and one with lower defects, as can be seen in this figure. In this Figure, Domains 1 and 3 (Fig. 4.11) have a relatively insignificant role, as the product in practice primarily works in the second phase. In such situations, the defects

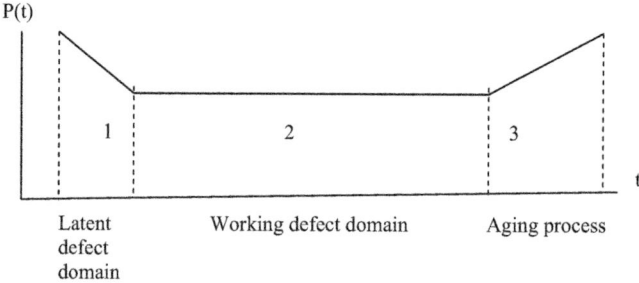

FIGURE 4.11 A product's defects by segments (domain) of its service life.

during the working period (second phase) can be approximated by the application of the exponential law.

If the product's life cycle defect pattern is similar to that shown in Fig. 4.11, then (b) $P(x + y) \approx [P(x) - P(y)]$.

If there is approximate equality in equation (b) above, the exponential law of distribution is satisfied and can be used in the prediction of product failures. This is an intersection of high-level processes with low levels of correlation.

If in aerospace research stations, such as satellites, if it does not collide with a meteorite in phase 1, it also should not collide during phase 2.

If adding a large number of independent streams, the summary stream— Poisson, i.e., ordinary without an aftereffect.

It happens when there are failures of complicated systems during a short part of time or events when certainly no exponential law occurs.

(a) If one considered region enough filling aging (Figs. 4.12 and 4.13).

(b) If there is an increasing intensity of failures at the beginning phase, the defects are called latent defects. This failure grouping is generally related to a large number or all of the products and is an indicator of low product quality. This can often be seen in farm and other machinery that is produced in undeveloped countries. Their products have many failures in the first phase shown in Fig. 4.11, because their quality is low and results in many latent defects (Fig. 4.14).

(c) If a product is produced by different suppliers or if details and units of the complete product are furnished by different suppliers, and these suppliers

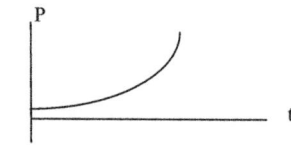

FIGURE 4.12 One variant of enough filling aging.

FIGURE 4.13   Second variant of enough filling aging.

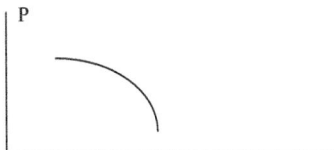

FIGURE 4.14   Illustration of the variant b.

have different quality cultures, some of them being suppliers with a low culture of quality manufacturing.

(d) Repair or maintenance time never has the characteristics of the exponential law of distribution. This is a situation when the exponential law of distribution can be used, although it is not actually occurring.

(e) Arriving Poisson's stream of failures, while the time of work has an exponential distribution, the time for repair does not have this distribution. The mean time of failures does not depend on the laws of distribution (Fig. 4.15).

If independent devices are connected sequentially, the mean characteristics of their reliability is not related to the laws of distribution.

For some models, the characteristics of failures (reliability) may exhibit a coefficient of variation.

Nevertheless, in most cases, the exponential law of distribution should not be used.

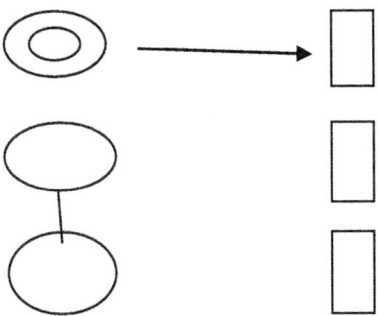

FIGURE 4.15   Illustration of 3a.

Let us demonstrate the result of using this law when it should not be used.

(a) Exponential law of distribution $P_1(t) = e^{-t}$

Weibull distribution $\qquad\qquad P_2(t) = e^{\pi} 4t^2$

$t = 0.01$

Probability of failures in the first case is $1 - P_1 = 0.01$

Probability of failures in the second case is $1 - P_2 = 0.000078$

So, in the above example, when there is actually a Weibull law of distribution, using the exponential law of distribution produces an error of 120 times.

(b) Estimating the number of spare parts necessary when there is a need for immediate restoration of the product to service **t**.

Given: t G

$\frac{t}{T_0} = 30; \quad \frac{G}{T_0} = \frac{1}{4} \lambda = 1/T_0; \quad \gamma = 0.99$

Using Poisson's law, the required number of spare parts is:

$N_0 = 41$

But by using the exact asymptotic formulas:

$N_0 = 33$

(c) There is a system consisting of 1000 equal components. The mean time to failure for each component is T_0. What is the mean time of system failure, if the components are combined sequentially?

If the life distribution for each component is estimated using the exponential law.

$T_N = T_0/N = T_0/1000$.

However, by using the Weibull law, $T_N = T_0/32$.

Using the Weibull law, the time to system failure will be 30 times greater than by using the exponential low.

In practice, Poisson's stream does not have constant parameters, so it cannot be considered to be the simplest method of approximation.

If a system is to be considered with recovery, the characteristics of reliability cannot be determinate in the isolated form (Fig. 4.16).

G (x) is an analysis of the maintenance shop, with the time needed for repair (maintenance).

The probability that a system of K devices needs repair:

$P_k \sim \dfrac{\lambda^k M h^k}{K!}$

If $k = 3$; $G(x) = 1 - G^{-\mu x}$;

$\dfrac{\lambda}{M} = 0.1$ (mean time of maintenance is in 10 times less than mean time between maintenances).

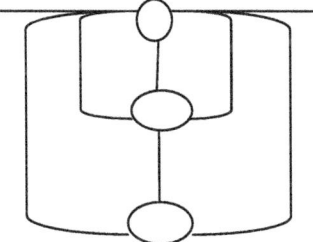

FIGURE 4.16   Illustration to 4c description.

Second variant.

Mean time of maintenance I/M.

In first variant $P_3 (1) = 10^{-3}$

In second variant $P_3 (I) = I/\sigma \cdot 10^{-3}$

If the variation coefficient $> I$, the error is on one side, but if $V < I$, the error will be on another side.

Conclusions

When the exponential law is not appropriate, as detailed in the proceeding analysis and if exponential law cannot be used, the methods of reliability evaluation described in the book Ref. [4.38] may be useful in such situations. When using the exponential law, it is necessary to validate the appropriateness of its usage statistically (through the acquisition of an adequate number of data points (typically in the hundreds) to verify conformance with this law) or to physically—justify the applicability of the exponential law from physical considerations, as was shown earlier in this subchapter.

If the applicability of the exponential law cannot be substantiated by either of the above methods, it may not be used, and alternative solutions must be found and applied.

4.3 Positive trends in the development of accelerated testing

4.3.1 Common positive trends in accelerated testing development

The development and implementation of accelerated reliability and durability testing technology (ART/ADT) is one of the basic positive trends in dramatically improving automotive and aerospace engineering.

As WILEY published in the back cover of the book Ref. [4.1] "...it is a key factor to accurately predict a product's quality, reliability, durability, and maintainability during a given time, such as service life or warranty periods. It covers new ideas and offers a unique approach to accurate simulation and integration of field inputs, safety, and human factors, as well as accelerated product development, as components of interdisciplinary systems engineering ..."

This direction of testing ART/ADT can be seen in the newest techniques in the testing field, and it provides many case studies of the implementation of this new direction. Its increasing use is solving previously unresolved product development problems, such as reducing product recalls and both the cost and time from design to market. It also is leading to higher levels of design, manufacture, and usage efficiencies.

Finally, ART/ADT leads to improved products, which is beneficial to both producers and users.

Fig. 4.17 shows the nature of ART/ADT.

A more detailed description of this accelerated testing technology can be found in Ref. [4.1].

The trends in the development of this technology are related to the combination of physical testing in the laboratory, including an electronic system of control, with field/flight testing, and leads to 21−22 Centuries product development. As has been shown, this also relates to entire testing machines with their interacted components (units and details). Fig. 4.18 depicts some of the current basic trends in the development of ART/ADT.

Greater information about the implementation of accelerated reliability and durability testing can be found in Chapter 6 of the book Ref. [4.2].

Fig. 2.9 (Chapter 2) depicts the basic steps on the path from traditional and accelerated testing to ART/ADT.

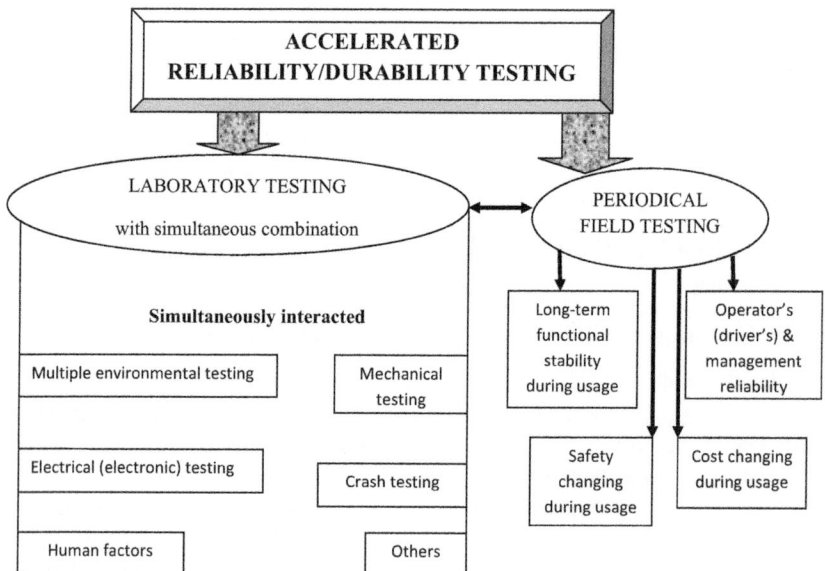

FIGURE 4.17 Depiction of accelerated reliability and durability testing (ART/ADT).

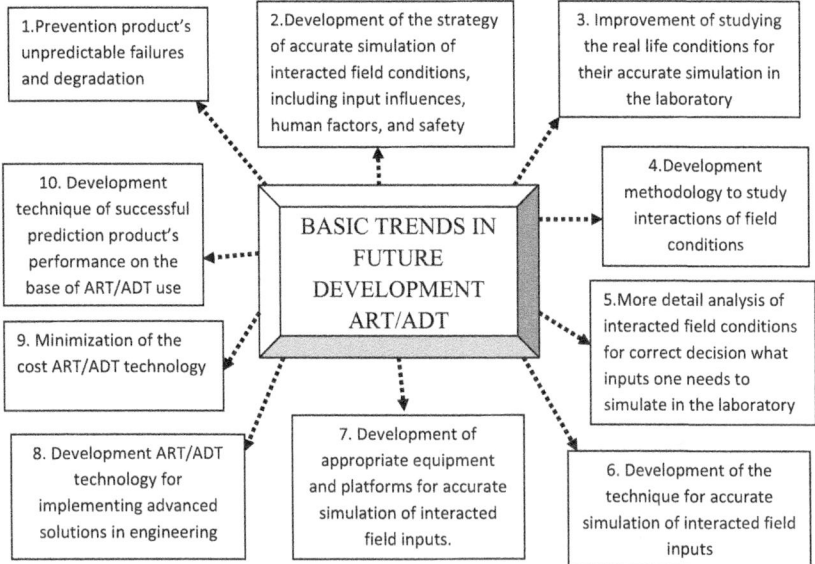

FIGURE 4.18 Schematic depiction of some of the trends in the future development of accelerated reliability and durability testing for automotive and aerospace engineering.

The path from traditional accelerated life testing (ALT), which uses separate (or two) field influences simulation to accelerated reliability and durability (ART/ADT) testing consists of:

1. Simulation using separate (or two) field inputs [ALT];
2. Increasing the accuracy of the field inputs;
3. "Modernized" testing (often wrongly called "durability" or "reliability" test, but, in fact, only simple accelerated testing [HALT, HAAS, AA, etc.,]). Most of this consists of increased (higher) loading simulation of temperature and vibration;
4. Testing with simulation using combined inputs;
5. Simulation of full-field conditions for true ART/ADT.

Antony James provided and published [4.21] an interview with Robert Rutledge, a fatigue testing expert at the National Research Council of Canada. In the interview, Mr. Rutledge said that fatigue testing had changed significantly in many aspects over the years. It went from using draftsmen to produce drawings to now using CATIA with three-dimensional layouts of the test and loading systems. The advances in computer control systems and data storage have allowed these tests to be performed with much more complexity and higher fidelity, while also being accomplished in shorter periods of time.

This has been predominately due to the National Research Council's (NRC) ongoing efforts to decrease project time.

But the above is only part of the positive trends that are being seen in the field of ART/ADT.

Schematically, the basic positive trends in future development ART/ADT can be seen in Fig. 4.18.

Next, consider the work of A and D Europe GmbH as detailed in Ref. [4.22]. A&D is involved in providing the hardware and software technology, as well as the application expertise and the commitment to help implement advanced and cost-effective solutions to the testing challenges faced by the industry.

One of their products is Automation iTest (DAC). This is Powerful Data Acquisition and Control Software for Power Train Testing.

The iTest system is a central control system that is useful for modern powertrain testing. It executes a predefined test schedule by coordinating the dynamometer and throttle systems, along with data acquisition from sensors on the test article or in-cell instruments. iTest achieves this level of functionality through an intuitive configuration editor and user-friendly operator GUI, while employing international standards for connectivity to the test cell instruments. The operator can select from a library of test cycles, which are based on the testing standards that have been established by North American, European, or Asian governing organizations.

The Automation Console software provides a powerful graphical user interface (GUI) that runs in a Microsoft Windows environment. The GUI provides access to the screen definition tools both in the test building environment and during run time.

A&D's iTest system has a modular system for adding test cell instruments or test cycles to its base application. Modules contain command sets specific to the instruments in use, with common commands used by common testing. This allows a single test to be used with a variety of instrument modules in different test cells, streamlining the test development process. iTest Application Modules are designed to integrate a wide variety of third-party devices and tests into a test cell. They contain the necessary application code to remotely control these devices.

Their integrated solutions combine the following elements:

- Industry-standard hardware/software for real-time control, data acquisition, and combustion analysis;
- Experience with power train test applications in diverse industries;
- Spans the full spectrum from upgrading a single facility to acting as prime contractor for multi-cell, turnkey installations.

While the above iTest system is used for testing under separate influences, it is not capable of true accelerated reliability or durability testing.

In Ref. [4.23] AB Dynamics wrote that they have grown steadily and have become major suppliers of automotive test systems.

Applications range from durability testing to precision control for critical new areas of technology development, such as active safety and autonomous driving, all with efficient data and protocol integration from the virtual to the physical to the real-world.

With fewer prototypes available to develop and more technologies and more derivatives, test systems must be quick to set-up and efficient to integrate seamlessly with the customer's development processes.

Their system includes:

- Driving simulators
- Kinematics and compliance testing:

Another product offered by AB Dynamics is the SPMM 5000 (Suspension Parameter Measurement Machine)

- The SPMM 5000 is a vehicle Kinematics and Compliance (K&C) test machine designed by suspension engineers for use by suspension engineers.
- It is designed to combine the most faithful possible simulation of on-road vehicle behavior, and as such, it would be the development tool of choice for suspension engineers around the world.

When equipped with the SPMM 5000 MIMS upgrade, it allows accurate measurement of Center of Gravity and Principal Moments of Inertia for full vehicles and for vehicle components.

4.3.1.1 A&D modular test system

A&D has developed a modular system for adding test cell instruments or test cycles to a base application. These modules contain command sets specific to the instruments in use, with common commands used by a common test. This allows a single test to be used with a variety of instrument modules in different test cells, streamlining the test development process. iTest Application Modules are also designed to integrate a wide variety of third-party devices and tests in a test cell.

But, again, this system also cannot be used for accelerated reliability or durability testing because it is not based on accurate simulation of full-field input influences, human factors, and safety.

4.3.2 Specific positive trends in accelerated testing that relate to any specific type of testing in automotive and aerospace engineering

4.3.2.1 Positives in reliability testing

To specific positive trends in the development of accelerated testing relate the different aspects of reliability testing. Unfortunately, these are seldom found in books or in articles and papers on reliability testing. For example, the annual

Reliability and Maintainability Symposium [4.39a] was held in 2019 in Orlando, Florida. This symposium provided over 100 tutorials, sessions, and workshop papers. But, of these, only the three papers listed below, focused on reliability testing:

- Martin Dazer, Berndt Bertsche, Alexander Grundber. "Cost and Time Effective Planning of Accelerated Reliability Demonstration Tests." This paper is an introduction to a new approach of comparing the expenditure of Success Run and End-of-Life tests as regards to time, cost, and the probability of success.
- Narasimman Sunderajan. "Replication of Field Scenario in Reliability Testing." This paper suggested a methodology of how to replicate the field scenario in testing by combining the stress in the test and effectively using Kurtosis in random vibration testing.
- Yuankai Gao. "Reliability Enhancement Test Evaluation Based on Accelerated Growth Model." This considered a quantitative evaluation method for reliability enhancement tests based on an accelerated growth model.

4.3.2.2 Positives in vibration testing

As can be seen in Ref. [4.24], the TEAM Corporation is responsible for some positive trends in the development of accelerated testing. As a supplier to the aerospace market in vibration test equipment, this company has developed and engineered vibration testing solutions for improved product quality.

TEAM has designed and manufactured testing systems for some of the largest platform systems, including for satellites, spacecraft, and tracked or wheeled military vehicles. Fig. 4.19 depicts one such large platform test table.

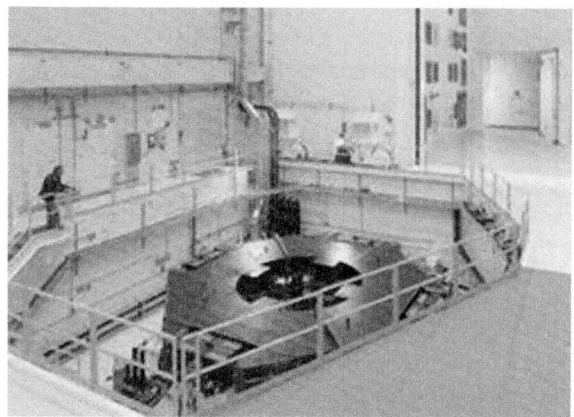

FIGURE 4.19 TEAM's large platform system [4.24].

Some of the TEAM system features include:

Rigid Table Design
Finite Element Modeling To Predict Table Response
Single or Multi-Axis Capability
Actuator Force Up to 1 million lb (4445 kN)
Variable Frequency Response for System Design
Available with Thermal Barrier for Use With Environmental Chambers

Applications

- Satellite Testing
- Spacecraft Testing
- Mil STD 810 Testing
- Tank Turret Combat Electronics Testing

Reference [4.25] provides the following information concerning accelerated corrosion testing.

The cost of corrosion to the USA is estimated to be $276 billion/year. This cost includes direct and indirect expenses associated with corrosion. This referenced corrosion web site was developed to inform and educate the public on issues involving environmental deterioration of materials. The corrosion engineering, research, and testing capabilities at the Kennedy Space Center (KSC) were presented as examples on how to develop corrosion control and detection technologies and to investigate, evaluate, and determine material behavior in various corrosive environments.

This Center includes the Corrosion Laboratory, Beachside Atmospheric Test Facility, Coating Application Laboratory, Accelerated Corrosion Laboratory, and Photo documentation Facilities.

Because the Kennedy Space Center's launch facilities are located within 1000 feet of the Atlantic Ocean, salt from the ocean, combined with a launch vehicles' acidic rocket exhaust, makes corrosion protection a high priority. For these reasons, KSC maintains state-of-the-art corrosion research and testing capabilities. The Corrosion Technology Laboratory is part of the Applied Technology Division, and any project involving corrosion may utilize this fully staffed and equipped corrosion laboratory as a resource.

To develop corrosion control and detection technologies and to investigate, evaluate, and determine material behavior in various corrosive environments, the Corrosion Technology Laboratory offers:

Facilities

Atmospheric Exposure Site, including:

- Cathodic Protection Compatibility Tank;
- Seawater Immersion;

On-Site Laboratories, including:

- Electrochemistry Laboratory
- Accelerated Corrosion Laboratory
- Coatings Application Laboratory
- Photodocumentation Laboratory

Capabilities include traditional salt spray techniques, as well as advanced cyclic and acidic methods.

Surface analysis

State of the art electron microscopes and experienced staff study corrosion mechanisms through surface chemistry, depth profiling, and composition mapping. Techniques available for surface analysis include:

- TEM (Transmission Electron Microscopy)
- SEM (Scanning Electron Microscopy)
- XPS (X-ray Photoelectron Spectroscopy)
- AES (Auger Electron Spectroscopy) Rutherford Backscattering Spectroscopy
- SIMS (Secondary Ion Mass Spectrometry)

Advanced in Fully Integrated Flight Test Systems From a Single Source [4.26].

Typically, flight test engineers must identify the various components needed to satisfy their test regime's unique requirements. This includes data acquisition units (DAU), gateways, transceivers, recorders, cameras, managers, and switches. These components then need to be sourced, often from multiple suppliers. It is not unusual for flight test component customers to acquire their DAU from one vendor and their recorders and switches from another. The engineers must then face the challenge of integrating these independent and disparate test components.

These design processes for the latter phase of flight test systems often create delays, which are a major source of risk to their programs. Such delays keep the test platform on the ground instead if in the air.

A better approach would be for the flight test components to be sourced from a single supplier. This would enable a fully integrated system providing a flight test solution that frees the customer from the need for integrating various components that were sourced from many independent suppliers. A single source vendor who would take on the responsibility for providing a complete set of interoperable products would be preferable.

The Curtiss-Wright Corporation, through the recent acquisition of Tele-tronics Technology Corporation (TTC) and its integration with its data acquisition product line for aviation and space platforms, is striving to make such a single-source fully integrated flight test system solutions a reality [4.27].

Curtiss-Wright now supports many aerospace flight test customers, platforms, and programs. Their combined product families have all of the component-level products needed to integrate a complete-level solution, including DAUs, gateways, transceivers, records, cameras, managers, and switches.

Some flight test engineers will have strong preferences for which setup software they prefer to use. For example, while TTC products use TTCW setup software, existing Curtiss-Wright data acquisition products use DAS Studio. Recognizing these very real preferences, Curtiss-Wright is pursuing selecting products and updating their ability to support both TTCWare and DAS Studio software. This will enable flight test engineers to select the development conditions that they are most comfortable with, while still providing them with a fully integrated system solution taking full advantage of the benefits of both product families. Curtiss-Wright miniature Axon DAU, which was scheduled for release in 4Q217, can be configured for both TTCWare and DAS Studio software setup.

An example of this single-source end-to-end-integration was demonstrated in various conferences in 2017. The typical flight test system demonstrated several inputs video, analog data, accelerometers, pressure sensors, bridge gauges and temperature sensors. They output bulk data to an Ethernet recorder and output serial PCM data typical of RF telemetry, with the data displayed graphically and in real time. The demo system seamlessly incorporated many diverse elements, including the Akra KAM-500 DAU, the high-performance TTC HBC-330 video camera, the next generation Axon DAU and the unique Axomite single module DAU that fits inside a bin (13tnm) and can operate 60 ft away from Axon. It also integrated the Curtiss-Wright NET/CWI/101 switch and the TTCADSR Ethernet recorder, which can record 788 GB of data at 150 Mbps.

Evolving flight test solutions from a complete-level approach to a system-level approach delivers a value that better addresses customer's problems. It reduces their risk and speeds up the time to deployment by delivering a fully integrated system based on proven, interoperable products.

This bringing together of the TTC and the Curtiss-Weight product families provides resources to accelerate the development of advanced testing technologies.

Reference [4.28] provides another example of how a company implemented advanced positive trends in the development of accelerated testing for automotive and aerospace engineering. The MTS and this author's analysis shown below demonstrates how a world leader in vibration testing—MTS Corporation—improved its products over a span of 33 years (Table 4.2).

MTS continues to develop its product line; the company is now building third-generation networks. They have built more than 28,000 3G base stations, with over 80% of them providing HSPA + Internet access at speeds up to 21 Mbit/s.

TABLE 4.2 MTS Corporation's Development of Vibration Testing Over 33 Years.

Date	Company's description
1962	MTS develops the first four-poster. This system was used for noise, vibration, and harshness testing of automobiles. The benefits of laboratory-based simulation soon became obvious.
1966	MTS develops the first multiaxis road simulator.
1974	MTS co-develops remote parameter control (RPC) with General Motors.
1980	MTS releases RPC2 with more advanced testing tools.
1982	MTS develops the first long-stroke multiaxis system to simulate rough road and maneuvering events.
1986	MTS engineers the first multiaxis test system for testing engine mounts.
1987	MTS introduces the first multiaxis simulation table (MAST) for the automotive industry.
1990	An MTS roads simulation system is the first road simulation system to be installed in an environmental chamber.
1994	MTS completes the flat-trac flat surface roadway, the first full vehicle flat-belt roadway system with vertical input and RPC.
1995	MTS introduces flex test 2 with adap trac, an adaptive controller that significantly simplifies the test setup process.

Their 3i strategy uses a special approach to each geographical segment. Through regionalization, the system allows an individual approach for each geographical segment. Through this regionalization capability, the distribution of resources between segments takes into account the specifics of each market. Called MTS OJSC, it effectively applies this feature to its businesses. This regionalization capability is a key Company advantage to be able to successfully withstand the constant pressures of the business need.

MTS has now introduced its Acumen Electrodynamic test systems. This product provides a versatile platform developed for and being implemented over the world. It provides precise road and motion control. Its streamlined design features include a rigid load frame and direct-drive linear motor to provide exacting control of force and displacement. The integration with MTS

Test Suite multipurpose software with the Acumed Electrodynamic Test System also enables automatic setup, smooth task flow, simplified actuator control, and intuitive limit setting, to reduce the risk of human error in setting these limits.

Another positive trend of accelerated testing development is described in Ref. [4.29]. This article stated that

> *...Compounding the issue is the fact that older detectors — with previous-generation Read Out Integrating Circuits (ROIC) — were nonlinear at low well fills. This caused the non-uniformity correction to break down, resulting in poor imagery and questionable temperature measurement accuracy. With the next-generation ROIC designs, detectors offer linearity to low well fill, allowing for accurate measurements at high speeds (short integration times) on colder targets. This is why it is critical for high-speed IR cameras to have a next-generation ROIC with linear response to low well fill.*

> *Product research and development on internal combustion engines, brake motors, tires, and high-speed airbags are just a few of the areas that truly benefit from high-speed, high-sensitivity thermal characterization testing. Unfortunately, traditional forms of contact temperature measurement such as thermocouples are not practical to mount on moving objects, and non-contact forms of temperature measurement such as spot guns − and even current infrared (IR) cameras − are simply not fast enough to stop motion of these high-speed targets in order to take accurate temperature measurements.*

> *Without the appropriate tools for adequate thermal measurement and testing, automotive design engineers can lose time and efficiency, and risk missing defects that lead to dangerous products and recalls.*

> *Next generation IR camera technologies may offer engineers a solution. These cameras incorporate 640×512-pixel high resolution detectors that can capture images at a rate of 1000 frames per second.*

> *Additionally, newer detector materials, such as strained layer super-lattice (SLS), offer wide temperature ranges with a combination of uniformity and quantum efficiency beyond that of earlier MCT and QWIP detector materials. These new technologies plus the ability to synchronize and trigger remotely, give engineers and managers the tools that they need to address the difficulties of high-speed automotive testing.*

By now having access to thermal imaging during the design and testing phases of automotive engineering, research and development teams can more readily identify weak points and improve overall product performance and safety. But access to this information depends on the type of camera and its features, which can have an impact on imaging success. Having access to a cooled thermal camera with the highest speed, sensitivity, and integration times available will allow researchers to accurately track temperature data over time at high-speeds. These cameras will also provide crisply detailed stop-motion frames. These capabilities will allow researchers to accurately measure temperature and thermally identify the exact time element in which a problem begins. This solution is similar to the solution which the author of this current book solved 30 years ago.

Another positive trend in the development of accelerated testing is the increasing availability of new force measurement solution that minimize force measurement variables and which protect tensile testing data. This development is related to computer software-based systems for force measurement and analysis. These systems offer test method flexibility, analysis, and reporting [4.30]. As was written in Ref. [4.30], these advances "require a higher level support than force gage-based systems. The new force measurement solutions provide, a range of application which can address everything from load limit and distance testing, to break limit, time average, cyclic count and duration testing, constant hold, and more. Optimized for production and quality control testing, the versatile, innovative architecture of these systems is designed for reliable, fast, repeatable, and easy operation. Users get the performance of a computer software-based system without the concerns and support requirements often associated with traditional systems."

Another positive contribution is [4.31] the use of the AUTOSAR standard, which includes numerous opportunities for improving embedded electronic systems in vehicles, including the early testing of software code. Predesign testing enabled by AUTOSAR can speed-up the development cycle, save time, and increase the level of functional design. This testing process is usually split into component and system-level testing.

The first key feature of AUTOSAR is the standardization of interfaces. Any function call, memory access, or hardware driver action that access a feature external to the current application component will be enabled in a standardized way. This has led to the creation of new tools and execution environments that can run simulations in a more realistic way. Greater detail on this is detailed in Ref. [4.31].

Finally, the author wrote that "...regardless of where you are along the path to using AUTOSAR there is a new light at the end of the testing tunnel of the form of virtual validation."

Ensuring data integrity is critical for automotive, aerospace, and other industries, where there are strict requirements for capturing data and maintaining testing records. Some force gages function as both force gages and controllers for digital force testers. With this technology, it is only necessary to set up the gage, and the gage has built-in testing templates for common force measurement test methods, such as load limit, distance limit, and break limit testing. For each test requirement, the user can specify the digital force tester functionality for the gage, and the user can specify the automatic return to zero position once the test has concluded.

As has been previously discussed, many tactical aspects and trends in the development of accelerated testing are connected with advanced statistical (theoretical) methods that are easy to use in the academic world, because it does not require expensive and complicated equipment for physical simulation, especially simulation of real-world conditions. Many scientists are involved in these theoretical tactical methods that increasingly are related to details and units of complete equipment.

Among these approaches, the two-parameter Weibull distribution is the most common distribution used to analyze test data and predict performance. A common method for estimating the Weibull shape and scale parameters is the Maximum Likelihood Method. As detailed in Ref. [4.32], the Maximum Likelihood method is known to have a positive bias for the estimation of the shape parameter [4.32].

With respect to this bias, there are ongoing efforts to estimate this parameter from a given data set, with minimal bias. An earlier paper entitled "Bias Correction for the Maximum Likelihood Estimates in the Two-parameter Weibull Distribution," written by H. Hirose, provided a formula for determining the unbiased estimate of the shape parameter. This paper also provides the results of a Monte Carlo simulation study using this methodology [4.32]. And, another example of the progress in this area was written in this author's publication [4.33]:

Automotive Centre of Excellence, University of Ontario (Canada) (ACE).This research and testing facility offers chambers and technology for climatic, structural durability and lifecycle testing. The wind tunnel simulates weather conditions, including severe wind, humidity, snow, ice and desert heat, to measure the safety of different vehicles (Fig. 4.20). Additional chambers include a climatic four- poster shaker, an anechoic chamber with a multi-axis shaker table within, and two large climate chambers.

FIGURE 4.20   Researchers and engineers can test product prototypes in the ACE Climatic Wind Tunnel in different types of weather scenarios [4.33].

Provincial economic development-driver for the Ontario Centre of Excellence was to establish a Regional Technology Development Site (RTDS) for controlled-environment driverless vehicle testing inside the university's three-story ACE Climatic Wind Tunnel.

At ACE, vehicles of any size can be tethered onto a giant platform in the Climatic Wind Tunnel, with the tires rotating at any driving speed. ACE can replicate climatic conditions observed anywhere in the world, simulating everything from the solar load of extreme desert heat to torrential downpours, freezing rain and blizzards.

The RTDS provides a close-to-home accelerated R&D environment where AV development companies and engineers will evaluate and validate prototypes.

Some other positive trends in Testing include the "Stepping Stone for Innovation" program.

The successful collaboration in the designing of a comprehensive testing program led to the innovative redesign of an aircraft qualified product [4.34].

Airlines and aircraft manufacturers are continuously working on new ways to improve services, however, having a great idea for something new, and then designing, manufacturing, and qualifying it in accordance with international aerospace standards, requires expertise, know-how, and skillful, innovative product development.

Dynamo Aviation (DA) is one of those companies. Peter Robadi, vice president of business development at DA explained [4.34] that

… a passion for aviation is vital for testing and qualification. Every new product we introduce to the market has to be tested in a way that guarantees safety and reliability in combination with our innovative design. Our partners should not only test and qualify product, they should understand how the aircraft works, which environmental impacts a given component experiences once it is installed and used in a flight situation and most importantly, they should be able to provide our engineers with detailed information if a unit fails during testing.

Trends in improving the efficiency of testing electrical machines and electrical inverter is critical to the development of advanced aircraft and future technologies [4.35]. Mitchell Marks wrote in Ref. [4.35] that the electrification of aircraft is heavily driven by the vision of making them more efficient, cheaper to operate, and more environmentally friendly in terms of exhaust gas and noise emissions. To transform today's conventional aircraft into electric ones, hydraulic and pneumatic actuators and systems will gradually be replaced by electric actuators and systems.

The main challenge is to reduce the weight and increase the efficiency, power density, and reliability of the components of such electric systems, i.e., generators, batteries, power controllers, and motors.

Testing the energy conversion efficiency of these components requires data acquisition systems that are able to measure the aggregate input and output power with high accuracy and reliability and deliver the underlying raw data necessary to understand and improve the energy efficiency in the R&D process.

A consequence of electrification is it will become more challenging to insure the stable operation of the aircraft's electric system. Individual components, as well as the entire aircraft electric grid, will need to be tested more extensively for power quality compliance. Data acquisition (A&Q) systems for such tests need to be able to continuously record the entire flight profile and switch to higher sample rates to capture singular disturbances. They should be able to perform harmonic analysis and power measurements with the same instruments.

What is needed to design an electric drive system for any application? Basically, there are three elements:

- Power source;
- Power converter;
- Motor.

Often these elements are powered by a battery acting as a DC bus—an inverter that changes DC power to AC or a motor that uses the power to convert electrical energy into mechanical power.

This is sometimes referred to as electromechanical power conversion.

Another positive trend relates to delivering product life-cycle performance [4.36].

Dramatic technological, economic, and social changes that have occurred over time have influenced the world of automotive testing, and now OEM and key component suppliers must design and test for an appropriate target life. Ideally, complex and varied usage profiles need to be well understood, and material properties fully characterized, which will help deliver high-quality components.

But more typically, tests are based on methods born of habit, not efficiency. Data is manually screened and then poorly indexed on local server file systems. This makes the analysis of the results unavailable for days and even weeks. Anomalous data may not be trapped, allowing it to ripple through the design process and introduce errors downstream.

By successfully integrating the design, test, and operation feedback loop, great improvements will be made to the quality and speed up the product moves from concept to market. While the price of this is considerable, there are now solutions that can help mitigate these costs. eCode International has developed [4.36] as a significant supplier of data acquisition, durability technology, and analysis software for the testing industry. This company has coined the phrase "product life-cycle performance" or PLP. Traditional life-cycle performance (PLM) has helped to optimize the flow of design and manufacturing information. However, PLM is not designed to manage or process giga-and terabytes of engineering test data, nor is it capable of the testing and operational feedback loops needed to ensure product performance. PLP works alongside PLM to help manage these large volumes of test data. It creates products designed to meet performance targets across the product life-cycle, by simplifying and automating the analysis needed to feed downstream engineers, encourage engineering collaboration, and enable data reuse.

PLM, which is now well established, although the birth of it is a software tool was initially problematic.

FDM software is typically a database application, configured to hold all the relevant design and product data. This computer-aided engineering analysis software enhances the design process by allowing the use of the data through computer simulation.

Description, definition, development, test, analysis, and validation form the design stages of PLM. But, existing PLM tools do not effectively manage test data, let alone the associated analysis and validation needed to design to a target life. To develop a more effective simulation of product durability (for example, by using DesignLife or FE Fatigue), one needs representative load data. Moreover, proving ground tests should be representative of real-world conditions.

But proving ground testing is rarely appropriate for durability testing because it does not take into account the changing multienvironmental influences experienced during a product's service life.

Examples were presented previously, showing how inaccurate simulation of real-world conditions made it impossible to use this approach for successful real-world durability and reliability prediction. This is another example of negative trends in accelerated testing development because of the use of underdeveloped software simulation.

We continue to see a continuous need for integrated simulation and testing throughout the development process. This is why professionals are working on developing faster and better simulation, and testing tools to provide more in-depth engineering insight to help OEMs, integrators, and suppliers design the best possible aircraft of the future. One example, for solving aircraft vibration problems utilized a system with 12 channels, which was attached directly to the pilot's body to acquire data from several microphones [4.37].

Reid Bollinger wrote [4.38] that there are two main approaches to testing materials for automotive or aerospace industries. He wrote that one is to apply a general-purpose test platform that is connected to a load frame created to a specific device. The second approach is to use test systems that are totally customized for a particular application. Wes Blankenship, president of Sumbrium Engineering company, favors the latter approach. "General purpose test machines are very expensive to be doing a simple test like exercising a gear tooth," he said. "By building a dedicated machine to do gear testing, we can increase the speed of the testing operation and enable testing of multiple devices in parallel" [4.38]. Similar approach for multiple-chambers was developed by the author (see Fig. 8.7 of this book).

In order to perform multiple testing operations simultaneously [4.38] described how the hydraulic motion controller that could control multiple motion axis simultaneously was used. Another example was using a controller with a very short control loop time so that subtle changes in how the device being tested responded to the test cycle and could be detected between test cycles. After building a spreadsheet containing different alternative motion controllers, the company selected the RMC200 motion controller from Delta Computer Systems. The RMC 200 is capable of running control loop times as fast as 250 μs.

The dynamic loading applied to the gear tooth is measured using a load cell connected to the motion controller. The position of the hydraulic axis is measured using a noncontact laser displacement sensor that is capable of sensing distance down to a tolerance of two microns in order to detect a break. As results, "Whereas other gear testers are capable of running 40Hz test cycles, purpose-build load frame is running tests for Fortune 500 powertrain companies at speeds up to 100 Hz. For the price of one off-the-shelf test frame, we can handle three test frames that each produce results three times faster.

With the success of the metal gear test system, Symbrium company has moved on to develop a smaller variant of the machine, producing 20 to 500 pounds of force that can be used to test the strength of gears made out of plastics." [4.38].

In conclusion, while the above information has been provided as a result of this author's analysis, it cannot include all publications regarding the negative and positive trends in the development of accelerated testing in automotive and aerospace engineering. The reader will find more information in Refs. [4.39—4.45], and other publications.

Finally, the following conclusions are presented from this chapter:

1. There are positive and negative aspects of trends in the development of accelerated testing. The negative trends lead to problems, such as increasing recalls, decreasing reliability, durability, safety, maintainability, and economic effects.
2. The chapter considered both strategic and tactical negative trends in the development of accelerated testing in a wide range of industries, including automotive, aerospace, electronics, and others.
3. The basic strategic negative trends are:
   - Virtual testing is increasing too quickly, and physical testing is developing too slowly;
   - There is less attention devoted to the development of both physical and virtual accurate simulation for testing;
   - There is an increasing gap of technical progress between design (and manufacturing) of new products and accelerated testing;

- Virtual simulation does not simulate accurately complicated real-world conditions, as well as complicated products, especially the interaction of units and details;
- More complicated equipment due to technical progress requires more complicated testing, but industrial companies avoid this largely because more complicated testing is more expensive;
- Organizations fail to take into account that saving money for testing often leads to increasing costs of subsequent processes, including design, manufacturing, and usage, and to increasing the total cost of the product;
- Narrow thinking management is too often averse to investing money in the more effective technologies for accelerated testing, especially ART and ADT;
- Continued separate simulations of real-world conditions and the corresponding accelerated testing do not take into account their interactions that exist in real life operation;
- There is continued reliance on use very old and ineffective theoretical approaches.

4. The chapter has provided examples of publications showing the above negative approaches.
5. The chapter also detailed positive trends in the development of accelerated testing, especially accelerated reliability and durability testing (ART/ADT) technology.
6. It considered the basic trends that can be seen in the development of accelerated reliability and durability testing.
7. It presented examples from different author's publications to demonstrated that in automotive, as well as aerospace, there are positive trends in the development of accelerated testing, especially pertaining to details and units.

Bibliography

[4.1]   Klyatis LM. Accelerated reliability and durability testing technology. Wiley; 2012.
[4.2]   Klyatis LM, Anderson EL. Reliability prediction and testing textbook. Wiley; 2018.
[4.3]   Meeker WQ, Escobar LA. A review of recent research and current issues in accelerated testing. International Statistical Review/Revue Internationale de Statistique Apr., 1993;61(1). Special Issue on Statistics in Industry.
[4.4]   Symposium "Autonomous vehicle test and development" (North America's only conference dedicated to test and validation of autonomous vehicles and self-driving technology). October 23–25, 2018. in Novi, MI, USA.
[4.5]   Klyatis LM, Klyatis EL. Accelerated quality and reliability solutions. Elsevier; 2006.
[4.6]   Klyatis L. Successful prediction of product performance. quality, reliability, durability, safety, maintainability, life-cycle cost, profit, and other components. SAE International; 2016.

[4.7] Wilson J. Why so many product recalls? Test engineering and management. April/May 2018.

[4.8] Stuart B. Mercedes' CLS: is most tested car. Automotive engineering. SAE International; October 2010.

[4.9] Hailey K. Poor test data costs industry billions per year. ITworld; May 15, 2015. https://www.itworld.com/article/.../poor-test-data-costs-industry-billions-per-year.html.

[4.10] King JP, Jewett WS. Robustness development and reliability growth: time, money, and risks. Prentice Hall. InformIT.; April 2010.

[4.11] Jayatilleka S, Okogbaa G. Accelerated life testing (ALT). 2014. Workshop on Accelerated Stress Testing and Reliability (ASTR).

[4.12] Autonomous vehicle test and development symposium in conjunction with autonomous vehicle international magazine. 5—7 June 2018. Messe Stuttgart, Germany.

[4.13] Automotive components fatigue and durability testing with flexible vibration testing table 10-02-01-0004. SAE International Journal of Vehicle Dynamics, Stability, and NVH-V127-10EJ.

[4.14] Kumar P, Prakaash J. Kumar P. Optimization of proving ground durability test sequence based on relative damage spectrum. SAE Paper 2018-01-0101.

[4.15] Presead S, Prakaash J, Dayalan P. Study the comparison of road profile for representative patch extraction and duty cycle generation in durability analysis. 2017. https://doi.org/10.4271/2017-26-0309. SAE Technical Paper 2017-26-0309.

[4.16] Sivash S., Hari Krishna SV, Mendez AN, Dodds CJ. Development of a specific durability test cycle for a commercial vehicle based on real customer usage. SAE Technical Paper 2013-26-0137.

[4.17] Lalanne C. Fatigue Damage. Mechanical Vibration and Shock, vol. 4; April 2002. ISBN:1903398066.

[4.17a] Radar Reliability Testing. The 14th institute of the China electronics technology group corporation. Aerospace Testing International June 2019:83.

[4.18] James T. The emerging role of physical test in product development. SAE International. Automotive Engineering 2008.

[4.19] Back to environmental basics. Aerospace Testing International April 2014.

[4.20] Barlow RE, Proschan F. Mathematical theory of reliability. Society for Industrial and Applied Mathematics; 1996.

[4.21] Rutledge R. Have you met...? Aerospace Testing International 2018. Showcase.

[4.22] A & D Europe GmbH: powertrain testing and vehicle development solutions.

[4.23] AB dynamics. https://www.abdynamics.com/.

[4.24] Team Corporation. Vibration testing manufacturer. www.teamcorporation.com.

[4.25] Marina Calle L. NASA Corrosion Technology Laboratory. Kennedy Space Center. http://corrosion.ksc.nasa.gov/.

[4.26] Fully Integrated Data Acquisition System for Flight Test Demonstration. https://www.curtisswrightds.com/news/press-release/cw-demonstrates-data-acquisition-system-for-flight-test-at-ettc-2017.html

[4.27] Albert B, Buckley D. From components to full integration. Aerospace Testing International 2018. Showcase.

[4.28] MTS Systems Corporation. www.mts.com.

[4.29] Flier Systems. Next generation infrared technologies solve high-speed automotive testing challenges. Test and Measurement August 2016.

[4.30]   Clinton JM. Selecting a force measurement solution to minimize variables and protect tensile testing data. Tech briefs Engineering Solutions for Design and Manufacturing; July 2018.

[4.31]   Fairchild J. Accelerated testing of embedded software code leverages AUTOSAR and virtual validation. Automotive Engineering September 2015.

[4.32]   Michael Tully J. Monte Carlo simulation of Two-parameter Weibull Distribution to determine unbiased estimate of the shape parameter. SKF USA Inc.; Bryan Dodson, SKF. SAE Paper. 19IDM-0008.

[4.33]   Automotive centre of excellence uoit (www.uoit.ca). maria.barrese@uoit.ca.

[4.34]   Shiffman D. Testing: a stepping stone for innovation. Twitter; August 9, 2018.

[4.35]   Marks M. Advanced testing of electric systems. Aerospace Testing International 2018. Showcase.

[4.36]   C. Mott. Delivering product life-cycle performance. Automotive Testing Technology Internationally. Report 2008.

[4.37]   Halle R. Listening post. Aerospace Testing International April 2015.

[4.38]   Reid B. Multi-axis motion controller accelerates gear testing. Tech brief. SAE; December 2018.

[4.38a] Gnedenko BV, Beliaev UK, Soloviev AD. (SC.D). About illegality wide use exponential law of distribution in reliability theory. Mathematical methods in reliability theory. Moscow, Russia. 1965.

[4.39a] RAMS. 2019. The 65th annual reliability & maintainability symposium. January 29–31, 2019. Orlando, FL.

[4.39]   Cheon S, Jeong H, Hwang SY, Hong S, Joseph D, Kim N. Accelerated life testing to predict service life and reliability for an appliance door hinge. Journal Procedia Manufacturing 2015;1.

[4.40]   Frank Murray S, Heshmat H. Latham, New York; Fusaro R. NASA Lewis Research Center, Cleveland, Ohio. Accelerated testing of space mechanisms. MTI (Mechanical Technology Inc.). Report 95TR29.

[4.41]   Autonomous vehicles testing methods review. 2016 IEEE 19th international conference on intelligent transportation systems (ITSC). November 1–4, 2016.

[4.42]   SAE International. Seminars. Accelerated test methods for ground and aerospace vehicle development C0316.

[4.43]   Klumpf M. Test and development engineer, Audi AG. Test development and execution system for autonomously performed scenarios.

[4.44]   Coo MS. DSD Testing, Austria. Challenges for testing with platform robots at high speed (>100 km/h).

[4.45]   Peter S, Millbrook Proving Ground Ltd. UK introduction the UK's controlled urban tested for connected and autonomous (proving ground).

Exercises

1. What are the right approaches to eliminating many of the problems seen in accelerated testing (AT) in automotive and aerospace engineering?
2. What are some examples of these?
3. Describe the basic negative aspects of the trends in the development of accelerated testing?

4. What are some of the real-world realities not being taken into account in the development of AT?
5. What are some of the strategic negative trends seen in the development of AT?
6. What are some of the wide-ranging factors in the strategic negative aspects of AT?
7. How is the gap between the level of design and manufacturing with testing changing?
8. What negative trends are happening in the development of virtual testing and physical testing?
9. Show the approximate curves pictorially for the above trends?
10. What publications documented some of the above trends?
11. What is the basic meaning of this phenomena?
12. How are recalls influenced by trends in the increased usage of virtual testing and decreasing physical testing?
13. Describe some examples of this dependence, from the published literature.
14. Show some examples of separate simulations of field input influences during design, manufacturing, and usage.
15. Show the difference between the use of a simulation for mechanical testing and for field situations.
16. What are some specific tactical negative trends in accelerated testing development?
17. Demonstrate some of the negative aspects when vibration testing is done in the laboratory?
18. Why is proving ground testing not appropriate for AT development?
19. Provide some examples of poor trends in testing development that have been described in publications.
20. Provide some examples of negative trends in Accelerated Life Testing.
21. Explain why fatigue failure testing is not appropriate for durability testing of a vehicle's components?
22. Why is ground testing a negative aspect for use in the evaluation and prediction of vehicle durability?
23. Why is the correlation methodology not appropriate for calculating durability after proving ground testing?
24. What is the problem with using the virtual simulation for accelerated testing of the product?
25. Why is the exponential law of distribution not appropriate for accelerated testing development?
26. Describe some of the basic positive trends in AT development.
27. Describe why accelerated reliability and durability testing (ART/ADT) are basic positive trends in the development of AT.
28. What is the definition of ART/ADT?
29. Demonstrate the scheme of ART/ADT pictorially.

30. Demonstrate the basic steps of the path from accelerated testing to ART/ADT.
31. Describe some of the basic trends in the development of ART/ADT.
32. Describe why the iTest system is a positive trend in the development of accelerated testing.
33. Describe the essence of the modular system developed by the A&D company.
34. Describe AB Dynamic's solution for test systems.
35. Describe the essence of the TEAM Corporation's solution to AT.
36. Describe some of the capabilities for corrosion control and testing available from the Kennedy Space Center.
37. Describe the MTS system of AT development.
38. Describe some of the solutions provided by the Automotive Centre of Excellence (ACE) in AT development.
39. Describe some positive approaches to improved efficiency provided by AT for electrical machinery.
40. Describe the development by eCode solutions in acceleration testing.

Chapter 5

The role of accurate simulation in the development of accelerated testing in automotive and aerospace engineering, and its connection with the engineering culture

Abstract

This chapter considers some of the aspects of accurate simulation and their influences on accelerated testing in the areas of automotive and aerospace engineering. In doing so, it discusses a number of key challenges in aircraft development, including:

- The role of engineering simulation approaches to ART/ADT and the related prediction methodology;

- Improvement of the engineering culture of management, which is necessary to understand and enable the interdisciplinary systems of approach to ART/ADT;

- Establishment of statistical criteria for physical simulation of input influences and accelerated stress factors;

- Determining the number and types of test parameters appropriate for analysis during ART/ADT;

- The author's approach to correctly choosing the areas of influence, which will account for all of the relevant factors of the product's field experience.

The chapter also considers improving the engineering culture for accelerated testing development, which directly influences the accuracy of the simulation.

5.1 Introduction

The basic concepts for achieving accurate simulation for the development of accelerated testing was published in some chapters of this author's books [5.1,5.2], as well as in his numerous articles and papers. Therefore, this chapter will only briefly include some specifics for obtaining accurate simulation for automotive and aerospace engineering, and the reader is directed to those references for a more detailed study of the subject.

Trends in Development of Accelerated Testing for Automotive and Aerospace Engineering.
https://doi.org/10.1016/B978-0-12-818841-5.00005-2
173

A key factor for successful accelerated testing is the accurate simulation of the field conditions. This is because the true objective is not just successful testing, but providing successful prediction of a product's performance in its real-life operation, as a result of the testing. It offers the possibilities for a successful development and improvement of product efficiency.

But, because a product's real-world operation includes a complex of many interacted factors too, often the testing protocols ignore many of these interactions in the interest of simplified or lower cost of testing. But as a result of not considering the importance of accurate real-world simulation, the testing may not reflect real-world operational results.

Therefore, the positive trends in the development of accelerated testing cannot be considered without paying the corresponding attention to the need for accurate physical or virtual simulation.

5.2 The role of accurate engineering simulation in the development of automotive and aircraft systems

One important positive trend in the development of accelerated testing can be seen with the increasing development of products whose completed version is closely correlated with the developmental testing methods and equipment. Shown below is an example of the evolution of engineering simulation for testing together with the evolution of the test subject.

In land, sea, and air applications, the use of Unmanned Aircraft Systems (UAS) has shown and continues to demonstrate explosive growth with no sign of slowing. But the UAS faces a number of key challenges, which must be addressed if it is to satisfy the future needs of UASs for the USA Departments of Defense (DoD), other governmental departments, nongovernment organizations and companies, and individual consumers.

There are two paths that can lead to either decreasing or increasing the technical and economic characteristics in automotive and aerospace engineering. The key role is accelerated testing (Fig. 5.1).

Both of these paths begin with the accurateness of the simulation of field/flight conditions for providing research, design, and manufacturing.

The benefits of accelerated testing are based on the level of real-world simulation for providing this testing. Fig. 5.1 relates the connection of accelerated testing level with its benefits, related to recalls and other problems solution.

As can be seen in this figure, accurate simulation and based on this accelerated reliability and durability testing (ART/ADT) leads to dramatic decreases in recalls and can save multibillions of dollars. Following the opposite path of traditional, accelerated life testing (ALT), which is used, although there is dramatic technological and higher speed progress in design and manufacturing, can lead to losses of multibillions of dollars, as shown in Chapter 2. Moreover, if this trend and situation continue, it will result in more costly and more dangerous future results.

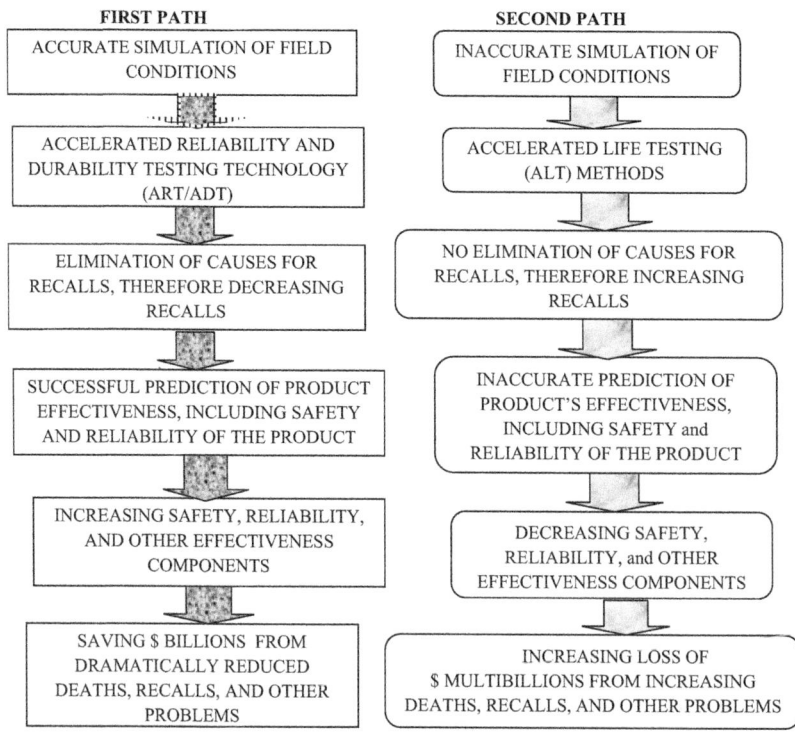

FIGURE 5.1 Two paths (ways) leading to dramatically reducing or increasing recalls and other problems solution.

From this the importance of stopping the development of negative trends in accelerated testing can be understood. The strategy and methodology of doing this are detailed in [5.1].

The challenges to achieving the first path (way) are included in [5.3]. These challenges related to numerous areas of automotive and aerospace engineering.

As an example, for Unmanned Aircraft System (UAS), these challenges include:

- Early inclusion of the accelerated testing program as a component of mission-specific platform design and development;
- Transitioning from multiple mission-specific platforms to a reduced number of common platforms that can serve multiple missions across and in conjunction with different domains;
- Increased platform capabilities, including all-weather flight, payload weight, speed, endurance, point to point navigation, and refueling;
- Reduced forward footprint to lower the staff required in the theater of operation;

- Development of effective micro-UASs capable of rapid tactical deployment;
- Expanded missions, including strike, cargo, and medical evacuation;
- Adaptability, especially in a fiscally constrained environment.

The evolving role of UAS

As the indispensable contribution of unmanned systems became clearer, the U.S. DoD published the first document.

Integrated unmanned systems roadmap in 2007 [5.3]

This document spanned all domains—air, ground, and marine. The roadmap was updated in 2009 to quantify how unmanned systems can be optimized to support a greater set of missions, pinpointing areas of technology maturation that can be shared across all domains and identifying technology that fosters collaborative operations. Some of the broader goals of this integrated roadmap were to identify opportunities for cost savings and to provide long-term strategic directions for the UAS contractor community.

Implication for UAS designers and suppliers

The rapidly expanding use of UASs demands an equally rapid integration of new technologies into existing platforms. The speed with which missions and capabilities are being developed means design and integration cycles must be very efficient and right the first time. In an increasingly competitive environment, the companies that succeed will be able to rapidly satisfy the needs of the end user. In the near- and medium-term, this will require customization of products to fit a variety of platforms and missions. In the longer term, these custom products will likely evolve into optimized, standardized, plug-and-play modules, and new capabilities will be developed to integrate in this way.

Further out, as products and solutions mature and become standardized, it will become more important to establish levels of reliability equivalent to those of manned aircraft and to extend the system life cycle. Reducing the required personnel forward footprint will demand improvements in system aerodynamics and system capabilities to support more autonomous takeoffs and landings.

The role of engineering simulation

Based on the historical trends in design and testing that have been observed and the UAS roadmap laid out by the major users, several key design constraints in the development of future UAS platforms and payloads can be expected, specifically, including:

- Very short development cycles;
- Short-term design customization with little design precedent;
- Medium to long-term design optimization for standardization;

- Increasingly complex missions with associated capability innovation and integration;
- Tightly controlled costs and a demand for right-the-first-time design.

Engineering simulation harnesses the power of computers and software to solve the fundamental equations of physics or those that are close approximations of these equations. This allows designers, test engineers, and others involved in the analysis to create virtual representations of complete UASs and their payloads for design analysis and optimization prior to physical testing.

Correct implementation of the technology has been verified and validated in a range of industry sectors, and the use of more accurate engineering simulation is, in some cases, mandated by regulatory bodies.

Research results [5.4] have shown that best-in-class companies

- Meet quality targets 91% of the time, compared with a 79% industry average;
- Meet cost targets 86% of the time, compared with a 76% industry average;
- On-time launch 86% of the time, compared with a 69% industry average.

The standout difference in strategy pursued by these best in class companies is the systematic use of engineering simulation (physical and software) regularly throughout the design and testing process. In essence, these companies consistently leverage engineering simulation throughout the design process, with accelerated testing as one of the basic components of this approach. They lead to improvements in quality, reliability, cost, time, and other components of performance when compared to companies that do not do this.

Research performed by the U.S. DoD revealed the staggering impact that engineering simulation can have [5.4,5.9].

A 3-year study reported that "... **for every dollar invested in accurate simulation the return on investment is between \$6.78 and \$12.92." There are recorded returns of between 678% and 1, 292%** [5.4].

The above simulation will add the most value when:

- It is applied to all aspects of design (pre-design and design with accurate simulation of field conditions, accelerated reliability, and durability testing, not just one or two influences that are taken in isolation);
- The interaction of the physics at a system level is included in the analysis (for example, analysis of the action of the full real-world input influences);
- The workflow is seamlessly integrated across the physical essence and with existing tools;
- Physics-based optimization is performed across the design envelope.

At an organizational level, it needs to be recognized that engineering complex accurate simulation tools need to offer more than just technical capability. The unique nature of the completed products with their lack of design precedent makes it critical to capture the whole design process and intent. By doing this, it can be systemized and scaled for future applications. Capturing and managing this engineering knowledge is best performed by using the successful prediction methodology and tools, as described in this author's publications, for example [5.1,5.2,5.6].

The ideal scenario for this is when the successful prediction, including accurate physical simulation of the real-world conditions, accelerated reliability and durability testing (ART/ADT), and prediction methodology (all in one complex tool), performs and improves the engineering culture of management and provides one entire system producing only the right type and needed information.

This approach relates to an Interdisciplinary System of Systems Approach.

The close collaboration between OEMs and suppliers required for the successful platform and payload integration demands the easy exchange of successful prediction while mitigating mutual intellectual property and data security concerns.

Having considered the growing complex needs of society and the benefits of accelerated testing technology as an interacted component of these needs, it is clear that advanced accelerated reliability and durability testing will be a fundamental enabler for the development of the next-generation systems.

Finally, the role of accurate simulation can be seen in Fig. 5.2. As it demonstrates, inaccurate field/flight simulation leads to a low level of testing.

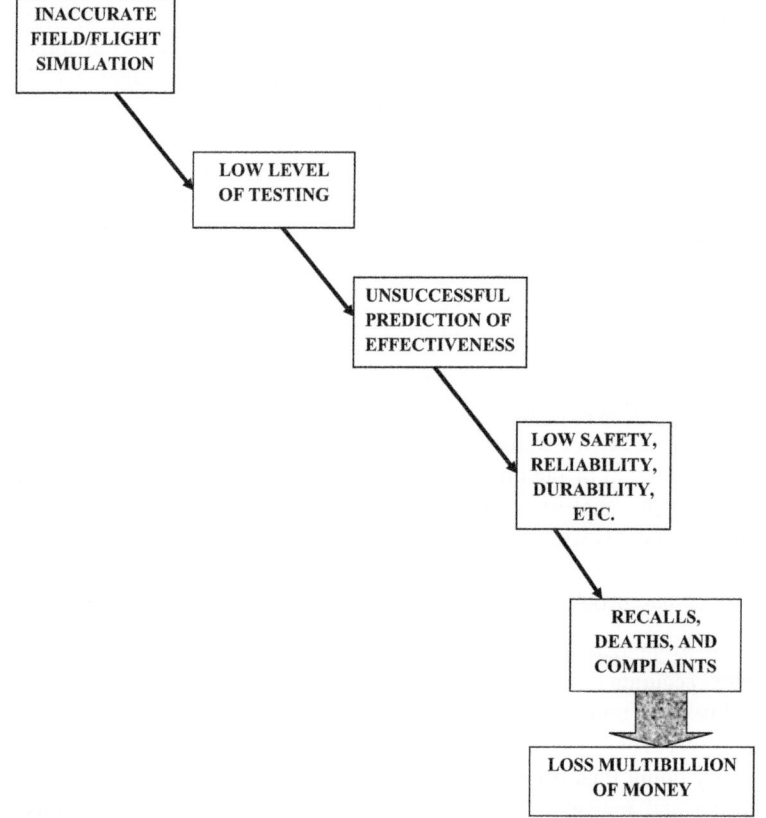

FIGURE 5.2 The path from inaccurate field/flight simulation to complaints, deaths, recalls, and economic losses.

The result of this testing is an inaccurate prediction of the effectiveness of the product or technologies, which then leads to low safety, reliability, durability, etc.

This is a basic reason for the decreased economic benefits obtained from the new equipment or technologies, and the reason for the unplanned costs of recalls, crashes, people's deaths, and other negative outcomes.

5.3 Establishing the concepts and statistical criteria for providing the physical simulation of the input influences on a product for accelerated testing

We have established that the field input influences must be accurately simulated in the laboratory, to provide the higher correlation necessary between the accelerated reliability or durability testing results and the actual field results. This is a fundamental requirement for obtaining a successful prediction of field reliability, durability, safety, and maintainability through ART/ADT. But exactly what are the needed parameters to provide this information? The answer to this question can be found in this author's books [5.1,5.2]. In general, it has been this author's experience that the most accurate physical simulation of the input influences processes occurs when each statistical characteristic—mean, standard deviation, and power spectrum and normalized correlation $[\mu, D, \rho(\tau,)$ and $S(\omega)]$ of all input influences differ from the field condition measurements by no more than 10%.

For each specific situation, one must calculate and use these statistical criteria to achieve this goal of obtaining an acceptable correlation between the reliability measured in an accelerated reliability testing as compared to field reliability. But all these analytic comparisons have testing acceleration coefficients.

The similarity of the degradation process in the field and in the laboratory will determine the practical limit of the applied testing stress and acceleration coefficient (Fig. 5.3).

The first method of AT is by increasing working time to the maximum possible. This type of acceleration can be used in testing if the product does not operate continuously (24/7), but can be tested 24 h a day, every day. Generally, it does not include idle time or operating time with minimum loading, etc. This method is based on the principle of reproducing the complete range of operating conditions and maintaining the proportion between heavy and light loads. The author's experience shows that this method of acceleration has the following basic advantages:

a) good correlation between field and laboratory testing results;
b) each hour of clear work performed by the product is faithfully reproduced in the stress schedule which is identical to the destructive effect of 1 hour of clear work under normal operating conditions;

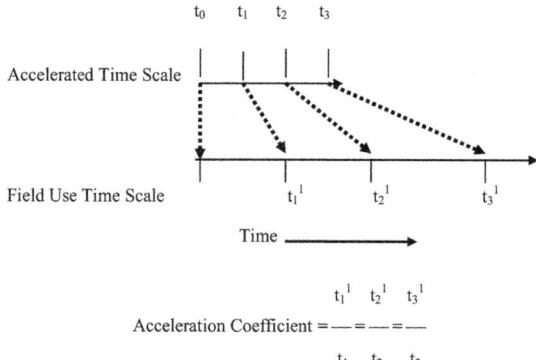

FIGURE 5.3 The accelerated coefficient described.

c) There is no need to increase the magnitude, size, or pace of the applied stress in the testing;

This is accelerated testing because the result of testing in the laboratory is faster than in the field by 10 to 18 and more times. This method is especially useful for a product, which works for short periods of time.

Specialists who use this method know that reproducing a complete range of operating conditions is not always easy, but it yields a more successful correlation of the accelerated testing results and the actual field results.

Test conditions are designed to include a full array of different parameters (temperature, humidity, vibration, radiation, and others) as experienced by the product in the field.

When it is necessary to obtain faster and simpler accelerated testing results, the increased stress methods of accelerated testing (Fig. 5.4) can be employed. However, when utilizing this method, the correlation between the laboratory test results and the field results will be lowered.

Higher stress means less correlation, which means less accuracy of the simulation and greater problems with accurate evaluation and prediction of product reliability and durability in the field. With lowered correlation there also are more problems in finding the true reasons for failures or product degradation, and the correct solutions for eliminating them.

Therefore, for any specific product, one must decide which stresses and how much increased stresses for products will be more effective.

Acceptance of how fully the method of simulating real-life input influences and their simultaneous combinations depends on the testing goal. If it is necessary to have independent simulation, for example, vibration testing without other influencing factors, the results of this testing will not accurately

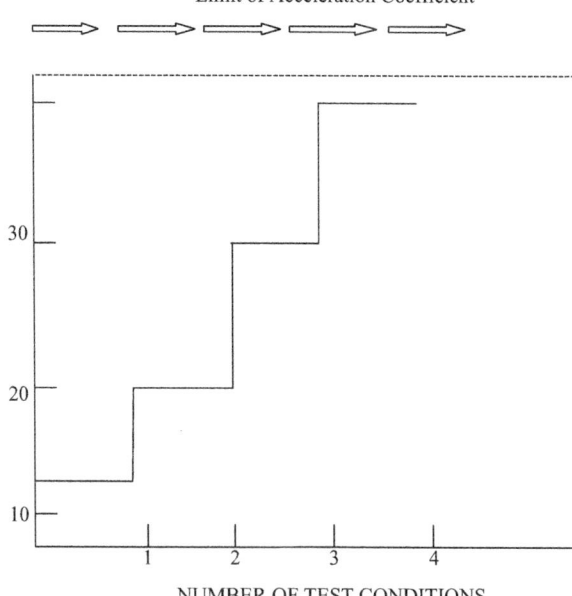

Limit of Acceleration Coefficient

NUMBER OF TEST CONDITIONS

FIGURE 5.4 Example of stress limit (acceleration coefficient).

predict the reliability or durability of the product. The same is true for the simulation of temperature testing. Temperature testing is not truly environmental testing, because the temperature is only one parameter out of the combination of possible field environmental influences that are experienced by the product.

There is a third popular use of the basic AT method—accelerated stress testing—which is often used for electronic products, aircraft, and airspace [5.8−5.10]. It has a smaller acceleration coefficient than HALT, and is more practically substantiated, but also has a high acceleration coefficient (usually around 25−30 and more). This method, while often used, does not directly find the initial information needed for successful reliability or durability prediction.

Acceleration (stress) factors are those factors that accelerate the product's degradation process as compared to that which would occur in normal use. There are many types of acceleration factors, such as the higher concentration of chemical pollution and gases; higher air pressure; higher voltage; higher temperature, exposure to fog and dew; a higher rate of change of input influences; shorter break time between work cycles, etc.

There is also a widely used AT method that employs simulation with only a minimum number of combinations of field input influences. One example would be a temperature/humidity environmental chamber for environmental testing [5.11]. As it is known that these are only some of the many environmental influences of the field on the product, these types of testing cannot produce accurate information about the reliability of the product. But it may provide at least a minimum correlation of accelerated testing results as compared to field results.

One must be cautious when there is a high level of acceleration, and/or a minimized combination of field influences and other aspects of AT as the accuracy of the prediction will be lessened.

5.4 Determining the number and types of test parameters for analysis during accelerated reliability and durability testing

Generally, the methods and types of simulation will depend on the specifics of the product's use and the limits of the facilities of the organizations performing the testing. As generally it is not practical to simulate all the various operating conditions, including input influences, in the laboratory, the objective then becomes determining the minimal number and types of test parameters to produce a comparison of the laboratory and field testing results, which are sufficient for accurate reliability, durability, and maintainability prediction.

In order to do this, it is necessary to establish the partial (basic) area of each influence, which can be introduced for each of the varieties of operating conditions and using this to set the minimum acceptable test conditions. To do this, the following approach is recommended:

$$E > N,$$

where: E is the number of field input influences $X_1 \ldots X_a$ [5.6];

N is the number of simulated input influences $X_1^I \ldots X_b^I$ [5.1].

And allowable error simulation input influences $M_I(t)$:

$$M_1(t) = X_1(t) - X_1^1(t)$$

where:

$X_1(t)$ are input influences of the field;

$X_1^1(t)$ are simulated input influences.

The author recommends the following approach for choosing the areas of influences which introduce all of the basic influences found in the field:

1. Establish the type of studied random process. For example, the stationary process is determined by the dependence of the normalized correlation using the difference in variables only.
2. Establish the basic characteristics of this process. For a stationary random process, we have the mean, standard deviation, normalized correlation, and power spectrum.

3. Define an area's ergodic, the possibility to make judgments about the process from one realization. This is when the correlation approaches zero if the time $\tau \rightarrow \infty$.
4. Check the hypothesis that the process is normal. Try the Pearson or other criteria.
5. Calculate the length of the influence area.
6. Select the size of divergence between the basic characteristics of different areas.
7. Minimize the selected measure of divergence and find the area of influence which introduces all possibilities of the field.

But, if it still remains true that if you cannot simulate the complete simultaneous combination of field input influences, the results of the laboratory testing may not be accurate initial information for reliability, fatigue, durability, and other problems solving.

The number and types of field input influences, which must be simulated in the laboratory depends on the results of the analysis of the action of the field influences on the product's degradation (failure) mechanism.

5.5 Improvement in the accelerated testing engineering culture

Development of the accurate simulation of field conditions need improvements in the engineering culture, because, as has been shown in this book, the current negative aspects of testing development are founded on a culture not embracing the engineering discipline. Many professionals, involved in the testing area, continue to work as they did 50–60 and more years ago. There is little effort devoted to improving the methods and details of accelerated testing and not thinking and implementing improvements to the basic methods and equipment for more accurate simulation.

As a result, there is a continuation of the use of the negative aspects of testing development, resulting in an inaccurate prediction of product quality, safety, and reliability, as well as money losses by producers and consumers, particularly in automotive and aerospace products. The details of this situation can be seen in Chapter 2 of this book.

Therefore, it is so important that there be improvements in the engineering culture for accelerated testing.

The subchapter discusses why it is so important for engineers and managers to develop the culture and how it relates to the System of Systems Approach, which provides an integrated composite approach, including people, products, and processes that provide the capability to satisfy the stated needs or objectives.

Cultural engineering is a conceptual approach that takes into account the changing concepts of the culture and applies them to practical strategies for dealing with issues and problems raised by the culture and the development of products in diverse contexts.

In other words, cultural engineering is about systems, processes, alternatives, and the formulation of creative solutions to challenges in the development of cultural institutions and the promotion of people's participation in cultural life.

Call it systems engineering, interface design, or whatever; it all means the same thing: something that works safely, efficiently, and economically.

The improvement of the culture inside and outside of companies and organizations means establishing positive relationships with everyone involved; reducing distress, creating a healthy work environment, improving communication with everyone involved; and influencing others to want to establish a working environment.

Because many engineers orient their identities and careers to their occupation rather than to their organizational communities [5.12], the adopting of the concept of an engineering culture is important.

For engineers, in particular, attention to their engineering culture may be more related to strictly engineering practices and values rather than examining being involved in the priorities and decision making processes of their organization.

This study also serves to advance our understanding of how and why engineers interact with one another and others in the organization. Much of the research on engineering teamwork has looked into the structure of teamwork and explains why engineers often do not work well with others. This is especially important for engineers who are involved in accelerated testing. As shown in this book, an engineering culture that embraces traditional simulation by separate input influences, using traditional methods and equipment, is the path to a low level of prediction that unanticipated costs. The same culture is also related to engineering managers who should be taught how to select team members, assign team roles, build cohesive unity, assess progress, and provide meaningful advice and guidance for the interconnected engineers who are involved in the development and the use of accelerated testing. Such a cultural transformation revamps the structure of organizational teamwork to help engineers work together. The present study concludes that organizational teamwork important for helping individuals to contribute meaningfully. However, understanding the present culture enables us to see better how the interactions between engineers and managers really work.

Scholars, practitioners, and employers are all noticing that although the scope and nature of most testing as a complex technical solution, engineers are not well equipped to work with other individuals to bring a project to fruition. Further, testing both inside and outside is such that it often requires multiple engineers working on them collaboratively and simultaneously.

Across a wide variety of literature, researchers in testing are identifying providing and developing consistently similar values and practices that characterize what we term an "engineering culture."

Engineers themselves are taught that they are members of a professional culture that sets explicit guidelines for what it means to be an engineer, and their ethical responsibilities not just their organization, but also to the general public. The amazing coherence and persistence of an engineering culture suggest that there is a certain mythos surrounding it. This professional culture of engineers and managers, who are involved in testing, should also include such components to dramatically improve the technology and methodology.

Gideon Kunda [5.13] offers a critical analysis of an American company's well-known and widely emulated "corporate culture." The company's management, Kunda reveals, uses a variety of methods to promulgate what it claims is a nonauthoritarian, informal, and flexible work environment that enhances and rewards individual commitment, initiative, and creativity while promoting personal growth. The author demonstrates, however, that these pervasive efforts mask an elaborate and subtle form of normative control in which the members' minds and hearts become the target of corporate influence. Kunda carefully dissects the impact this form of control has on employees' work behavior and on their sense of self. In conclusion, written especially for this edition, he reviews the company's fortunes in the years that followed publication of the first edition, reevaluates the arguments in the book, and explores the relevance of corporate culture and its management today.

The engineering culture consists not only of the relationship of the people but also the consideration for products and processes in one complex structure. This is very important to professionals who are involved in accelerated testing. Failure to understand this leads to narrow thinking at all levels, beginning with the president or CEO of an organization and continue to all levels of the top, middle, and first-level management, as well as engineers and other specialists in the company. All of these people should be sharing in one final goal of making their company successful in the market.

Too often the basic problem of management is the lack of strategic thinking. Many managers do not appreciate that their level of engineering culture depends on their product's long-term success, and this is highly dependent on successful accelerated testing to produce highly efficient products.

Moreover, specialists in the company that are involved in design, reliability, durability, safety, human factors, life cycle cost, profit, recalls, and other processes have to work interconnectedly through their management. The same relates to the various models of the product, which are currently produced by a company. These various models are interconnected, because they are like the links of a chain, and poor performance or quality issues with any of the products impacts negatively on all of their products.

It also relates to companies that produce the product and to ones that obtain parts of their products from independent suppliers.

But too often, beginning from top management and throughout the organization, people are concerned about their narrow areas of responsibility, and less about their responsibilities are connected in the real-world with other areas of the companies activities, as well as their suppliers. For example, often a technical director who is responsible for the human area does not think about how his area is connected with other areas of the company. Therefore, they do not address the human resource problems of the company, because they are considered secondary to the technical problems, and separately from the problems of other interconnected areas of the company.

The above situation is similar to that found in testing, where, for example, a person responsible for corrosion testing may think only about corrosion testing. The engineer may not consider that corrosion of the product also depends on vibration, deformation, dust pollution, etc. (see Figure 4.9 in Chapter 4).

Moreover, these engineers often do not realize that degradation, and finally, failure of the product, depends on the interacted wear, cracking, solar radiation, human factors, and many others in addition to just the chemical-induced corrosion.

In this case, improvement in the engineering culture entails understanding and taking into account all of the above factors.

Another problem founded in the need for an improvement in the engineering culture of organizations relates to testing. A common method of comparing different testing approaches is based on presenting the cost of these approaches and selecting the cheaper testing approach.

In this case, there is a direct comparison of the costs of the testing approaches. But this is a demonstration of a low engineering culture because the direct cost of the testing approach is the evaluation of providing the testing. While sometimes one thinks that the testing approach does not have an influence on subsequent processes costs, often, the testing approach selected is based on inaccurate simulation of the real-world, influences which lead to unpredicted degradation, failures, and, finally, recalls and unplanned expenses. In such cases, the improvement of the engineering culture will lead to using life cycle cost for determining which of the different testing approach should be used. In practice, using accelerated reliability/durability testing technology (ART/ADT) leads to decreasing life cycle costs, especially through reducing product changes, recalls, and other negative results from poor testing.

Therefore, the development of engineering culture has to go through the development of the interacting of many other subsequent factors, not only testing, during design and manufacturing, and the costs of these factors.

Similar to the above is current poor predicting of product performance, which based on poor results of accelerated testing, based on inaccurate simulation, as initial information for this prediction and leads to increasing recalls and decreased profit.

Successful accelerated testing cannot be done without the development of the engineering culture, which connects the many interacting components along with understanding their importance. This testing of the system requires knowledge of all the components in the chain of interacting various input influences, human factors, and safety (Fig. 5.5).

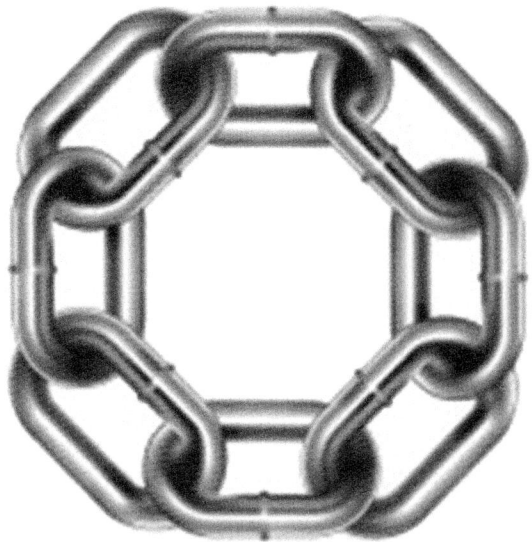

FIGURE 5.5 Schema of successful accelerated testing, which is similar to a chain of interacted links.

This is a basic reason why automotive and other industries are faced with increasing recalls.

The simulation of the real-field conditions necessary for accelerated testing is inaccurate. This is because there is no widely implemented methodology and equipment, and resources devoted to this accurate simulation and successful predicting of the product's performance.

The author's methodology for successfully predicting the performance of the product consists of the following basic components [5.6]:

- Determine the criteria for successful prediction of the performance of all components after testing; the mathematical dependences for calculating the various components of performance;
- Develop mathematical descriptions of the dependences between the quantitative indices and factors that influence the product's performance;
- Develop mathematical descriptions of the connections of the influencing factors with the testing results for the product specimens during the design process;
- Define the coefficients for the recalculation of the future, including correlated and uncorrelated factors, and quantitative indices of performance for all components during and after manufacturing.

This methodology includes both qualitative and quantitative levels of performance.

When a prediction is based on the traditional approaches of ALT data, and when the degradation (failure) process differs substantially from the product's degradation process during service life under real-world conditions, the prediction is inaccurate. Such predictions are not based on the available technology of accelerated reliability and durability testing (ART/ADT), which should be the source of initial information and a key factor for successful predicting. Why is this so? There are two basic groups of causes why ART/ADT is not often used for this situation:

- Cultural
- Technical

Basic cultural issues:

- Many professionals, especially top managers who control investments, erroneously think that ART/ADT is more expensive than the traditionally used types of accelerated life testing. They, as well as many other managers involved in this area, think so, because they consider testing as a separate procedure, and do not take into account the cost of failures in prediction due to poor testing that is done on subsequent procedures during design, manufacturing, and usage during the life cycle. Of course, for example, vibration testing or vibration/temperature testing is cheaper than full ART/ADT, which provides a full simulation of the real-world conditions. But vibration/temperature testing only simulates a portion of the part of the real-world conditions and ignores other parts of these conditions.

It is, therefore, an inaccurate simulation of the real-world conditions and cannot offer the possibility for studying the real-field product use, and as a result, do not provide the initial information necessary for successful prediction of the product's performance. Therefore, one has to know the example that, as was written earlier in this chapter, "… **for every dollar invested in accurate simulation the return on investment is between \$6.78 and \$12.92." There are recorded returns of between 678% and 1, 292%** [5.4].

- This narrow thinking by many managers slows the development, especially the implementation of ART/ADT.
- Many remain skeptical that one can accurately simulate in the laboratory in conjunction with periodical field testing and can actually duplicate real-world conditions. But through the above simulation, one can quickly study the nature of the product during its service life, with only a minor difference between the ART/ADT results and real-world results.

But the above simulation requires transferring the field environment into the laboratory with a high degree of accuracy. This technology of testing is often not used in the accelerated development of products in comparison with traditional methods.

- The important role the inertia in thinking presents to a company's development.

- Many professionals in their organization do not want to be responsible for carefully analyzing and having to report:
  - why their companies profit is 4–5 (and more) times less than was predicted during design and manufacturing;
  - what role was played during the testing process that facilitated this situation;
  - why the comparison between testing results compared so poorly with real-world conditions.

The basic technical causes of the above include:

- Inadequate development of the theory of accurate simulation of the field conditions.
- Inadequate development of the strategy of accurate simulation of the field conditions.
- Poorly developed methodology of accurate simulation of the field conditions, which includes the full input of influences, including safety and human factors, in the laboratory for the actual product.
- Equipment that is not suited for accurate simulation of the world conditions in the laboratory.
- The knowledge of advanced testing, including durability testing, obtained from the literature, is often inaccurate or misleading.

These are the basic reasons why recalls and corresponding cost losses have increased from year to year (see Chapter 2). Improving the engineering culture in testing has been lacking in the following ways:

- First. Use of laboratory accelerated stress testing or traditional, accelerated life testing (ALT). This approach began many years ago. In this testing, the simulation has mostly been of a single factor of the field influences (temperature, humidity, vibration, pollution, etc.), and multiple parameters were seldom used simultaneously.
- Next are the developments that improved the accuracy of these simulations, but still separate field factors.
- In the 1950s, engineers began to understand that the separate simulations of one field input influence were not an accurate approach to replicate the field conditions where many factors are acting in combination. From that time, engineers began to study and use combined accelerated stress testing, such as temperature, humidity, and vibration in special test chambers. Combined test chambers first began to be used for electronic products. But now, some companies continue to use separate simulations.
- Then, in the 1990s, the more advanced companies, beginning with electronic products testing, developed test chambers that employed simultaneous combinations of multiple factors. Typical combined factors test chambers in electronics most often included combined simulation of temperature, humidity, vibration, and input voltage. From this time

forward, the improvements in the engineering culture in the development of ALT moved in two directions.

The one direction moved to highly accelerated life testing (HALT), highly accelerates stress screening (HASS), and accelerated adding (AA). The basis of these types of accelerated testing was the simulation of two factors (commonly temperature and vibration), but with the level of each of these factors being much higher than the level seen in the field. For example, for the automotive industry, where the maximum field temperature is typically 70°C, the temperature in the test chambers was raised as high as 120°C. Also, while the typical minimum field temperature is −40°C, the temperature in the test chambers was as low as −150°C.

A similar situation can be seen with vibration testing. But this is a blind approach of testing because the physics-of-degradation process has been altered, and therefore, does not accurately simulate the real-world conditions. By changing the physics-of-degradation process in the laboratory, as compared with the field, the corresponding time to failure (and for different details of a vehicle acceleration coefficient differs) changed. As a result, the acceleration coefficient can only be determined after the experimental results are compared to the actual results during the service life for each test subject, and second, it is not possible to accurately evaluate the acceleration coefficient for the whole vehicle or for units that consist of many details.

The fundamental drawback to this approach is the inability to successfully predict the product's quality, reliability, safety, durability, life cycle cost, profit, recalls, and other performance components during the product's warranty period or service life.

This direction is sometimes called *modernized simulation* of field inputs, but the strategic level of these types of testing is not much better than the traditional (earlier) level of testing.

While this approach does place some distance toward the development level to HALT as compared with the traditional separate testing for each separate input influence, this type of testing is popular among testing providers and users because it is simple and inexpensive. But once again, this simplicity and inexpensiveness do not consider the costs to subsequent processes.

Moreover, this approach does not take into account the costs of the recall process, especially the costs involved with crashes or personal injury issues. For example, as can be seen in Chapter 2, it is projected it cost billions of dollars due to the crashes of Boeing 737 Max commercial jetliners in 2018−2019. Therefore, this advancement in accelerated testing is not the solution for successfully predicting the quality, safety, reliability, durability, and other components of the product's effectiveness.

Another mistake is people often incorrectly call combined types of testing, or vibration testing, or proving ground testing, as "durability testing." This is not correct, because durability testing, like reliability testing, needs an accurate simulation of real-world conditions, i.e., simulation of the full input influences of the real world, plus the human factors, and the maintenance, and safety issues.

The second way is moving to Accelerated Reliability/Durability Testing (ART/ADT), which is described in [5.1,5.2,5.6,5.14], and others. One can read the definition of this term in Chapter 1.

There is one more problem, which is related to improving the engineering culture, and as a component, the accelerated testing culture. During the past 20 or 30 and more years, there has been an increasing effort to move product manufacturing to the lowest cost locations. As a result, in the case of many complex products and their related supply chains, this leads to less than optimum of the products supplied by the supply chain system. When the supply chain relies primarily on the lowest priced provider, it actually adds costs to the whole vehicle.

A Systems Engineering approach to analyzing each component in the supply chain to determine the value added and not just the pursuit of the lowest cost suppliers can help to optimize a supply chain and result in overall cost savings and product improvements in each and every component.

To understand the capabilities of supply chain vendors, it is necessary to evaluate many characteristics of the supplier. Eight characteristics have been identified for products and services in a lean environment—known as the eight rights—and must be evaluated and understood [5.15,5.20]. These eight rights include:

- The right product;
- The right quantity;
- The right conditions;
- At the right place;
- At the right time;
- From the right source;
- At the right price;
- With the right service provided.

While there is no one-to-one relationship between the eight rights, the supply chain wastes collectively can be seen as root causes producing poor performance with the eight rights.

The most common tool used to address the eight rights is a plan for every part (PEEP). PEEP is used in the planning for all new parts and suppliers. It is a holistic tool in which all supply chain performance characteristics of a purchased component are documented. PEEP allows an organization to drill down into details of the supply chain and determine optimal methods to manage suppliers such that complexity can be driven out.

The above eight rights are actually a subset of a PEEP in that they allow for the measurement of critical performance components of purchased parts for every shipment received. Each right is measured by the percentage of successful executions. The "perfect execution score" is derived by multiplying together the percentage of the successful executions for each of the eight rights.

More detail regarding this approach can be found in [5.20].

Improvement of the engineering culture is also found in the related article "The Future of Quality. In fifteen years, will you recognize your organization?" [5.15]. The authors consider Quality in a broad sense (including reliability, durability, maintainability, and others). Some quotes from ([5.17,5.21]):

- One future-looking trend in today's leading organizations is the renewed interest in customer focus.
- Buzzwords are already emerging for the new quality world, as predicted by quality researchers and futurists.
- The world is changing faster than management realizes.
- In 2020, we will need bifocal leadership: clear, short-range thinking and sharp action to steer through the downturns, as well as accurate vision and steady nerves, to see well into the future.
- Know and go:
  - Quality organization in 2020 will be dominated by the acronym FUTURE which stands for fast, urban, universal, revolutionary and ethics;
  - quality professionals will continue to be motivated by factors other than money; the evolution of information technology and the formation of virtual companies demand new approaches to managing quality;
  - forget Six Sigma, tomorrow's focus will be on error-free performance; Quality efforts in the United States will focus less on manufacturing, because that industry will represent just a fraction of our gross national product.

An important interview with quality legend Joseph M. Juran [5.18] relates to the improvement of the engineering culture. This interview discusses the impact of manufacturing on the United States.

This interview is actually very current. Below are some components of Juran's answers [5.18]:

- Some people think that high quality costs more. That confusion exists in many different companies.

The word "quality" has two very different meanings. One meaning is the features of the product that enable it to sell. There, higher quality generally costs more. It takes more product research, more product development, and so on. People don't even call it a cost; they call it an investment, which will bring back higher returns. That's quality on the marketing side or the income side.

- Quality on the cost side is quite different. The cost of failure, the internal failure — scrap, rework, slow deliveries, failure to deliver on time — and the external failures — field failures, law suits, safety problems.

- A lot of CEOs believe that they are too busy to lead the quality charge, and so they delegate it. This hasn't worked very well. Leadership by the top people is an essential ingredient in getting out of that steep slope.
- A lot of companies believe that getting certified to ISO 9001 solves their quality problems. That simply is not true.
- The different members from companies of different standardization bodies are not going to agree to standards, that their companies are not able to meet. They are starting to change the standards, but that's at a glacial pace. It takes a long time to change an international standard."

Today's world is fraught with risk. A failure mode and effect analysis (FMEA) is an analytical methodology used to ensure that potential problems have been considered and addressed throughout the product and process development cycle.

An offshoot of Military Procedure MIL-P-1020, titled Procedures for Performing, Effects, and Criticality Analysis, issued in 1949 [5.17]. FMEA was first used as a reliability evaluation technique to determine the effect of system and equipment failures. Failures were classified according to "their impact of mission success and personnel/equipment safety" [5.17]. FMEA was further developed and applied by NASA in the 1960s to improve and verify the reliability of space-program hardware.

Today, the procedures called out in MIL-STD-1629A are the most widely accepted methods used throughout the military and commercial industry.

FMEA is a prevention-based, risk management tool that focuses the user or team on systematically:

- Identifying and anticipating potential failures;
- Identifying potential causes for the failures;
- Prioritizing failures;
- Taking action to reduce, mitigate, or eliminate failures.

The real value of the FMEA is reflected in its use as a long-term, living document. It is essential that the document is owned and updated as changes are made to the design or process. There are two types of FMEAs:

- DFMEA—an analysis process used to identify and evaluate the relative risk associated with a particular hardware design.
- PFMEA—a similar analysis, which relates to process design.

FMEA was first developed and used by reliability engineers in the 1950s to study malfunctions of military systems. As such, it has been a worthy and valuable technique in quality and other areas.

As common, as the tool is, however, it is often used incorrectly.

FMEA is a type of reliability and quality analysis, which can be used to anticipate and prevent problems, shorten product development times, and achieve safe and reliable products and processes.

But simply using FMEA does not indicate a strong engineering culture and its component—an accelerated testing culture. For success using this approach, it must be performed on the parts by the correct team, during the correct timeframe and with the correct procedure.

For obtaining improvements in the engineering culture, one needs to learn the following top FMEA [5.19] mistakes and avoid using them (Fig. 5.6). These mistakes are continuing even today.

But the above research work did not consider the accuracy of the simulations with the flight inputs.

One other important aspect of the engineering culture is correctly understanding the meaning of terms and definitions. The book [5.6] demonstrates

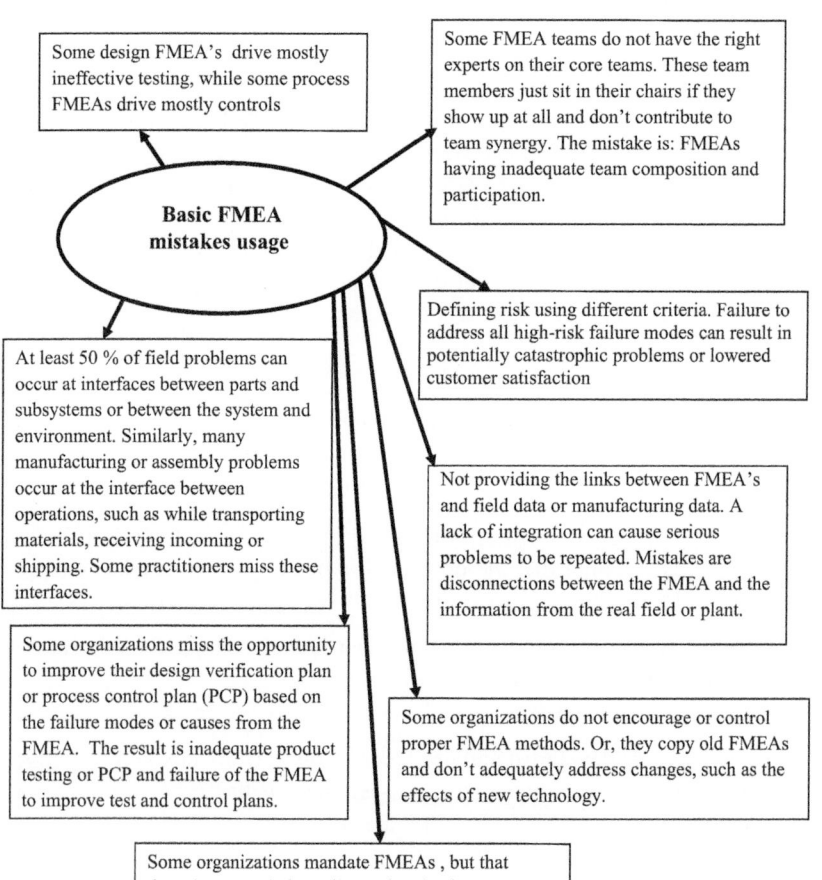

FIGURE 5.6 Which FMEA mistakes are you making?

how the inaccurate understanding of the term "durability testing" leads to poor testing, evaluating, and predicting of product performance.

5.5.1 An organization's aspects of culture as a component for improving the engineering culture

Often, improvement practitioners must deliver successful products and nudge their organizations forward on the improvement journey. Finding the right path involves dealing with uncertainty because an organization's culture and past experiences are unique. With uncertainty in an organization, it is difficult or impossible to get clear answers. As a result, initial efforts may not be successful, and the team can easily take missteps. In such situations, a frequently seen response to the team's misstep is a criticism of the team and its processes, but in such situations, the correct response to inevitable setbacks is to use them as the key to guiding improvement on the journey forward.

Direct confirmation, tactics to overwhelm the opposition, and hard-selling solutions or uncompromising positions often create negativity that leads to stonewalling, disengagement, and painful memories of the improvement initiative. In other words, all pain, no gain.

What does one do in this situation? Before selecting a test method and equipment, gauge the organization's pulse. Interact with all levels of management and staff in the company, as well as high-level professionals in accelerated testing outside the company. Get a sense of what they do and identify the key challenges that exist. Study advanced literature in accelerated testing and try to adopt the most advanced methods and equipment.

Identify any initiatives launched in the past, the reason for undertaking these initiatives, how they were managed and supported, and how different parts of the organization adapted to the changes [5.13].

Professionals often recommend this approach for gauging an organization's culture before fully committing to a testing solution. Has the organization or other organizations undertaken similar test solutions in the past? What was their experience and outcome? Developing these insights before starting work on a lean test method and equipment help to craft a better strategy by addressing three key issues:

- Address the past bad experience;
- Use early engagement, problem solving, and negotiations to win over employees;
- Frame it right.

For more details on organization culture, the reader is referred to [5.22].

Bibliography

[5.1]   Klyatis L. Accelerated reliability and durability testing technology. Wiley; 2012.

[5.2]   Klyatis L, Klyatis E. Accelerated quality and reliability solutions. Elsevier; 2006.

[5.3]   Nome F, Hariman G, Sheftlevich L. The challenge of pre-biased loads and the definition of a new operating mode for DC-DC converters. In: Power electronics specialists conference. IEEE; 2007.

[5.4]   Harwood R. The role of engineering simulation in the evolution of unmanned aircraft systems. Engineering solutions for military and aerospace. In: DEFENSE Tech Briefs. SAE International. Supplinaire to NASA Tech Briefs, vol. 8; December 2012.

[5.5]   Determining the value to the Warfighter, a 3-year ROI study. DoD HPCMO.2010.

[5.6]   Klyatis LM, Anderson EL. Reliability prediction and testing textbook. Wiley; 2018.

[5.7]   Research Brief. The impact of strategic simulation of product profitability. Aberdeen Group; June 2010.

[5.8]   Chan HA, Parker PT. Product reliability through stress testing. In: Annual reliability and maintainability symposium (RAMS) tutorial notes; 1999. p. 1−26.

[5.9]   Brecher BI. Accelerated testing experience with avionics. In: The 54th annual quality congress proceedings; May 8−10, 2000. Indianapolis.

[5.10] Morelli M. Fundamentals of HALT & HAAS. IEEE workshop on accelerated stress testing. October 1999. Boston, MA.

[5.11] Thermotron. Environmental stress, test simulation, and screening solutions.

[5.12] Whalley, Barley. Technical work in the division of labor stalking the wily anomaly. University of Texas; 1997.

[5.13] Kunda G. Engineering culture. Philadelphia: Tampe University Press; 2009.

[5.14] L. Klyatis. About trends in development accelerated reliability and durability testing technology. SAE 2012 World Congress. (Paper 2012-01- 0206).

[5.15] James Harrington H, Frank V. The future of quality. In fifteen years. Will you recognize your organization? Chico, CA: Quality Digest; 2005.

[5.16] Bailey DD, Alter H. New weak links. Use lean and quality tools to strengthen global supply chain performance. Milwaukee. Wisconsin: Quality Progress; 2014.

[5.17] U.S. Department of Defense. Procedures for performing a failure mode and effects and critically analysis. 1949.

[5.18] Scot M, Juran P. A lifetime of quality. An exclusive interview with a quality legend. Chico, CA: Quality Digest; 2002.

[5.19] Carlson C. Which FMEA mistakes are you making? Milwaukee, Wisconsin: Quality Progress; 2014.

[5.20] Sharma M, Sharma V. Discovering the right path. Milwaukee, Wisconsin: Quality Progress; 2015.

[5.21] Abe S, Hirokawa M, Zuitzu T. Stability design consideration for on-board distributed power system consisting of full-regulated bus converter and POLs. Power electronics specialists conference. 2006. 37th IEEE.

[5.22] Klyatis L. Successful prediction of product performance. SAE International 2016.

Exercises

1. Why is accurate simulation a key factor for accelerated testing development?
2. What is the meaning of the term (UAS) as applied in aircraft systems?
3. Describe several of the key challenges for the future roadmap for UAS development.
4. Describe the role of engineering simulation.
5. What is the estimated return for every dollar invested in accurate simulation.
6. When is accurate simulation most valuable?
7. Describe the statistical criteria used in the physical simulation of the input influences.
8. Describe the basic advantages of accurate simulation of the field conditions.
9. Describe how establishing the basic area of each influence can be introduced for all varieties of operating conditions.
10. Describe the author's approach for choosing which areas of influences should be included in the field testing decision.
11. Why does the development of accurate simulation of the field conditions depend on improvements of the organizations engineering culture?
12. What are the indicators of a good engineering culture?
13. Describe the basic cultural and technical components of the author's methodology for the successful prediction of a product's performance.
14. Describe the basic causes of why ART/ADT is not often used in practice.
15. some of the ways of improving a lacking engineering culture.
16. Describe the eight rights that must be understood and evaluated for the supply chain to assure product effectiveness.
17. Describe some of the quotes for considering in a broad sense quality development.
18. Describe the interview with quality legend Joseph. M. Juran as it relates to improvements in the engineering culture.
19. Describe the top FMEA mistakes.
20. Describe an organization's aspects of its culture as a component for improving the engineering culture.

Chapter 6

Implementation of basic positive trends in the development of accelerated testing

Abstract

Chapter 6 describes the positive trends that are occurring with new concepts of accelerated testing (AT). It will consider the different areas of implementation of accelerated testing development, including:

- Better understanding of the (AT) science and approach by professionals in the field;
- The actual advances in accelerated reliability and durability testing (ART/ADT);
- The published citations by this author's publications, with examples;
- Some of the many official written solutions that relate to organizational aspects of wide implementation new technologies of accelerated testing and recognition of the author's expertise in this area;
- Information from some of the many published reviews of the author's previous books that relate to ART/ADT;
- Some strategic aspects and principles of these trends in advanced AT from author's presentations to International Congresses and Symposiums;
- Some results of the implementation in different forms in United Nations papers, Institutes Academy of Sciences the Soviet Union, and world-class industrial companies, including Nissan, Jatko (Japan), Carl Schenck and Hereus Vothch (Germany), Iscar (Israel), DaimlerChrysler, Marcel Dekker, and others;
- Implementation of positive trends through seminars for the American Society for Quality, the Ford Motor Co., and others;
- Implementation of reliability testing and risk assessment through international standardization in Aerospace and Automotive engineering.

Trends in Development of Accelerated Testing for Automotive and Aerospace Engineering.
https://doi.org/10.1016/B978-0-12-818841-5.00006-4

6.1 Introduction

Chapter 6 includes the results of the implementation of the basic positive approaches to accelerated testing development—Accelerated Reliability and Durability Testing Technology—with which, the author is familiar. Of course, no one can be familiar with all aspects and value of implementation of this technology worldwide, especially those described in languages with which the author is not familiar. This is known to be occurring, because Elsevier, Wiley, SAE International, as well as Amazon and other companies have sold the author's books in many countries, and second, this information is in addition to the information that was included in the author's previous book [6.1].

As was previously detailed in Chapter 4, the basic positive trends producing the development of Accelerated Testing in the Automotive and Aerospace Engineering fields is the increasing recognition that accelerated reliability and durability testing (ART/ADT) provide the following benefits:

- It offers the possibility for the successful long-term prediction of the product's quality, reliability, safety, durability, maintainability, life-cycle cost, profit, and other efficiency components over the service life;
- Reduces the time from design to market;
- It reduces expenses in product design, manufacturing, and usage;
- It reduces recalls and customer complaints through the true analysis and results of ART/ADT, which provides the actual failure causes. This alone has the potential of avoiding multibillion-dollar losses;
- It helps to reveal where best to invest for increased profitability;
- It offers new possibilities for increasing the quality, safety, reliability, durability, maintainability, supportability, profit while decreasing the product's life-cycle cost;
- Globally employing it can improve the identification of many other problems and opportunities during design, manufacturing, and usage of the product.

For all of these reasons, the implementation of ART/ADT is very important in the engineering and development of products for an organization.

6.2 Some aspects of implementation of accelerated reliability and durability testing, including citations from other authors' publications

This author's new scientific-technical direction for the development of accelerated reliability and durability testing (ART/ADT) has been adopted for various types of sciences and products in many countries throughout the world. Some results of this implementation have been detailed in Chapter 4 "Implementation of Successful Reliability Testing and Prediction" in the book *Reliability Prediction and Testing Textbook* [6.1]. That chapter consists of 94 pages.

Additional descriptions and demonstrations of some of the important positive trends relating to the development of ART/ADT, as well as additional information, will be presented in this chapter.

6.2.1 Areas of ART/ADT implementation presented by other authors

Besides the examples presented in this chapter, as well as in other chapters of this book, instances of the implementation of this author's aspects of ART/ADT can be seen in the changing approaches of many professionals. This is usually a result of understanding the causes of negative traditional trends in the development of testing, as presented by this author in his other publications.

This evolution in the thinking of these professionals can be seen in their work. Some examples of this implementation have been included in other authors' publications, including:

- Implementation of different aspects of accelerated reliability lifetime test methods;
- ART/ADT used for improving various product reliability aspects in transport refrigeration;
- ART/ADT use for marine energy converters;
- ART/ADT use in model aircraft and launcher controllers;
- Employing statistical processing of accelerated life data;
- ART/ADT being used in the development of product models;
- Durability testing of various products;
- Aggressive efforts in improving quality testing and reducing customer complaints;
- Improving the wear and life characteristics of composite materials products;
- Accelerated testing being used in the development of various aspects of reliability and reliability/durability testing;
- Developments in reliability assessments;
- The expanding role of ART/ADT in the improvement of electronic systems reliability;
- Developments in the fields of renewable and sustainable energy;
- And others.

6.2.2 Examples with some citations from publications other than those published in the book *Reliability Prediction and Testing Textbook*

The following are some examples:

1. Markus W. Kemmner. Reliability demonstration of a multi-component Weibull system under zero-failure assumption. A Dissertation approved on May 9, 2012. Doctor of Philosophy Department of the Industrial Engineering University of Louisville. Louisville, Kentucky, USA.

"A method to calculate the LCL of reliability of a system with Weibull components was presented by Klyatis, Teskin and Fulton (2000). This method is

not applicable for zero failure but works with complete and censored data. It is intended for situations in which the different components of the system were tested individually with different sample sizes and test durations. Based on the rate of censoring different factors have to be determined table-based. They are used to calculate a LCL of the reliability of the components and finally the system".
From Reference:
Klyatis LM, Teskin OI, Fulton JW. Multi-Variate Weibull Model for Prediction System-Reliability from Testing Results of the Components, Proceedings of the Annual Reliability and Maintainability Symposium (RAMS). Los Angeles, CA, 2000.

2. Popa Ionut, Lupescu Octavian, Popa Valica & Scurtu Popa Ramona. Researches regarding the reliability assessment using the boxplot method. International Journal of Modern Manufacturing Technologies. ISSN 2067−3604, Vol. I, No. 1, 2009.

"Analysing the special literature (Billinton & Allan, 1992; Klyatis & Klyatis, 2004; Mărăsescu, 2004), the reliability represents the probability in which the component parts, products and systems perform their functions without faults for what they were designed, in the specificated conditions, for a certain time period and with a given confidence level. Some product or technological equipment reliability".
From Reference:
[2] Klyatis LM, Klyatis EL, Accelerated reliability testing problems solutions, Reliability and Robust Design in Automotive Engineering, 2004 SAE World Congress, Detroit, MI, March 8−11, 2004, pp. 283−290, ISBN−13: 978-0-08-044924-1.

3. Janne Kiilunen. Development and evaluation of accelerated environmental test methods for products with high reliability requirements. Tempere Institute of Technology. Tampere 2014. Publication 1242. ISBN 978-952-15-3359-4.

"However, a challenge with this kind of testing is that environmental stresses rarely occur alone or consecutively in service conditions and often their combined effect may be substantially more severe than their individual effects. Therefore the specific stresses applicable for the actual operating conditions of a particular product should also be applied simultaneously during reliability testing. This makes it possible to study the interaction effects of various stress factors. Consequently the reliability test may be improved and made to better represent the real use environment and the potential reliability risks.[Kly06][Oco05]"
From Reference:
[Kly06] Klyatis LM, Klyatis EL. "Accelerated quality and reliability solutions", Elsevier Inc., United States of America, 2006, p. 544.

4. Zaharia SM, Martinescu I, Morariu CO. Statistical processing of accelerated life data with two stresses using monte carlo simulation method.
Eighth International DAAAM Baltic Conference Industrial Engineering —
19—21 April 2012, Tallinn, Estonia.

"... Estimating reliability is essentially a problem in probability modeling.
A system consists of a number of components. In the simplest case, each
component has two states, operating or failed. When the set of operating
components and the set of failed components is specified, it is possible to
discern the status of the system. The problem is to compute the probability that
the system is operating - the reliability of the system [1]."
From Reference:
[1] Klyatis LM. Accelerated reliability and durability testing technology.
Wiley, New Jersey, 2012.

5. Sebastian Marian Zaharia, Ionel Martinescu. Improving product reliability
under accelerated life testing using monte carlo simulation. Transilvania
University of Braşov B-dul Eroilor, nr. 29, Romania.

"Quantitative ALT consists of tests designed to quantify the life characteristics of the product, component or system under normal use conditions and
thereby provide reliability information [4,5]."
From Reference:
[4] Klyatis, L.M. (2012), Accelerated Reliability and Durability Testing
Technology, Wiley, New Jersey.

6. Ni—Mn—B, Seung-Hwan Ma, Young-tai Noh, Gun-ik Jang. The study on
accelerated life-time reliability test methods. Korea Conformity Laboratories. Department of Materials Science and Engineering, Chungbuk National University. Journal of the Korea Academia-Industrial Cooperation
Society, Vol. 16, No. 5 pp. 2993—99, 2015. https://doi.org/10.5762/KAIS.
2015.16.5.2993. ISSN 1975—4701/eISSN 2288—4688 2993.

Ni-Mn-B 삼원합금도금 가속수명 및 신뢰성 평가에 대한 연구
(The Text is in Korean).
From Reference:
[5] Lev M. Klyatis "Accelerated Reliability And Durability Testing
Technology" WILEY, 2012.

7. Alejandro Romo Perea-1, Javier Amezcua -2, *and* Oliver Probst1,1. Validation of three new measure-correlate-predict models for the long-term
prospection of the wind resource.
  1. Department of Physics, Instituto Tecnológico y de Estudios Superiores
  de Monterrey, CP 64849, Monterrey, Nuevo León, Mexico
  2. Department of Atmospheric and Oceanic Sciences, University of
  Maryland, College Park, Maryland 20,742-2425, USA

Publisher: American Institute of Physics. Journal of Renewable and Sustainable Energy 3, 023105, 2011.

"While it has been discussed that a linear regression emerges naturally if a bivariate joint-normal distribution of the variables X and Y is assumed, in the case of wind speed prospection this approach is questionable. A bivariate joint-normal distribution implies normal marginal distributions. Hourly wind speed averages, the ones typically used in wind energy prospection, do not follow normal distributions but are described rather by Weibull distributions. It is therefore plausible to look for a bivariate pdf whose marginal probability density functions are Weibull. This type of bivariate pdf's have been proposed by several authors [13—17]."

From Reference:

[16] Klyatis LM, Teskin OI, Fulton JW, "Multi-variate Weibull model for predicting system-reliability, from testing results of the components", RAMS, pp. 144—149, 2000.

8. Rajkumar K., Aravindan S., Kulkarni MS. Wear and life characteristics of microwave-sintered cooper-graphite composite. Journal of Materials Engineering and Performance. Indian Institute of Technology. Delphi.2396—Vol 21(11) November 2012. JMEPEG (2012) 21:2389—2397. ASM International https://doi.org/10.1007/s11665-012-0161-z 1059—9495/$19.00. (Submitted March 17, 2011; in revised form January 12, 2012).

"In actual field, a product/component usually fails under the combination of multiple stresses (Ref 12)".

From Reference:

[12] Klyatis LM, Klyatis EL, Accelerated quality and reliability solutions, 1st ed., Elsevier, Amsterdam, 2006.

9. Sebastian Marian Zaharia. Lifetime estimation from accelerated reliability testing using finite elements analysis postdoctoral research Department of Manufacturing Engineering, Technological Engineering and Industrial Management Faculty, Transilvania University of Brasov, Romania,

Prof. dr.ing. Ionel MARTINESCU, Department of Manufacturing Engineering, Technological Engineering and Industrial Management Faculty, Transilvania University of Brasov, Romania. Fiabilitate si Durabilitate - Fiability & Durability No 1/2013 Jiu, Editura "Academica Brâncuşi", Târgu Jiu, ISSN 1844 — 640X.

"Reliability accelerated tests (ART) are statistically based sampling tests performed to approximate the long term reliability of a product before products are mass produced in manufacturing. These tests also provide acceleration factors that are used to estimate the mean time to failure, life expectancies of a product under test. There are different types of accelerated experiments plans in use, which include subjective, traditional, best traditional, and statistically optimum and compromise plans. Accelerated reliability models relate the failure rate or the life of a product to a given stress such that measurements taken during accelerated testing can then be extrapolated back to the expected performance under normal operating conditions. The implicit working assumption here is that the stress will not change the shape of the failure distribution. The most significant acceleration models are: Arrhenius, Eyring, Inverse Power Law; Life - Thermal Cycling, Life - Vibration, Life - Humidity [2,3]".

From Reference:

[3] Klyatis LM, Accelerated Reliability and Durability Testing Technology, Wiley, New Jersey, 2012.

10. Higashi-Fuji Technical Center

From Wikipedia, the free encyclopedia.

Higashi-Fuji Technical Center (東富士研究所 Higashi-Fuji Kenkyūjo?) is a Toyota research and development facility in Susono, Shizuoka, Japan. The facility was established in November 1966 [2].

Notably, the center contains an advanced driving simulation housed inside a 7 m (23 feet) diameter dome with an actual car inside [6]. The simulator is used to analyze driver behaviors in order to improve safety [6]. Higashi-Fuji also includes a crash test building [7].

References

1. "Japanese Facilities". Toyota. Retrieved December 27, 2013.
2. "Higashi-fuji Technical Center: Facility Overview" (PDF). Toyota. 2010. Retrieved December 27, 2013.
3. "Toyota Develops World-class Driving Simulator" (Press release). Toyota. November 26, 2007. Retrieved December 27, 2013.
4. Kageyama, Yuri (November 12, 2012). "Toyota tests cars that communicate with each other". Associated Press. Retrieved December 27, 2013.
5. "Design and R&D Centers". Toyota. Retrieved December 27, 2013.
6. Klyatis LM. Accelerated Reliability and Durability Testing Technology. February 3, 2012. John Wiley & Sons. pp. 58−59. ISBN 9781118094006.
7. Abuelsamid, Sam (30 July 2010). "Autoblog gets seat and simulation time with Toyota's newest safety technology [w/video]". Autoblog. Retrieved 27 January 2014.

11. Ringgold, Inc., Portland, OR. 2016. The Free Library. Ringgold, Inc., Portland, OR. 2016. December 23, 2018.

"Rather than considering safety aspects of failures in automotive, aerospace, and commercial products, Klyatis focuses solely on the scientific and technical aspects of predicting the performance of such products accurately enough to avoid failure. There are no published books on predicting product performance, only books on predicting reliability, as though reliability were a separate component that did not interact with other performance components.

Among his topics are an analysis of current approaches in simulation and testing, methodological aspects as the first basic component of successful prediction of product performance, integrated equipment for the physical simulation of interacting real-world conditions, and improving the standardization of accelerated reliability and durability testing".

From Reference:

Lev Klyatis. Successful Prediction of Product Performance: Quality, Reliability, Durability, Safety, Maintainability, Life-Cycle Cost, Profit, and Other Components. SAE International, 2016.

6.3 Some citations from published reviews to this author's previous books related to the implementation of the basic positive trends in the development of accelerated testing

These are reviews to this author's development of accelerated reliability/ durability testing (ART/ADT) that were previously published in his books, and were published in journals and magazines worldwide. While some are published in English, some others are not yet published in the author's books.

For example, Bryan Roggles from Boeing, Everett, WA in November 2013 published in ASQs Magazine *Quality Progress* a review to the author's book *Accelerated Reliability and Durability Testing Technology*, which was published by John Wiley & Sons in 2012. In this review he wrote: "... Anyone interested in high-quality products, particularly in the areas of reliability and durability, will find this book a valuable resource. This book is an excellent resource for teachers and students."

Another example is James Rodenkirch's book review, published by *The Journal of Reliability, Maintainability, and Supportability in Systems Engineering*, Summer 2012 (DoD). The reviewer analyzed this author's book published by Wiley. In this two-page review, James Rodenkirch wrote:

> ... *is an interesting and eye opening excursion on how to conduct, properly, accelerated reliability or durability testing (ART/ADT). New concepts and ideas are centered on close simulation of the conditions that equipment(s) are exposed to out "in the field ..."*

And

> ... *Mr. Klyatis' book has a wide range of applicability. His holistic approach to reliability and durability testing will resonate with testing professionals as well as aspiring test engineers. His highlights of the drawbacks associated with current testing practices offer a sound basis for the improved testing alternatives he proffers. The reader walks away with cost effective arguments that can be offered to decision makers to explain how changes to testing procedures and training, along with the purchase and use of the proper test equipment, can reduce cost and improve system reliability and safety.*

And one more citation from this review: "... Equally important is the need for the test community to grasp a fundamental concept: those performing the testing must be well trained and experienced".

One more citation comes from Susan Fingerman's review of this book. The review was published by Jefferson, American Public University System. Sci-Tech Book News Reviews, Volume 66, Issue 2, Article 14, April 2012, page 41. In this review she wrote

... Klyatis demonstrated several applications of the methodology and equipment that provide physically accelerated reliability testing and accelerated durability testing on an actual product in a way that represent the real world's many interactions that influence products. Previous testing approaches focused on one factor in isolation, he says, rather than looking at the product as a whole. His topics include accelerated reliability testing as a component of systems approach, accelerated reliability and durability testing methodology, financial and design advantages of accelerated reliability/durability testing, and standardization.

6.4 Some strategic aspects of the implementation of the positive trends in accelerated testing for successful prediction of a product's efficiency

The basic strategic aspects of the implementation positive trends in accelerated testing in the automotive and aerospace engineering fields are related to the adoption of accelerated reliability and durability testing. These can be found in this author's other published books [6.1−6.5], and in the over 80 other publications in English in numerous journals, world congresses, international symposiums, workshops, and other forums (Figs. 6.1−6.12).

Some strategic aspects can be seen in the following slides from this author's presentations (converted from Power Point format to Word) at National and International meetings:

Some slides from this presentation are provided below:

First two slides:

2007 RAMS

The Reliability/Maintainability Integrated with Quality

Lev Klyatis
Eccol, Inc.
Habilitated Dr.-Ing., Sc.D., PhD

FIGURE 6.1   Title page of presentation Lev Klyatis at the 2007 RAMS (Annual Reliability and Maintainability Symposium (The International Symposium of Product Quality and Integrity) [6.6].

ACCURATE PHYSICAL SIMULATION OF FIELD
INPUT INFLUENCES

1. Physical simulation of real life input influences (chemical and mechanical [dust] pollution, radiation, features of the road, etc.) means physical imitation of the above influences on the natural test subject.

Usefulness of this simulation depends on its accuracy.

2. The concepts of accuracy include:

-Maximum number of simulations of field conditions;

-Accurate simulation of each group of field conditions (multi-e nvironmental, electrical, mechanical, etc.) simultaneously and in combination;

-Simulation of simultaneous combination of each type of complex influence [for example, pollution = mechanical (dust) + chemical].

3. Simulation of whole real influences, but not including:
    -idle time (breaks, etc.);
    -time with minimum loading which does not cause failures.
4. Degradation mechanism as a basic criterion for accurate simulation;

5. Consideration of system (unit) interactions among components (details).

6. Reproduction of the complete range of field schedules and maintenance (repair)

7. Corrections of the simulation system after an analysis of field degradation & failures, and comparison of this with degradation and failures during ART/ADT.

FIGURE 6.2   Two slides from presentation at the 2007 RAMS.

Basic Principles of Accelerated Reliability Testing

1. Improvement of the technique and equipment for accurate physical simulation   of field input influences;

2.Entire technology of STEP-BY-STEP;

3. Providing accelerated laboratory testing (ALT) in a combination with periodic field testing;

4. Providing ART as a simultaneous combination of multiple-environmental (including corro sion), mechanical (including vibration), electrical, and   other components of ART.

FIGURE 6.3   Third slide from Lev Klyatis's presentation at the 2007 RAMS.

(cont.)

As a result:

1. One hour of pure workpe rformed by the product under accelerated testingconditions is identical in itsdegradation ef fect to one hour of pure work in the field.

2. One can easily calculate the acceleration coefficientfor tes ting;

3. Therefore, one can:
   a) predict the dynamics of future degradation & the time to failures in the field;

   b) find and eliminate causes for degradation & failures of product quickly and at lower costs.

FIGURE 6.4   One more slide from 2007 RAMS presentation.

Below are three slides from the presentation at the SAE 2012 World Congress (Detroit, 2012).

About Trends in the Strategy of Development Accelerated Reliability and Durability Testing Technology

Lev Klyatis
Habilitated Dr.-Ing., Sc.D., Ph.D
Sr. Advisor
SoHaR, Inc.

FIGURE 6.5   The title page of the presentation at the SAE 2012 World Congress.

FIVE BASIC STEPS of ART/ADT STRATEGY COMPONENTS

1.Study of the field conditions   (integrated input influences, human factors, and safety problems) to determine parameters to be simulated in the laboratory.

2. **Accurate simulation** of the field conditions.

3. **ART/ADT performance**, including **analysis** and **management** of degradation (failures) and their **causes**.

FIGURE 6.6   Slide from the presentation for SAE 2012 World Congress.

FIVE BASIC STEPS (cont.)

4. Development of recommendations for elimination of these causes.

5. If these recommendations are used, there are:

- **accurate prediction** of quality, reliability, safety, durability, maintainability, life cycle cost, and others.

- **accelerated development** of the product.

FIGURE 6.7   Slide from the Lev Klyatis's presentation for SAE 2012 World Congress.

The presentations at the SAE World Congresses, ASQ Congresses, and RAMS symposiums provided only 20–30 min that included time for questions and answers. Therefore, it was developed as a more detailed presentation, as a tutorial for the RMS Partnership (Reliability, Maintainability, and Supportability) workshop/symposium. Attendees included professionals from the Department of Defense, the Department of Transportation, and industry. The author presented a tutorial for this event. Below are some slides related to the strategic aspects of ART/ADT from that tutorial.

ACCELERATED RELIABILITY/DURABILITY TESTING. RELIABILITY & SAFETY ACCURATE PREDICTION and SUCCESSFUL PROBLEMS PREVENTION

Lev Klyatis
Dr.-Ing. Habilitated, Sc.D., Ph.D.
Senior Advisor
SoHaR, Inc

TUTORIAL

FIGURE 6.8   The title page of the tutorial for RMS Partnership workshop/symposium.

MORE STRESS means

GREATER ACCELERATION and

LOWER CORRELATION of **AT** RESULTS WITH FIELD RESULTS.

FIGURE 6.9   Axiom of Stress Testing.

FIVE NON-TRADITIONAL PRINCIPLES
OF ART/ADT TECHNOLOGY

1. The **integration** of Reliability Testing components in each other – from the study of field conditions to **Reliability & Safety accurate prediction and problems successful prevention.**

System of Systems (SoS) approach is used for this.

2. The **accurate simulation** of field conditions **is one complex**, including **integrated** full field input influences, safety, and human factors.

FIGURE 6.10   The slide from the above tutorial.

FIVE NON-TRADITIONAL PRINCIPLES (cont.)

3.   **Accurate physical simulation** of all **interconnected** components of field conditions.

The **fully-integrated test equipment** is used for this purpose.

4. **ART/ADT**   is based on this simulation and is a **keyfactor** for **accurate prediction** of product safety, reliability, maintainability, **and their problems successful prevention.**

FIGURE 6.11   The slide from the above tutorial.

FIVE NON-TRADITIONAL PRINCIPLES (cont.)

5. a). **Accurately moving,**
   through accurate physical simulation of field conditions,
   the field to the laboratory.
   b). Then one can use in the laboratory the
      scientific-technical apparatus for studying the causes of product **degradation** in the field
      (and ways for elimination them).
   Used ART/ADT Equipment will be **useful:**
   - **not only for design, but also**
   - **for manufacturing and usage**
   - **for the following models of the product.**

One **cannot** do this using current test approaches, including field "reliability" testing.

FIGURE 6.12   Five non-traditional principles from the above tutorial.

Greater details on the implementation of the strategy of reliability and durability testing was presented at the RMS Partnership (DoD) seminar and symposium/workshop in San-Diego, as well as at other world forums. Some slides from this event related to ART/ADT strategy, including new and effective nontraditional principles are presented below (Figs. 6.13–6.19).

When we provide the above strategy, we have to take into account the divergences that occur during design and manufacturing that affect the testing level (Fig. 6.20).

Most accurate simulation of input influences occurs when each statistical characteristic (mathematical expectation, variance, normalized correlation, and power spectrum) of all input influences differs from the field s by no more than given limit:

$$Y_{1\ FIELD} - Y_{2\ ART} \leq \text{Given Limit (3 \%, 5 \%, or 7 \%)}$$

FIGURE 6.13   Accuracy of the simulation as key factor for accelerated reliability and durability (ART/ADT) testing.

The reliability function distribution after **ART** is $F_A(x)$, in the **real world** is $F_0(x)$.
The measure of their difference is:
$$\Delta[F_A(x), F_0(x)] = F_A(x) - F_0(x).$$

On the $\Delta[F_A(x), F_0(x)]$ one gives limitation Δ_A.

If $\Delta[F_A(x), F_0(x)] \leq \Delta_A$ it is possible to determine the reliability by **ART** results with high accuracy.

But if $\Delta[F_A(x), F_0(x)] > \Delta_A$
it is not recommended.

FIGURE 6.14   Statistical criteria of comparison reliability as one from ART results and real-world results.

If it is unknown the functions Fa(x) & F0(x), one can construct
the graphs of the experimental data:

$F_A(x)$ **and Fo(x)** and determine:
$$D_{M,N} = [F_{AE}(x) - F_{0E}(x)]$$

where: F_{0E} and F_{AE}

are empirical distributions of the reliability function as

results of the testing of machinery under operating conditions
and by **ART/ADT** .

FIGURE 6.15   If one does not know the function $F_A(x)$ and $F_0(x)$ (often in practice).

One needs to calculate the accumulated parameter's function and the values of the confidence coefficient found in the equations:

$$\hat{Y}(x) = \sum_{m=k}^{n} C_n^m p^m (1-p)^{n-m}$$

$$\underline{Y}(x) = \sum_{m=0}^{K} C_n^m p^m (1-p)^{n-m}$$

and evaluate the curves that are limited to the upper and lower confidence areas.

where: $C_n^m p^m (1-p)^{n-m}$ is the probability that based on an event will be in **n** independent experiments **m** times. The values of **Y** are found in the tables of books on the theory of probability if the confidence coefficient is

$$\gamma = 0.95 \text{ or } \gamma = 0.99.$$

FIGURE 6.16 Comparison parameter's function with predetermined accuracy and confidence area.

CONSIDERED ART/ADT INCLUDES 8 NON-TRADITIONAL PRINCIPLES (1)

- The integration of components in each other in one complex from study of field conditions for accurate prediction and accelerated development.
  For this goal is used System of Systems (SoS) approach.

- The **accurate simulation** of field conditions **as one complex**, including **integrated full** field input influences, safety, and human factors.

FIGURE 6.17 The slide from RMS Partnership seminar.

8 NON-TRADITIONAL PRINCIPLES (cont.)

- Since field conditions act simultaneously and in combination, **accurate simulation** of them also means **acting all integrated field conditions**.
  For this purpose is used fully-integrated test equipment.

- **ART/ADT**, which based on this simulation, **is a keyfactor** for **accurate prediction** of product quality, reliability, maintainability, safety, durability, life cycle cost, **and accelerated product development**.

FIGURE 6.18 The slide from RMS Partnership seminar.

8 NON-Traditional Principles (3)

- **Accurately moving**, through accurate simulation of field conditions, **of the field to the laboratory**, that offers possibility for using in the laboratory the apparatus and other scientific-technical solutions for studying the causes (and ways for eliminate them) of product <u>degradation in the field</u>.
- **We cannot do this if use other current test approaches.**

- As a result, quickly obtaining the information (for any time of usage) for all parameters **accelerated development and accurate predictionwith minimal costand time**

- **Not description** of Quality, Reliability, Maintainability, Supportability, and Durability and other problems, **but demonstrationhow** one can solve their problems with considering them in interconnection (integration).

FIGURE 6.19 The slide from RMS Partnership seminar.

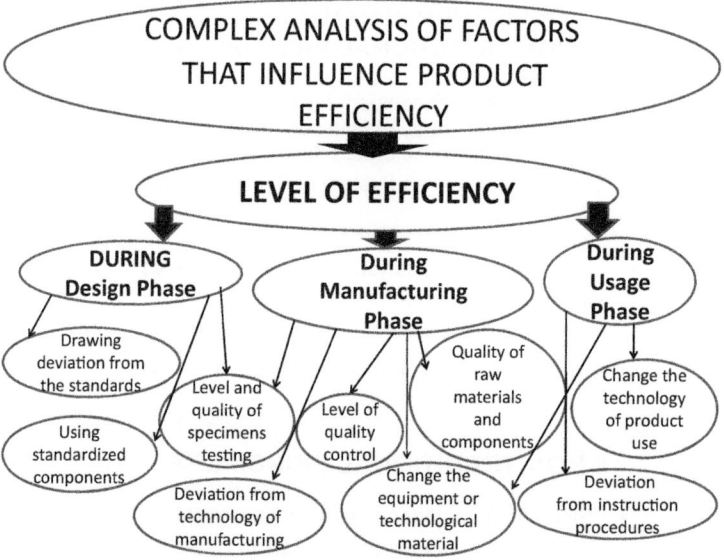

FIGURE 6.20 Scheme of factors that one needs to study during design, manufacture, and usage phases.

The above slides demonstrate some of these strategic principles. Besides the above, specific real-world input influences that are specific to the type of test subject and corresponding conditions of the subject's use. Also, one needs to provide complex analysis, during design, manufacturing, and usage phases, of factors that influence product efficiency, as shown, for example, in Fig. 6.20.

ART/ADT was partly implemented in many countries and by many companies (see Ref. [6.1]). Fig. 6.21 shows the official letter how Iscar Ltd. was involved in implementation process, and obtained the benefits from implementation during design and manufacturing. As can be seen from this letter, the development and implementation of new system for defect analysis, organizing interdisciplinary teams for emergency improvement, including quality assurance, design, manufacturing, marketing, and selling was accomplished. The result for the 3 years was an increase of over 40% for the company's product sales, defects reduced from 2.6% to 1.7%, and customer complaints were greatly decreased. As one result, American billionaire Warren Buffet bought the Iscar Ltd. company.

Letter of recommendation

September 19, 2006

To whom it is concern

RE: Eugene (Evgeny) Klyatis

Iscar, Ltd. was recently acquired by Warren Buffet, the world's leading expert in company value, operational competence and production highlighting our company's rapid rise to preeminence among the precision manufacturing companies of Israel. Our company is research and development intensive, engaging almost 200 researches.

In the past three years sales of each of our products have increased no less than 40% and defective finished goods reduced from 2.6% to 1.7%. Favorable feedback from our wholesalers and ultimate product users have increased exponentially, while complaints have greatly decreased.

From April 2001 Eugene Klyatis was engaged by our company as chief researcher and developer of Quality Assurance, critical elements of our operations. His contributions to our improved results were signal, consisting in his research, analysis, research and developments in developing entirely new approaches for product quality, increasing the engineering culture of our company.

Mr. Klyatis developed new systems for defect analysis and inter-disciplinary teams for emergency improvement, which included professional from all our departments, including quality assurance, marketing, sales, design and manufacturing.

As the Iscar's Product Manager, Milling since 2000 based upon my personal knowledge, it is my professional opinion, that Eugene Klyatis was indispensable to the growth of our company his research in the fields of quality assurance is among the world's most important.

Shmuel Weinshtein   *Woinshtein*

Product Manager Milling
Iscar Headquarters.
Tel: + 972(4)9970508
Fax: + 972(4)9970113
Mobile + 972(50)5302615
E-Mail: Shmulikw@iscar.co.il

FIGURE 6.21 Some results of ART/ADT partial implementation by Iscar Ltd.

6.5 Some of the author's patents in accelerated testing improvement that were actually implemented

Below is a list of some of the over 30 patents held by the author in accelerated testing (AT) improvement, and that reflect positive trends in the development of AT.

Partial list of patents

By Lev Klyatis

1. Method of Machinery Accelerated Testing. Patent 1794320. USSR Governmental Committee for Inventions Discoveries for Science and Engineering.

2. Method of Machinery Accelerated Testing. Patent 386308 (USSR).

3. Generator of Nonstationary Random Processes. Patent 430370.

4. Equipment for Accelerated Reliability and Durability Testing of Machines. Patent No. 180840 (Russia).

5. Equipment for Accelerated Reliability and Durability Testing of Machines. Patent No. 173465 (USSR).

6. Equipment for Accelerated Reliability and Durability Accelerated Testing. Patent No. 134490.

7. Equipment for Harvesters Testing. Patent 104645 (USSR).

8. Multi-Channel Generator of Random Processes. Patent 466511 (For system of control of testing equipment).

9. Equipment for Accelerated Reliability and Durability Testing of Farm Machinery with Pull Flax Working Heads. Patent 365003 (USSR).

10. Method of Accelerated Testing. Patent No. 365003 (Russia).

11. Equipment for Accelerated Testing. Patent No. 477686 (Russia).

12. Equipment for Testing Working Heads of Fertilizer's Applicators. Patent No. 1667680 (USSR).

13. Equipment for Testing Fertilizer Applicator. Patent No. 366681 (Russia).

These patents were implemented at the Kalinin State Test Center (Russia), research institute VNIIZIVMASH (Ukraine), and other countries and companies. As an example, two (title pages) from these patents can be seen in Figs. 6.22 and 6.23.

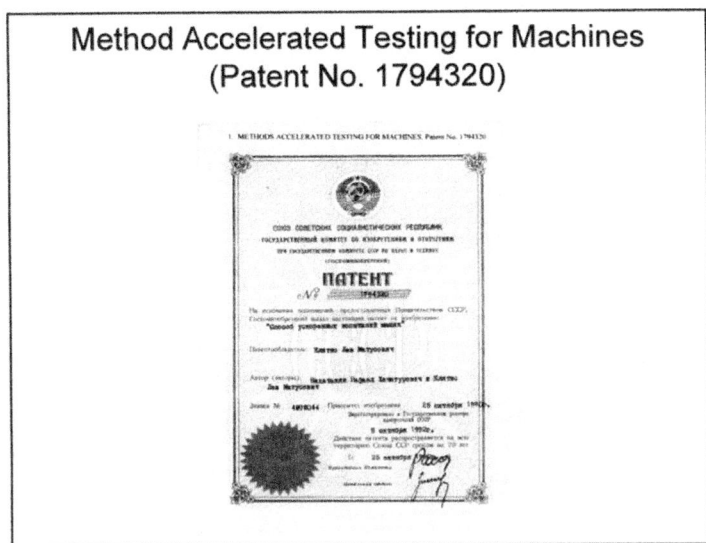

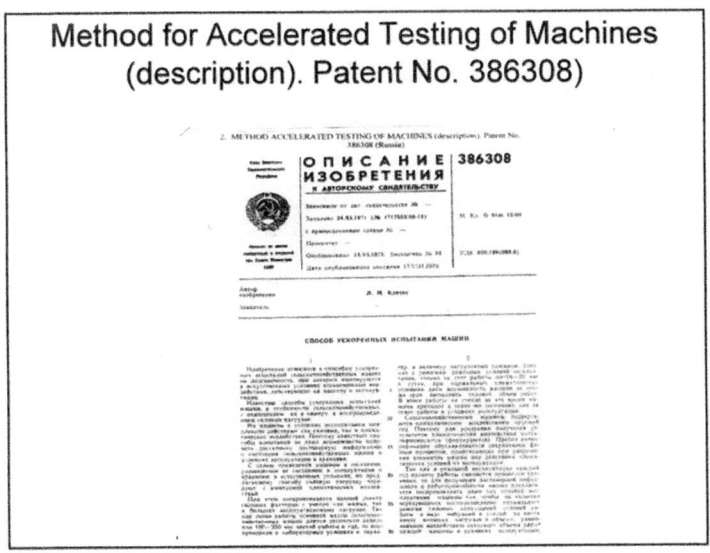

FIGURE 6.22 Example of author's patents for new methods of accelerated testing of machines.

United States Patent [19]

Minenko et al.

[11] Patent Number: 4,852,927

[45] Date of Patent: Aug. 1, 1989

[54] DEVICE FOR GRIPPING NECKS OF PACKED SACKS

[76] Inventors: Leonid P. Minenko, ulitsa Narimanova, 72/7, kv.6.; Mikhail I. Osher, ulitsa Lenina, 93/4, kv.28.; Evgeny V. Belikov, pereulok Pronsky, 106a, kv.35., all of Rostov-na-Donu; Lev M. Klyatis, ulitsa Bestuzhevykh, 10, kv.152, Moscow, all of U.S.S.R.

[21] Appl. No.: 183,186

[22] PCT Filed: Jun. 20, 1986

[86] PCT No.: PCT/SU86/00065

§ 371 Date: Feb. 19, 1988

§ 102(e) Date: Feb. 19, 1988

[87] PCT Pub. No.: WO87/07883

PCT Pub. Date: Dec. 30, 1987

[51] Int. Cl.⁴ B66C 1/42
[52] U.S. Cl. 294/86.4; 294/103.1
[58] Field of Search 294/86.4, 103.1, 88, 294/68.1, 68.3; 414/607, 608, 621

[56] References Cited

U.S. PATENT DOCUMENTS

4,226,458 10/1980 Achelpohl et al. 294/88

FOREIGN PATENT DOCUMENTS

262347 5/1970 U.S.S.R. .
740152 6/1980 U.S.S.R. .
802162 2/1981 U.S.S.R. .
931654 5/1982 U.S.S.R. .
1009970 4/1983 U.S.S.R. .
1039858 9/1983 U.S.S.R. .

Primary Examiner—James B. Marbert
Attorney, Agent, or Firm—Fleit, Jacobson, Cohn, Price, Holman & Stern

[57] ABSTRACT

A device for gripping necks of packed sacks comprises a housing (1) with a horizontal support (2), and a frame (3) one side (6, 7) of which is pivotably connected to the housing (1), whereas the other side has a horizontal rod member (8) forcing the neck (9) of the packed sack (10) to the horizontal support (2) of the housing (1). The pivot (11) for connecting the frame (3) to the housing (1) is arranged below the horizontal support (2) of the housing (1). Side bars (12, 13) of the housing (1) have ears (14, 15) horizontal axes (16, 17) of which are offset relative to the horizontal support (2) of the housing (1) and arranged in a vertical plane (18) tangent to the horizontal rod member (8) of the frame (3) in a position, when the horizontal rod member (8) is brought in contact with the support (2) of the housing (1).

8 Claims, 4 Drawing Sheets

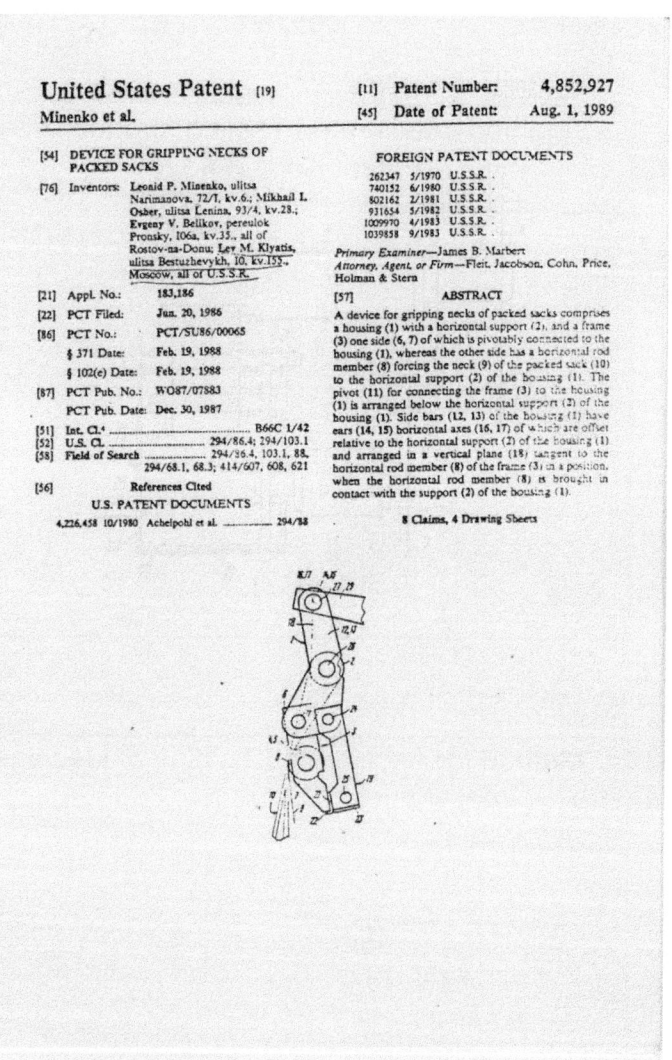

FIGURE 6.23 Author's patent in the United States.

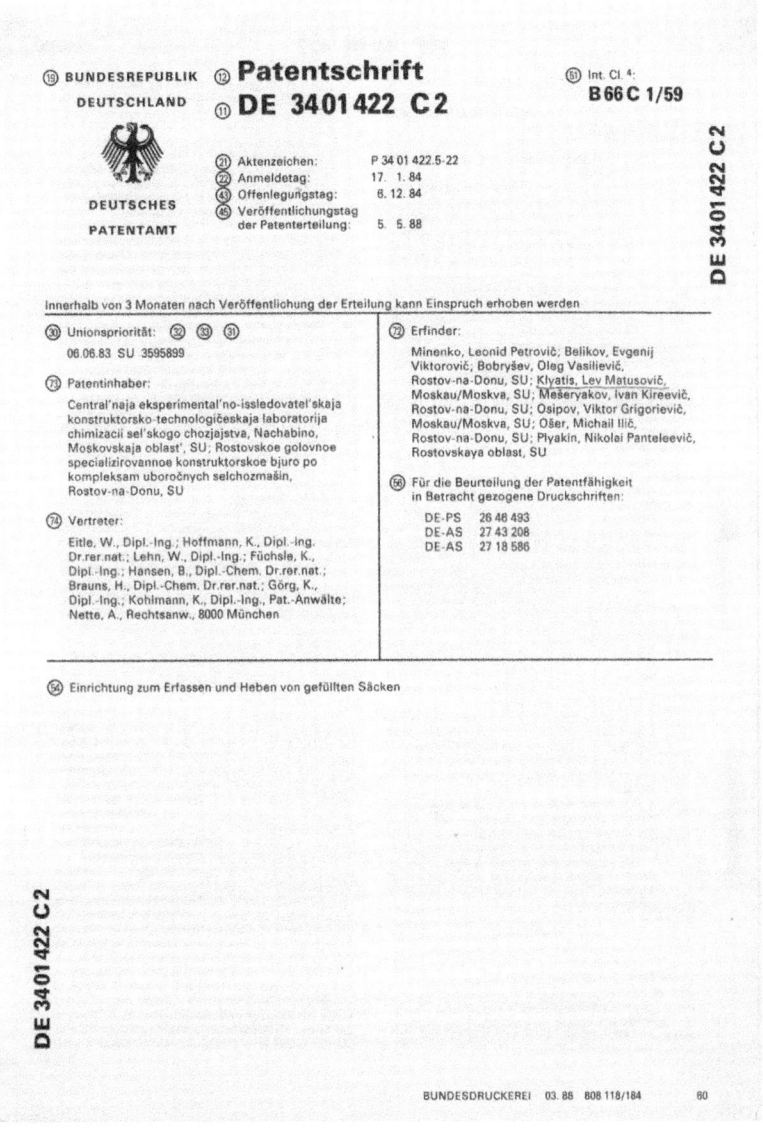

FIGURE 6.23 (A) Author's patent in Germany.

6.6 Implementation of the new concepts of accelerated testing improvement

New concepts of accelerated testing improvement were reported at the United Nations Economic Commission for Europe. It was conducted over 2 years in Geneva, and the final report by the United Nations was published in New York.

Two pages (Figs. 6.24 and 6.25) from the above mentioned report in Geneva [6.7] are presented below. The concepts contained in this report were first implemented for agricultural machinery, and then implemented in the USSR for automobiles. Following these successful applications it was implemented in many other countries throughout the world in aerospace, electronic, and other highly technical areas.

Distr.
RESTRICTED
AGRI/WP.2/11?
16 January 1969
ENGLISH
Original: RUSSIAN

ECONOMIC COMMISSION FOR EUROPE
COMMITTEE ON AGRICULTURAL PROBLEMS
Working Party on Mechanization of Agriculture

ACCELERATED TESTING OF AGRICULTURAL MACHINERY

Note by the Secretariat

In accordance with the decision adopted at the fourteenth session of the Working Party on Mechanization of Agriculture (see document AGRI/298, AGRI/WP.2/111), Mr. A.F. Kononenko and Mr. L.M. Klyatis, experts from the USSR, have prepared a final draft of the report on the accelerated testing of agricultural machinery. This report, the text of which is reproduced below, will be discussed at the Working Party's fifteenth session, with a view to its approval for publication in the AGRI/MECH series.

GE.69-762

FIGURE 6.24 The first page of the report Accelerated Testing of Agricultural Machinery in United Nations Economic Commission For Europe (Geneva).

In the 21st Century, these trends in accelerated testing strategy development, either completely or partially, were widely implemented in countries, where languages other than English are used, and references and citations are more difficult to obtain.

As these new author's, as well as other researchers, adopt these strategies in the areas of successful prediction of a product's efficiency, it inevitably leads to quicker and better development to improve civilization throughout the world.

AGRI/MECH/43
page 8

Table 1

Variable stress amplitudes in components of the
load-bearing system of the T-4 tractor, recorded
at the testing-ground (speed of tractor, without
plough, 5.8 km/h)

Strain gauge No.	Pass over 20 pairs of obstacles 160 mm high on concrete track			Pass over 40 pairs of obstacles 140 mm high on cobbled track		
	Stress amplitudes, kg/cm^2			Stress amplitudes, kg/cm^2		
	A max.	A av.	A min.	A max.	A av.	A min.
1.	x			x		
2.	720	645	520	870	705	610
3.	375	320	215	-	-	-
4.	505	435	375	-	-	-
5.	780	645	545	900	725	630
6.	785	650	445	800	675	575
7.	750	600	460	815	610	435
8.	910	775	615	710	610	435
9.	680	585	470	-	-	-
10.	1,240	1,020	850	1,450	1,076	760
11.	530	420	360	535	420	370
12.	1,190	855	640	-	-	-
13.	720	520	455	840	815	770
14.	695	560	410	800	600	515
15.	970	770	615	1,030	835	740
16.	805	705	590	850	660	525
17.	930	655	505	765	540	445
18.	690	550	490	600	490	385
19.	995	795	665	1,000	730	630
20.	570	485	387	615	490	420
21.	x			x		
22.	830	690	575	435	350	300
23.	935	620	435	-	-	-
24.	70	540	465	-	-	-
33.	1,270	1,120	975	795	655	520
34.	1,180	960	675	1,210	935	715

Notes:
1. "x" = stress less than 300 kg/cm^2
2. "-" = test not recorded for technical reasons.
3. The numbers of the gauges selected for detailed tests are underlined; the numbers of the extra gauges selected for the schedule of 180° turns are double-underlined.
4. The oscillogram analysis took into account the greatest curve-span corresponding to passage over two obstacles.
5. The average amplitudes (A av.) are shown in the table as parameters of the amplitude distribution series and cannot be used as an exhaustive characteristic in analysing the level of strain in the components.

The results of analysis of the oscillograms recorded in the detailed studies
made on the testing-ground and in field operation are given in tables 2 and 3.

FIGURE 6.25 The Table 1 from report Accelerated Testing of Agricultural Machinery in Economic Commission for Europe of the United Nations (Geneva).

These trends, including, as an example, two types of test chambers for accelerated corrosion testing that were also implemented in the Research centers, Academy for Sciences of the Soviet Union & Russia. Fig. 6.27 presents these documents of the implementation of this in the Physical Chemistry Institute, Academy of Sciences of the Soviet Union.

As result of the positive trends obtained by this implementation, the Soviet Union's Government organized a special organization TESTMASH (Fig. 6.26). Based on the research obtained by the department, where the author worked, he was invited to become the chairman of TESTMASH. The solutions developed there were directed to wide scale implementation of the new Accelerated Reliability and Durability Testing (ART/ADT) technology first in the USSR, and then later in Russia.

ЗАМЕСТИТЕЛЬ ПРЕДСЕДАТЕЛЯ

ГОСУДАРСТВЕННОГО КОМИТЕТА СССР

по НАУКЕ и ТЕХНИКЕ

103905, Москва, ул. Горького, 11
Тел. 229-11-92 Телетайп 417531 „Триод"

от _06.02.90г._   № _МК6-33/19_

на № _6/24-1911 от 27.12.89г_

Об организации хозрасчетной
структурной единицы Инженерный центр "Тестмаш" с права-
ми юридического лица в составе НИИАЭ НПО"Автоэлектроника"

Заместителю Министра автомобильного и сельскохозяйственного машиностроения СССР

т.Морозову В.П.

Государственный комитет СССР по науке и технике рассмотрел
предложение министерства об организации самостоятельной хозрасчетной структурной единицы Инженерный центр "Тестмаш" в составе
научно-исследовательского и экспериментального института автоэ-
лектроники и электрооборудования (НИИАЭ) НПО "Автоэлектроника"
Министерства автомобильного и сельскохозяйственного машиностроения СССР и сообщает.

ГКНТ СССР неоднократно рекомендовал Минавтосельхозмашу СССР
создать такой инженерный центр в г.Москве на правах юридического
лица на базе отдела ускоренных испытаний ВНИИКОМЖа и поэтому не
имеет возражений против предложения министерства.

Однако при утверждении положения об инженерном центре счита-
ли бы целесообразным возложить на него выполнение следующих глав-
ных направлений научно-исследовательской и проектно-конструктор-
кой деятельности:

- машины и оборудование для проведения испытаний полнокомп-
лектных тракторов, автомобилей, сельскохозяйственных машин, а
также их узлов и деталей;

- диагностическое оборудование;

- климатические камеры;

- автоматизированные системы управления стендово-испытате-
льным и диагностическим оборудованием на базе микропроцесссорных

Зак. 550—10.000

FIGURE 6.26   The Soviet Union Government's letter about organization TESTMASH.

The goals, specifics, and contents of this wide implementation were described in detail in an interview with the author and published in the All USSR Journal *Tractors and Farm Machinery*.

These trends were later implemented in other countries (see Ref. [6.1]).

Several German companies, and companies from other countries, expressed interest in implementing the author's strategy and ideas in

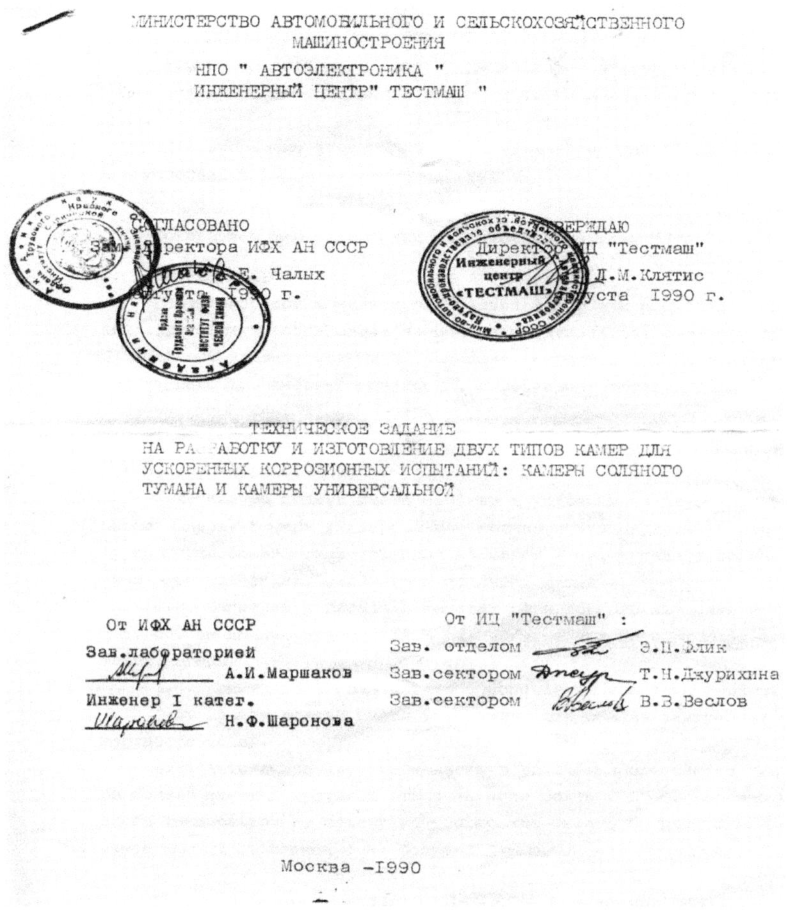

МИНИСТЕРСТВО АВТОМОБИЛЬНОГО И СЕЛЬСКОХОЗЯЙСТВЕННОГО
МАШИНОСТРОЕНИЯ

НПО " АВТОЭЛЕКТРОНИКА "
ИНЖЕНЕРНЫЙ ЦЕНТР" ТЕСТМАШ "

СОГЛАСОВАНО
Зам. директора ИОХ АН СССР
Е. Чалых
июня 1990 г.

УТВЕРЖДАЮ
Директор ИЦ "Тестмаш"
Л.М.Клятис
августа 1990 г.

ТЕХНИЧЕСКОЕ ЗАДАНИЕ
НА РАЗРАБОТКУ И ИЗГОТОВЛЕНИЕ ДВУХ ТИПОВ КАМЕР ДЛЯ
УСКОРЕННЫХ КОРРОЗИОННЫХ ИСПЫТАНИЙ: КАМЕРЫ СОЛЯНОГО
ТУМАНА И КАМЕРЫ УНИВЕРСАЛЬНОЙ

От ИФХ АН СССР

Зав. лабораторией
А.И.Маршаков

Инженер I катег.
Н.Ф.Шаронова

От ИЦ "Тестмаш" :

Зав. отделом   Э.П.Элик

Зав. сектором   Т.Н.Джурихина

Зав. сектором   В.З.Веслов

Москва —1990

FIGURE 6.27 The document about implementation TESTMASH's new solution of corrosion test chamber in Physical Chemistry Institute of the USSR Academy for Sciences.

accelerated testing development. One such example is presented in Fig. 6.28, where Hereus-Votsch GmbH wrote about their special interest in joining a continuation of the work developed by this author's department in developing a new complex test chamber for corrosion testing.

The following are some examples of this author's direction in the development of ART/ADT which can be seen in Fig. 6.29 (first page of the invitation) of US-USSR Trade and Economic Council, Inc. (organized by both US and USSR Governments), where William D. Forrester, President of this Council, invited Dr. Klyatis, member of this Council, to the XIV meeting of this Council, which was held in the USA.

Heraeus
VÖTSCH

P R O T O K O L L
über
Besuch der sowjetischen
Delegation Minavtoselchozmasch
bei Heraeues-Vötsch GmbH,
D-7460 Balingen v. 30.6. - 3.7.89

ПРОТОКОЛ
переговоров советской делегации
Минавтосельхозмаша   с фирмой
Хераус-Фёч Гмбх Д-7460 г.Балинген
от 30.6.89.

Die Firma Heraeus befindet sich
z. Zt. In einer Bestandsaufnahme
bezüglich der Möglichkeiten, die
gegenseitigen Geschäftsbeziehun-
gen in der Sowjetunion zu vertie-
fen. In diesem Rahmen fanden die
o.g. Gespräche in Balingen statt.

Фирма "Хераус-Фёч" в настоящее
время готова к углублению взаимных
с СССР комерческих отношений. В этих
рамках состоялись организационные
переговоры в г.Балинген.

Zwischen Minavtoselchozmasch und
Heraeus-Vötsch wurde folgendes
vereinbart:

Между Минавтосельхозмашем и "Хераус-
Фёч" достигнута следующая договорен-
ность:

1) Bis September 1989 Klärung der
Möglichkeiten seitens Minavtoselchozmasch für die Schaffung
einer Service-Organisation zur
Betreuung der Heraeus-Produkte,
evtl. auch auf Basis eines Joint
Ventures.

1)До сентября 1989 года последует
разъяснение со стороны Минавтосельхозмаша о возможности создания сервисногс
центра по обслуживанию продукции
"Хераус" в рамках совместного предпри-
ятия.

2) Bei der Vorstellung der Möglichkeiten seitens Minavtoselchozmasch
sollte auch aufgezeigt werden, in-
wieweit gemeinsame Entwicklungs-,
Montage- und Produktionsprojekte
realisiert werden können.

2)При представлении возможностей для
этого со стороны Минавтосельхозмаша
должно быть указано,в какие сроки
могут быть реализованы совместные кон-
структорские,монтажные и производствен-
ные работы.

3) Insbesondere besteht seitens
HVB Übteresse ab der gemein-
samen Fortführung eines bei
NPOVNJIKOMG und Firma GSPKTB
Bubrujsk begonnenen Projektes
einer komplexen neuen Korro-
sionsprüfkammer.

3)Особый интерес для фирмы представляе
совместное продолжение работы начатой
ВНИИКОМХем и ГСПКТБ г.Бобруйск разрабо
кой новой комплексной камеры для ускоренных коррозионных испытаний.

4) Sich daraus evtl. ergebende
weitere Vereinbarungen müßten
folgende Punkte berücksichtigen:

4)Вытекающие отсюда дальнейшие догово-
ренности должны учитывать следующие
пункты:

FIGURE 6.28 Protocol results of negations delegation of the USSR Automotive Department with Heraeus-Votsch GmbH (Balingen, Germany), where Germany's company demonstrates especial interest in join continuation of work, which the author developed.

During this Council meeting, Dr. Lev Klyatis made a successful presentation to the engineering group about his company's development of ART/ADT. As a result of this presentation, the author received:

- A proposal from N. Bekh, Chairman of KAMAZ, Inc. (large Russian company in truck design and manufacturing with over 100,000 employees) to implement the author's strategy in the development of equipment for accelerated testing, as well as for testing systems in the company. This proposal later resulted in a signed contract of implementation by KAMAZ, Inc. with TESTMASH, with this author as chairman;

US-USSR Trade and Economic Council, Inc.
3 Shevchenko Embankment, Moscow 121248, USSR

№ 533
"20" ноября 1990 г.   Директору Инженерного центра
                                        "Тестмаш", члену АСТЭС

                                        г-ну Клятису Л.М.

Уважаемый Лев Матусович,

Я находился в Москве в середине октября с.г. в связи с поездкой в СССР представительной группы американских бизнесменов под эгидой Национального комитета республиканской партии США, программа пребывания которой была организована АСТЭС. В ходе этого визита я имел встречи с лидерами советских деловых и финансовых кругов, с представителями союзных и республиканских органов.

Я хотел бы сообщить Вам о тех изменениях в нашей деятельности, которые мы производим для того, чтобы обеспечить дальнейшую активизацию нашей работы в плане удовлетворения возрастающих требований членов АСТЭС в условиях продолжающейся экономической децентрализации, вызванной реформами, проводимыми в СССР, а также в результате быстрого расширения американо-советского делового сотрудничества.

Для того, чтобы эффективно оказывать содействие почти 700 советским и американским членам АСТЭС, мы увеличиваем количество руководителей проектов в нашем Московском представительстве примерно на 10 человек и соответственно расширяем штат советских и американских сотрудников в нашей штаб-квартире в Нью-Йорке.

С этой же целью мы решили компьютеризировать наши офисы в Нью-Йорке и Москве, что повысит эффективность работы по обеспечению оперативной связи между американскими и советскими партнерами и облегчению поиска партнеров среди американских и советских членов АСТЭС.

Американо-Советский Торгово-Экономический Совет
СССР, Москва, Набережная Шевченко, 3, 121248 ● Телекс 413212 ● Факс 230-2467

FIGURE 6.29 Invitation Dr. L. Klyatis from USTEC President W. D. Forrester to attend the USTEC Council meeting in the USA.

- A proposal with the American company Steptoe & Johnson (see Fig. 6.30) to organize implementation in U.S. industrial companies to provide positive developments in accelerated testing for automotive and other industrial areas (see Chapter 4);
- Received additional invitations from other American companies in the automotive industry for implementation under his direction of ART/ADT development.

The positive trends in development accelerated testing (see Chapter 4) developed by this author, and reflected not only by their use in many research and testing facilities, and in publications, but are apparent to the author of this

STEPTOE & JOHNSON
ATTORNEYS AT LAW
1330 CONNECTICUT AVENUE, N. W.
WASHINGTON, D. C. 20036-1795

(202) 429-3000
FACSIMILE NO. (202) 429-9204
TELEX: 89-2503

July 16, 1991

SPA "Autoelectronika"
Engineering Center "Testmash"
Dr. Lew M. Klyatis
Director
Box 73
103055 Moscow, USSR

Phone: 250-8020
Fax: 258-7504

Dear Dr. Klyatis:

Thank you for your letter of July 3, 1991 and also for your organization's charter and other background materials. We are very pleased that Testmash and Kamaz wish to employ Steptoe & Johnson in a legal and business capacity in an effort to produce and market the new type of vibration test system developed by Testmash.

Based on Mark Davis's communications with us (you met Mr. Davis in Moscow several weeks ago), we understand that there are three types of services that you would want Steptoe & Johnson to perform. First, Steptoe & Johnson would help your organization to find a Western investor willing and capable of becoming a partner in the joint venture with Testmash and Kamaz. Second, Steptoe & Johnson would represent Testmash (and possibly Kamaz) in negotiations with the Western partner, provide tax and other legal and business advice, draft the necessary documents (such as a joint venture agreement, charter, technology licensing agreement, etc.), and help to register the joint venture. Third, if the parties to the joint venture wish, Steptoe & Johnson would provide legal advice to the joint venture or, if Testmash wishes, exclusively to Testmash. We agree that the above division of work is appropriate and believe that we can be of great use to you at every stage of the project.

As you probably realize, each of the above undertakings will require significant time and effort on our part. The second and third stages will be the most time consuming. However, even the first stage, in order to be success

FIGURE 6.30   First page of the letter from Steptoe & Johnson (US) in an effort to produce in the Western market the new type of vibration test system developed by TESTMASH.

book. Examples can be seen in German companies who have expressed interest in collaborating with the author (Figs. 6.31−6.33).

Fig. 6.31 demonstrates one such moment in Germany with the company Carl Schenk during a visit by the USSR Automotive Department delegation. After the author's presentation, the Carl Schenk representatives wrote a protocol, which included special interest of this company in the implementation of equipment for accelerated testing, which was developed by the author (see protocol in the book [6.1]). After that Carl Schenk sent to

FIGURE 6.31 In In Germany: from left Dr. Lev Klyatis, two Carl Schenk representatives, Dr. Y. Zukov, chair of USSR delegation.

the author a reminder that they wanted to come to Moscow for decision on the details of this collaboration for implementation (Fig. 6.32).

The positive trends in the development of accelerated testing in the Soviet Union (and later Russia) (see Chapter 4) were included in the Government Programs with financial support for implementation (there are corresponding documents available in Russian).

After the author came to the USA, his research results in accelerated testing were developed and implemented, first, for farm machinery, and then later for other automotive and aerospace engineering applications.

In the beginning of this part of his career, his work relied heavily on relationship with colleagues from Western countries.

But most of his previous publications were in Russian and were not available in English. Because of this he was virtually unknown as an expert in the field of testing in the USA. In order to meet expenses, Dr. Klyatis began working in the USA by delivering fish (Fig. 6.34). This was physically demanding hard work that had to be done in *any* weather. Whether it was

```
20 16 24
412116 SHMA SUT
20 16 32
412116 SHMA SU
78422 3Z NEMA DD      FS-NR. 104    20.03.92    14. 31      BU
04


TESTMASH MOSKAU
Z. HD. HERRN PROF. DR. KLYATIS


SEHR GEEHRTER HERR PROF. DR. KLYATIS'

- IM INTERESSE DER ZUSAMMENARBEIT MIT ''HERAEUS-VOETSCH'' PLANEN
  WIR EINE REISE ZU IHNEN NACH MOSKAU.
  HIERMIT BITTEN WIR SIE PER TELEX AN UNSERE ADRESSE EINE EIN-
  LADUNG FUER DIE AUSSTELLUNG DES EINREISEVISUMS ZU SCHICKEN.

    - REISENDE: HERR MICHAEL TERMOELLEN, GEB. AM 23.05.1959
                HERR JUERGEN KAISER,      GEB. AM 07.02.1944
    - REISETERMIN: AB 05.04.1992
    - REISEDAUER:  MAXIMAL 10 TAGE

- BITTE ANTWORTEN SIE PER TELEX BIS ZUM 25.03.1992'
- WIR VERSUCHEN LAUFEND HERRN DR. FLIK ANZURUFEN, LEIDER BE-
  KOMMEN WIR KEINE VERBINDUNG (AM 19.03.92 WURDE DAS TELEFON-
  GESPRAECH NACH 2 MINUTEN GESTOERT). WIE KOENNEN WIR HERRN
  DR. FLIK IN MOSKAU TELEFONISCH ERREICHEN (DATUM, UHRZEIT)?


MIT FREUNDLICHEN GRUESSEN
J. KAISER
ABT. SERVICE
HERAEUS-VOETSCH-NEMA GMBH
NETZSCHKAU


412116 SHMA SU
78422 3Z NEMA DD
```

FIGURE 6.32 Copy of the fax to the author from Hereus-Voetsch company (Germany) about collaboration.

raining all day, or was very hot in the summer, the author still needed to bring the fish to the customers on time and by walking.

But while doing this, it also provided the opportunity to further his development and to implement new approaches to accelerated testing through presentations at national and international meetings, attending standardization committees, direct contacts with colleagues, and other ways. By doing this he was able to implement new approaches, first, in America, and then worldwide.

Another example of how the author raised funds for travel to further his new ideas, can be seen in Fig. 6.35 acknowledging his Research Grant Award of grant monies. Considering that the American Society for Quality has over 100,000 members, and that they gave only eight of these research grant awards each year, this was a significant accomplishment. The author sent

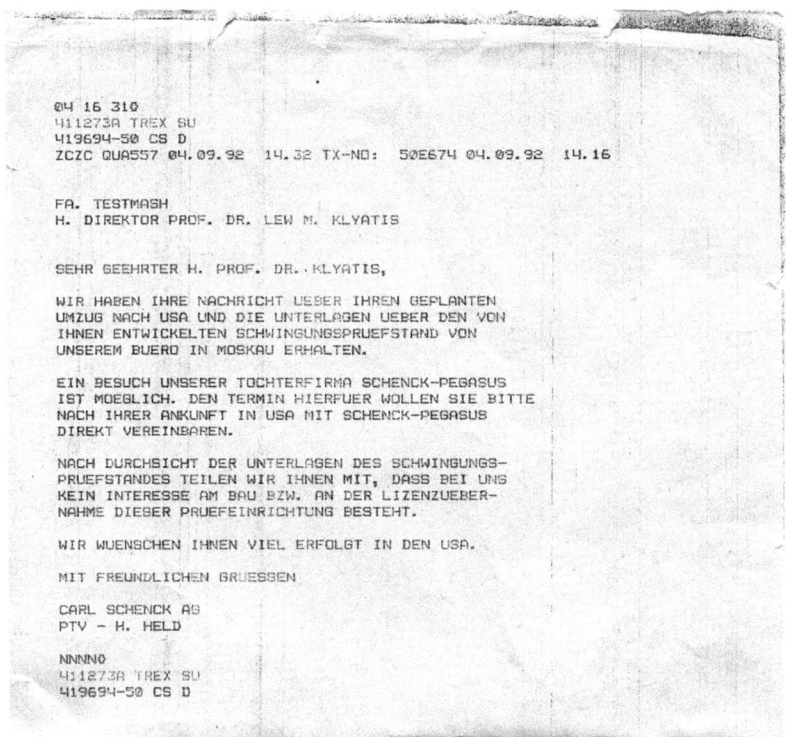

```
04 16 310
411273A TREX SU
419694-50 CS D
ZCZC QUA557 04.09.92  14.32 TX-NO:  50E674 04.09.92  14.16


FA. TESTMASH
H. DIREKTOR PROF. DR. LEW M. KLYATIS


SEHR GEEHRTER H. PROF. DR. KLYATIS,

WIR HABEN IHRE NACHRICHT UEBER IHREN GEPLANTEN
UMZUG NACH USA UND DIE UNTERLAGEN UEBER DEN VON
IHNEN ENTWICKELTEN SCHWINGUNGSPRUEFSTAND VON
UNSEREM BUERO IN MOSKAU ERHALTEN.

EIN BESUCH UNSERER TOCHTERFIRMA SCHENCK-PEGASUS
IST MOEGLICH. DEN TERMIN HIERFUER WOLLEN SIE BITTE
NACH IHRER ANKUNFT IN USA MIT SCHENCK-PEGASUS
DIREKT VEREINBAREN.

NACH DURCHSICHT DER UNTERLAGEN DES SCHWINGUNGS-
PRUEFSTANDES TEILEN WIR IHNEN MIT, DASS BEI UNS
KEIN INTERESSE AM BAU BZW. AN DER LIZENZUEBER-
NAHME DIESER PRUEFEINRICHTUNG BESTEHT.

WIR WUENSCHEN IHNEN VIEL ERFOLGT IN DEN USA.

MIT FREUNDLICHEN GRUESSEN

CARL SCHENCK AG
PTV - H. HELD

NNNN0
411273A TREX SU
419694-50 CS D
```

FIGURE 6.33 The copy of the fax from the Carl Schenck AG (Germany) to this book author about collaboration with Schenck Pegasus in the USA.

FIGURE 6.34 First job professor Lev Klyatis and Doctor Boris Ganelin in America.

American Society for Quality

611 East Wisconsin Ave
· P.O. Box 3005
Milwaukee, WI 53201-3005
414-272-8575
Fax 414-272-1734
800-248-1946

May 29, 1998

Dr. Lev Klyatis
International Assoc. of Arts & Sciences, Inc.
17 Battery Place
7th Floor North, Rm 772
New York, NY 10004

Dear Lev:

It gives me great pleasure to inform you that the review committee has approved a grant for your ASQ Research Advisory Committee Research Fellowship proposal. Enclosed is a check in the amount of $ 5,000 which represents 50% of the total award. Another 25% will be paid upon receipt of your six-month progress report, and the remaining 25% upon receipt of the final report.

On behalf of the ASQ Research Advisory Committee, I offer my congratulations on your selection as a recipient and look forward to the result   our work with great anticipation.

Sincerely,

Lawrence S. Aft
Chair
Research Advisory Committee

FIGURE 6.35   Information about the approval of a grant for the author's research fellowship proposal.

reports about his research work to ASQ over a span of 2 years, and used this award money during his early years of living in America to cover the cost of attending World Congresses, International Symposiums, Workshops, and advanced industrial companies and universities, international and national committees. At these events, he educated participants on the implementation of his new approaches through presentations at these meetings and seminars providing, and began collaboration with colleagues in America and in other countries.

The first implementation in the USA of these new ideas and strategies (see Chapter 4) was through the American Society of Agricultural Engineers (ASAE), T-14 Quality, Reliability, and Test Committee.

Fig. 6.36 shows that the author was invited as a coordinator for reviewing the standard EP-456 "Test and Reliability Guidelines" which needed to be updated for improvement in testing and reliability techniques. The work in updating this standard was completed, and this updated standard was successfully balloted and approved.

Minutes of the ASAE T-14
Quality, Reliability, and Test Committee
Tuesday, 13th December 1994
11:30 am - 1:00 pm

Attendance:

Lawrence Ellebracht	Visitor	AGCO
David Jones	Member	Univ of Nebraska
Lev Klyatis	Member	New York
Ronald McAllister	Member	New Holland
Steve Newbery	Member	John Deere
John Posselius	Member	New Holland
David Sandfort	Visitor	J I Case

1: Meeting called to order at 11:30 am.

2: Minutes from December 1993 Winter meeting approved as submitted (noted corrections on spelling and wording on the first copy of minutes sent out last April. Corrected minutes are what was submitted to ASAE headquarters)

3: Old Business

David Jones led a discussion about T-14 sponsoring a book or monograph. It was decided that there is a need but details need to be worked out. Possible contributors may be authors making technical presentations at a technical session that T-14 may sponsor at the November Power and Machinery Conference. Immediate action was tabled for the time being.

T-14 is not sponsoring a technical session at the 1995 Annual Meeting (June 95). It was noted that there is going to be a technical session "Testing and Reliability" at the Machinery Conference to be held in Ceder Rapids, Iowa, in May 1995, further information about the program or who to talk to about getting on the program can contact committee member Steve Newbery. It was suggested that T-14 sponsor a session at the P&M conference to be held in Chicago, November 1995 by Ron McAllister (a coordinator for the program). John Posselius agreed to coordinate a session at the November meeting providing each committee member try to get a person to present a technical paper or report. If a paper is presented it would be considered for our monograph.

Lev Klyatis agreed to be the coordinator for rewriting EP-456. He will include discussions on additional testing and reliability techniques. John Posselius agreed to get help from headquarters in the form of rewrite procedures and a copy of the standard on disc to aide in the rewrite.

FIGURE 6.36 The first pages from minutes of the ASAE T-14 Quality, Reliability, and Test Committee, where one could be see that the author was invited as coordinator for rewriting the EP-456 standard.

He was invited to be the chief of the Reliability Department at ECCOL, Inc. (USA), where he worked until 2009. During this time, he developed and implemented ART/ADT technology, including the following:

- Developed the theory and strategy of ART/ADT, including new ideas. These theories and strategies were followed with implementations through presentations, seminars for industrial companies, journal articles (*Journal of Quality Engineering*; *Journal of Reliability Review*; *The Journal of*

Reliability, Maintainability, & Supportability in Systems Engineering (DoD); and others);

- Prepared first two English language books on these theories and strategies that were published by Mir Collection in New York (2002) and Elsevier in UK (2006) [6.4]. The Elsevier's book included the subchapter "Trends in development of useful accelerated reliability testing technology"(eight pages), as well as Chapter 8 "Basic Concepts of Safety Risk Assessment" developed from experience as an expert of the International Electrotechnical Commission (IEC) with International Organization for Standardization (ISO) "Joint Study Group in Safety Aspects of Risk Assessment";

- Consultant work for Black & Dekker, Co, Ford Motors Co., Daimler-Chrysler, and others;

- Practical step-by-step help for American industrial companies in implementing ART/ADT technology.

Several years after coming to the USA, the author made a presentation "Principles of Accelerated Reliability Testing" at the IEEE Workshop on ACCELERATED STRESS TESTING, in Pasadena, California. The photo below is from this workshop (Fig. 6.37).

Fig. 6.38 demonstrating that the expertise in the area of reliability and accelerated testing were recognized by his appointment as Reliability Division session manager at the American Society for Quality's (ASQ) Annual Quality Congress. This was after only 6 years after immigrating to the United States.

FIGURE 6.37 From left: Paul Parker, chairman of IEEE Workshop Accelerated Stress Testing; Lev Klyatis, speaker; Kirk Gray, vice chairman of this workshop, Pasadena, CA.

American Society for Quality

June 8, 1999

Mr. Lev M. Klyatis
72 Montgomery St., #1311
Jersey City, NJ 07302

Dear Lev,

This letter is to confirm that you are appointed session manager for the Reliability Division's session at the 54th ASQ Annual Quality Congress, May 8-10, 2000 in Indianapolis, IN. Your task will be to find speakers for a one and one-half hour session on reliability at the conference. Speakers and yourself will receive complimentary registration to the conference, but no further expenses will be paid by ASQ or the Reliability Division.

I have attached copies of the tentative timeline for the conference. Any further information will be sent to you by ASQ headquarters.

Thank you for your participation.

Ronald L. Akers. P.E.
Chair – Reliability Division

Cc: T. Gurunatha

Headquarters   611 East Wisconsin Avenue   P.O. Box 3005   Milwaukee, Wisconsin 53201-3005   414-272-8575   Fax 414-272-1734   800-248-1s

FIGURE 6.38 Confirmation letter that the author has appointed session manager for the Reliability Division session at the 54th ASQ Annual Quality Congress.

At this Annual Quality Congress, Lev Klyatis has meeting with the Presidents of the American Society for Quality and the China Society for Quality (see Figs. 6.39 and 6.40).

Later he developed a group of six standards under the common title "Reliability Testing" for the SAE International G-11 Committee, which were focused on aerospace reliability standardization. The draft versions of these

FIGURE 6.39 Dr. Lev Klyatis with Presidents of Society for Quality of the USA and China. 56th Annual American Society for Quality Quality Congress. Denver.

FIGURE 6.40 Nellya Klyatis who for many years helped Lev Klyatis and Harold Williams, Executive Editor Engineering Journal "Reliability Review" at the 56th Annual ASQ Quality Congress, Denver. Picture taken at the American Society for Quality Reliability Division booth at that Congress.

standards were discussed several times during SAE G-11 meetings. One outcome from the discussions of these draft documents was the invitation to this author to lecture in Honeywell's office in Washington D.C. Figs. 6.41 and 6.42 show this book's author with a group of this committee's experts at this site, following these discussions.

FIGURE 6.41   Dr. Lev Klyatis, one of the participants of SAE G-11 Reliability Committee for standardization in aerospace, during the meeting at the Honeywell International, Inc., Washington D.C.

FIGURE 6.42   Group of experts in aerospace reliability and maintainability (SAE G-11 members) in Washington DC. Second from left − Lev Klyatis.

For this and other work in aerospace engineering testing development and implementation in practice, the author received several awards that can be seen in Figs. 6.43−6.45.

FIGURE 6.43   The first page of brochure about SAE Aerospace Awards recipients.

During this time, his colleagues from TESTMASH, Dr. Yakhya Abdulga-limov and Dr. Yuriy Piatine (seen in Fig. 6.46 discussion) continued discussing his new ideas and ART/ADT strategy development after coming to the USA.

Terry C. Kessler
TK Consulting LLC

Terry Kessler is acknowledged for his work on Committee J and Chair of the OEM Subcommittee. He has led and coordinated OEM's and Manufacturers of Maintenance materials in efforts to develop improved products, standardize specifications and introduce new technologies. He has provided expertise and guidance to airline customers on many aspects of engine maintenance and inspection as a GE staff member for 30+ years. He has more recently been a member of the Engine Titanium Consortium with Iowa State University, investigating the relationships between overhaul and repair processes and is currently a member of the Centre for Aviation Systems Reliability project continuing that work.

Dr. Lev M. Klyatis
Eccol, Inc.

Lev Klyatis is recognized for his work on the Reliability, Maintainability, Supportability and Logistics (RMSL) Division (G-11) and the US TAG for IEC TC56 (Dependability) (Expert of the USA Technical Advisory Group for International Electrotechnical Commission), update and development of international standards in Dependability (Reliability, Maintainability, Availability). Klyatis developed a new approach to Accelerated Reliability Testing of mobility vehicles, accurate physical simulation of the entire complex of real life input influences on the actual automotive system, simultaneously and in a combination and accurate physical simulation of interactions among different field input influences as well as different components of the system. He also developed 11 basic steps of accelerated reliability testing technology, the methodology of specific accelerated testing such as vibration, corrosion, environmental and IEC international standards in dependability, including group of standards "Equipment Reliability Testing."

Stephen C. Lowell
Defense Standardization Program Office

Stephen Lowell is recognized for his work on the Aerospace Materials Division. Lowell took over chairmanship of the Aerospace Materials Division at a time when the committees were struggling with the proper assimilation of military specifications canceled as a result of Mil Spec reform. He led the committees through this difficult transition period, providing guidance and establishing policies to help facilitate the rapid word-for-word conversion of the documents and the subsequent adoption and full conversion to SAE standards. The new standards built on those old canceled Mil Specs reflect the latest technology and the technical excellence that is the hallmark of SAE Aerospace standards.

12

FIGURE 6.44   The second page from brochure about SAE Aerospace Awards recipients.

This close cooperation and discussions on implementing these new ideas in accelerated testing development, were initially created in the USSR and then further developed in the United States. They have since been implemented throughout the world. The forms of this implementation were reviewed in the beginning of this subchapter.

Technical Standards Board

2003
OUTSTANDING CONTRIBUTION
AWARD

In recognition of

DR. LEV M. KLYATIS

Nominated by the

Aerospace Council

for outstanding contributions in furthering the goals of the Technical Standards Board and this Council, which are to serve the public, government and industry, through standardization, documentation and dissemination of information aimed at improving and advancing transportation safety and technology.

alan R. Dohner

Dr. Alan R. Dohner
Chair, Technical Standards Board

SOCIETY OF AUTOMOTIVE ENGINEERS, INC.

SAE The Engineering Society For Advancing Mobility Land Sea Air and Space
INTERNATIONAL

FIGURE 6.45 The Outstanding Contribution Award Lev Klyatis from SAE International aerospace Council for outstanding contributions in furthering the goals for serving the public, government, and industry, through standardization, documentation, dissemination of information aimed at improving and advancing transportation safety and technology.

FIGURE 6.46 Dr. Yuriy Piatine, his colleague from TESTMASH, who came from Canada to Dr. Lev Klyatis in the US to discuss evolving trends in accelerated testing development. They are shown looking at his awards for first years work in the USA.

As a U.S. expert in accelerated testing, he was appointed as the USA Delegate to the International Electrotechnical Commission, Technical Committee TC-56 — Dependability (Fig. 6.47).

As a result of this work, some of his ideas and methods in accelerated testing development were implemented through the adoption of international standardization, and through meetings with experts in different countries.

During his work with the TC-56 Technical Committee this work was adopted in the USA, China, Australia, Germany, and other countries. China's representatives were particularly interested in implementing this author's ideas.

United States National Committee of the International Electrotechnical Commission
A Committee of the American National Standards Institute
25 West 43rd Street Fl. • New York, NY 10036 • (212)642-4936

FAX: (212)730-1346
(212)302-1286 (Sales Only)

24 September 2001

Mr. Lev Klyatis
ECCOL Incorporated
72 Montgomery Street, #1311
Jersey City, New Jersey 07302

Subject: *Delegate Accreditation Letter*

Dear Mr. Klyatis:

The U.S. National Committee of the IEC is pleased to confirm your appointment to the USNC delegation for the announced IEC/TC 56 meeting. An Accreditation and Identification Card is enclosed for your use. Also, a USNC/IEC logo pin will be mailed to you shortly.

As you know, at this meeting you will represent the USNC/IEC. The positions on the technical agenda items will have been determined by the US Technical Advisory Group. Positions on policy and administrative matters should be developed in consultation with the USNC office. This will assure a unified US position at all levels of the IEC organization. For your information also please find the website link to the booklet "Guide for U.S. Delegates to IEC/ISO Meetings."
http://web.ansi.org/public/library/intl_act/default.htm

The USNC is grateful to you and your employer for agreeing to support the voluntary standards system and for your willingness to present US positions to IEC.

Sincerely,

Charles T. Zegers
Charles T. Zegers
General Secretary, USNC/IEC

CTZ:dn

Copy to: J.A. Miller, TA
N. Criscimagna, DTA
P. Kopp, GM

FIGURE 6.47 Accreditation letter for Dr. Klyatis as US Delegate as an expert representing USA in Dependability in IEC, where his approaches were implemented into international standards.

The author's development of accelerated testing included discussion with colleagues in different countries.

By visiting the test centers of many companies for consultant work and by observing their testing technologies, especially those of large industrial companies, such as Detroit Diesel, Ford Motor, Carl Schenk, Martin Lockheed, Honeywell, and many others, he was able to refine and improve his theories and strategies. He played an important role in analytical work during the SAE World Congresses as session chairman, the ASQ session manager, as well as a reviewer to Quality Press Publisher, and work with technical literature. In this book, the author is able to present the analysis of the speed of the technical progress in present testing technologies and to compare it with the technical progress in design and manufacturing technologies over a span of dozens of years, and of the trends and the development of accelerated testing in different countries and companies.

These activities helped him to better understand the trends and to analyze them, and as a result, to create and implement new ideas for the further development of accelerated testing. The results of these ideas on the implementation of accelerated testing can be found in the English language books in Refs. [6.1–6.5], as well as, in his over 30 patents in different countries, articles, and papers in English [6.8–6.32] and other languages (totaling over 300 publications).

One of the more effective means of implementing accelerated testing, especially accelerated reliability and durability testing, is through the author's seminars with industrial companies (Ford Motor, Black & Decker, etc.), and the ones organized by the American Society for Quality (Figs. 6.57 and 6.58). These seminars were specifically designed to analyze the methods and equipment used by these companies for accelerated testing, and to show how these companies could increase the effectiveness of their testing.

During the IEC Conference in Beijing, this author received an award from the Chinese government for outstanding achievements in accelerated testing development. This can be seen in Ref. [6.1]. There was some controversy, over this award, as he was the only person from the American delegation to receive this award, and the chief of the U.S. delegation questioned why he was the only one to receive this award, and no one else, including the chief of the delegation. The answer became apparent when the Chinese government representative Dr. MEI Wenhua (The Chinese representative in IEC T-56, involved in aerospace) asked the author to present his book *Successful Accelerated Testing* for implementation in China, which he did. The next year, in Australia, Dr. MEI Wenhua presented his new book *Reliability Growth Test* through his colleague to Dr. Klyatis (Figs. 6.48 and 6.49).

FIGURE 6.48 Lev Klyatis in Beijing (China) during IEC Conference, where presented his first book in English "Successful Accelerated Testing"

This book was written in Chinese, but included an English table of contents. The English table of contents is provided below:

INTERNATIONAL ELECTROTECHNICAL COMMISSION TECHNICAL COMMITTEE No. 56 DEPENDABILITY Working Group 2: Methodology	56/WG2/PT 2.8 January 2001

Project 2.8: Revision of IEC 60300-3-1

Date: January. 20, 2001 in Philadelphia

Location   Philadelphia Marriott
            1201 Market Street
            Philadelphia, PA 19107 , USA
            Phone: +1-215-625-2900
            Fax: +1-215-625-6101
            Room: 406 (10:00 – 17:00)

Preliminary agenda

1   Opening of the meeting

2   Approval of the agenda

3   Current membership

Prof. Bobbio, Andrea	Italy	T. +39 01 1670 6742	bobbio@di.unito.it
Dr. Braband, Jens	Germany	T. +49 531 226 3231	Jens.Braband@vt.siemens.de
Mr. Jones, Jeffrey A.	United Kingdom	T. +44 / 1203 572 616	jones_j@wmgmail.wmg.warwick.ac.u
Mr. Kleinrath, Peter	Austria	T. +43 5 1707 35918	peter-viktor.kleinrath@siemens.at
Dr. Klyatis, Lev	USA	T.	lev-nellya-klyatis-1-assoc@worldnet.att.net
Ms. Krasich, Milena	USA	T. +1 508 766 1134	Milena_Krasich@bose.com
Mr. Lohmar, John William	USA	T. +1	lohmarjw@navsea.navy.mil
Dr. Masuda, Akihiko	Japan	T. +81 554 63 6956	amasuda@ntu.ac.jp
Mr. Midgette, William H.	USA	T. +1 301 443 2536 ext 12	whm@cdrh.fda.gov
Mr. Miller, John A.	USA	T. +1 714 842 4776	millerja@earthlink.net
Mr. Sateesh, Kris	USA	T. +1 / 941 739 4229	kris.sateesh@edwards.spx.com
Mr. Schwarz, Erich	Germany	T. +49 89 636 48696	erich.schwarz@mchp.siemens.de
Dr. Turconi, Giorgio	Italy	T. +39 02 4388 8195	giorgio.turconi@italtel.it
Dr. Walls, Lesley	United Kingdom	T. +44 141 548 3616	lesley@mansci.strath.ac.uk
Dr. Wild, Antonin	Canada K2H 7X7	T. +1 613 820 0410	tonywild@echelon.ca
Mr. Loll, Valter	Denmark	T. +45 20 73 01 15	Valter.Loll@nmp.nokia.com

4   Confirmation of minutes 56(WG2/PT2.8)03 (LA meeting)

5   Further work on the revision of IEC 60300-3-1

Document "Revision IEC 60300-3-1 Draft Jan-2000 c.doc"
considering the comments in 56(WG2/PT 2.8)06 and 56(WG2/PT2.8)05

6   Time schedule for the revision, allocation of work

7   Date and place of the next meeting

8   Close of the meeting

- 1 -

FIGURE 6.49 First page of agenda the IEC TC-56 working group Methodology meeting for revision of the international standard IEC 60300-3-1.

The IEC technical committee T-56 meeting in Sydney, Australia, was a very interesting meeting as it involved experts from various other countries (see Figs. 6.50 and 6.51). It was especially important when the chief of the T-56 Committee Mr. Loll proposed that the author of this book prepare a draft of the IEC standard in reliability testing. Dr. Klyatis agreed to draft a standard, which was implemented by the SAE G-11 Committee in Aerospace Equipment Reliability Testing (see Chapters 7 and 8 of this book and [6.2]).

FIGURE 6.50 Group of IEC TC-56 experts from different countries (Sydney, Australia). Lev Klyatis from right.

Another form of advancing the successful implementation of the positive trends in the development of accelerated testing (see Chapter 4) was through the author's seminars at large industrial companies and professional societies, as well as with consulting meetings with Ford Motor Co, NISSAN, TOYOTA, and other companies' testing professionals and top management.

For example, Mr. Takashi Shibayama, Vice President Jatko Ltd. (Japan) met Dr. Klyatis after his presentation at the SAE 2013 World Congress. Having read his books and papers, he was very interested in implementing accelerated reliability and durability testing (ART/ADT) in Jatko Ltd.

Showing the author's book *Accelerated Reliability and Durability Testing Technology*, Wiley 2012, he said that both he and his experts in testing had studied this book and wanted to learn more with some detailed descriptions and answers to their questions, related to this type of testing. He agreed with the author that they would meet at next year's SAE World Congress, and as Dr. Klyatis could not travel to Japan, he would do a special seminar for Jatko Ltd. As they had agreed at next year's Congress, several professionals from Jatko, who had come to this congress as speakers (one of whom can be seen in Figure 4.28 in Ref. [6.1]), along with Mr. Shibayama (Fig. 6.56) along with two senior managers from Nissan attended the seminar-meeting to discuss testing improvement and ART/ADT implementation in their company. This seminar took place during the Congress (Fig. 6.56).

One more from many examples of the practical implementation of this author's knowledge of accelerated testing development for industrial companies is through his preparation and presentation of seminars for senior management, engineers and managers.

Main Identity

From:	<mjc@iec.ch>
To:	"Siegfried RUDNIK" <siegfried.rudnik@siemens.com>; "G C ALSTEAD" <alstead@gca-con.demon.co.uk>; "Yoshinobu SATO" <yoshi@e.kaiyodai.ac.jp>; "Lev KLYATIS" <lklyatis@agoron.com> "Charles SIDEBOTTOM" <Charles.Sidebottom@Medtronic.com>; "Ron BELL" <ron.bell@hse.gsi.gov.uk>; "Friedrich HARLESS" <Friedrich.Harless@siemens.com>; "Hartmut KROSIGK, VON" <hartmut.krosigk@siemens.com>; "H HUHLE" <huhle@zvei.org>; <roger.david@cramif.cnamts.fr>; <nicola.worsell@hsl.gov.uk>
Cc:	<#TC44@iec.ch>; <#TC56@iec.ch>; <#TC62@iec.ch>; <#TC65@iec.ch>
Sent:	Thursday, July 29, 2004 9:29 AM
Attach:	acos-jsg-iso-da-final.doc; acos-safety-risk-assessment-experts.xls
Subject:	ACOS - Safety aspects of risk assessment - Joint study group with ISO - Draft agenda

Subsequent to my e-mail of 2004-06-11 inviting you to participate in the first meeting of the above mentioned joint study group with ISO organised for 2004-09-08 (for further information see document ACOS/312/AC), please find attached the draft agenda.

As the meeting documents are a series of publications, I will be sending them to you in a zip file in a second e-mail because the size of the file at 11.9 MB might block your e-mail system.

If this is the case, then a ftp server site has been created as an alternative route for downloading the meeting documents.
Details on the ftp server site are given below :
Server name : ftp.iec.ch
Members :
- username: memacjsg
- password: brighton
The procedure for using the ftp site is available in IEC's Bits&Bytes vol. 22
http://www.iec.ch/support/bbyte/vol2201/bb_vol2201.htm

You will also find attached an Excel spread sheet with the members invited to participate in the ACOS joint study group, please could you :
a) confirm your participation and obtain your NC's approval, if this has not been already done. Noting that for the IEC members, NC's approval should be sent to me directly as unfortunately ACOS groups are not incorporated in the recently, created Experts Management System.
b) confirm your participation at the first meeting again for those of you who have not done so.

If you have any questions, then please do not hesitate to contact me.

Thanking you in advance for your co-operation and looking forward to meeting you in Frankfurt on 2004-09-08.

Regards,

Michael J. Casson
ACOS Secretary
IEC CO

(See attached file: acos-jsg-iso-da-final.doc)(See attached file: acos-safety-risk-assessment-experts.xls)

8/4/2004

FIGURE 6.51   The letter inviting Lev Klyatis to become an expert member of the IEC/ISO Joint Study Group in Safety Aspects of Risk Assessment.

During the SAE Congress, this author after his presentation, did a seminar with Kazuhiko Oishi, General Manager, Vehicle Performance Development, Toyota Technical Center, USA, and a group of managers from Toyota, who were involved in testing and improvement of long-term quality and reliability (Figs. 6.52—6.54).

In his practice, this author experienced many similar examples.

One such was a seminar for Ford Motor Company's professionals (Fig. 6.55). This seminar invitation was a result of his presentation at the Reliability and Maintainability Symposium (RAMS).

Aerospace Council
Request to Continue WIP > 5 Years

Date of Request:	[04/29/2013]
Committee:	G-11, Reliability Committee
Document Number:	JA 1009
Document Title:	Reliability Testing
Date WIP Initiated:	1998
WIP Type:	New
Document Sponsor:	Lev Klyatis
Committee Chair:	Chris Sautter

Current WIP Status

Question	Answer/Description
Summarize the most recent activity (include dates)	At the 2007 Fall G-11 meeting Dr. Lev Klyatis proposed to prepare JA1009 Reliability Testing as a complex of six standards, as often doing IEC and ISO with complicated standards. Dr. Klyatis proposed at the 2011 Fall G-11 meeting the abstracts of the following six standards: 1. JA1009/A – Definitions (mostly specific definitions for below 5 standards) 2. JA 1009/1 – Strategy (common and specific components) 3. JA 1009/2 – Procedures (step-by-step the design of Reliability Testing technology) 4. JA 1009/3 - Equipment (requirements to equipment and its components) 5. JA1009/4 - Statistical criteria for comparison Reliability Testing results with real world results during service life 6. JA 1009/5 - Collection and analysis of Reliability Testing data, development recommendations for improvement of test subject reliability and life cycle cost
Has document been balloted?	Yes

FIGURE 6.52 The balloted approval of proposed by the author complex of six standards with common title "Reliability Testing".

And, in addition to these known events, the author cannot be familiar with all of the implementations of his technologies. One basic reason that these implementations cannot be known is that when he came to the USA, he did not patent any of these new inventions, but included them in open form in his over 80 English language publications. Anyone who was read these publications was free to use these innovations. This is very different from the situation during the author's previous life in the Soviet Union, Ukraine, and Russia, where he had obtained over 30 patents. These patents were internationally recognized by the United States, Germany, France, Bulgaria, and other countries.

 **Gmail**   **lev klyatis< klyatislm@gmail.com>**

Minutes from G-11R Breakout session
1 message

Lesmerises, Alan< Alan.Lesmerises@standardaero.com>   Thu, Dec 6, 2012 at 4:46 PM
To: "Bartlett, James K AMRDEC" <james.k.bartlett@us.army.mil>, "McGill, Jennifer AMRDEC" <jennifer.r.warren@us.army.mil>, lev klyatis <klyatislm@gmail.com>, "Gorelik, Michael (MCOE)" <Michael.Gorelik@honeywell.com>, dan.k.fitzsimmons@boeing.com
Cc: Kevin Thompson <kevin.thompson@honeywell.com>

I drafted some notes on what we did during the G-11R breakout session so Kevin can compile a set of minutes for the meeting. Please take a look and let me know if I missed something significant or if you see any errors:

The G-11R Breakout Session

Attended by:
Mike Gorelik (Chairman)
Dan Fitzsimmons
Lev Klyatis
Alan Lesmerises
Jennifer McGill

The G-11R team met to discuss the latest set of edits for JA1009, prepared by Jim Bartlett & Jennifer McGill (a marked-up copy of the document had been provided prior to the start of the meeting). Before discussing specific changes made in the latest revision of the document, the team debated the overall approach for the Reliability Testing related documents being planned, including JA1009. Dr Klyatis explained his vision for having a set of related documents, each addressing different aspects of Reliability Testing (starting with a glossary, then moving onto requirements for testing, methods for accurate prediction, etc.). However, after considering a number of factors relating to the glossary (its utility to users, the cost of having a document separate from the existing list of R&M terms, the potential overlap of terms defined between the two glossaries, etc.), the team eventually agreed that folding the terms currently shown in JA1009 into a single glossary (ARP5638) would make the most sense.

The team then went through the edits in the marked-up copy (provided by Jim) and discussed the proposed changes. The team was able to review and approve the changes through paragraph 3.4 before we had to wrap up on the second day. Before we concluded, Dr Klyatis said he accepted the proposed changes in the remainder of the document (at least a few of the other team members had not yet reviewed all of the other changes).

Thanks.

Alan L. Lesmerises, MS, CRE
Reliability & Life Cycle Management Engineer
StandardAero Engineering Services
3523 General Hudnell Dr.
San Antonio, TX, 78226
United States of America
Office: +1.210.334.6187
Fax: +1.210.334.6181

https://mail.google.com/mail/?ui=2&ik=eb6487a3ec&view=pt&search=inbox&th=13b72...   06.12.201:

FIGURE 6.53   From G-11R Minutes. Standardization in Reliability Testing.

When preparing this book, the author learned from searching the Internet (Klyatis Lev M., WorldCat Identities) that the publications were present in 1913 large libraries, and included:

- The book *Accelerated Reliability and Durability Testing Technology*, (Wiley. 2012); by Lev M Klyatis. 14 editions published, held 645 WorldCat member libraries worldwide;

DEC 17 '01 10:35 FR SMALL CAR PLATFORM   248 576 2017 TO 612012000942   P.01/01

DAIMLERCHRYSLER

DaimlerChrysler Corporation

December 14, 2001

SAE G-11 Probabilistic Methods Award Committee
400 Commonwealth Drive
Warrendale, PA 15096

Dear Sirs:

I would like to nominate for the Probabilistic Methods Award the work of Dr. Lev Klyatis as described in the paper *METHODOLOGY OF SELECTING INPUT INFLUENCES PROBABILITY CHARACTERISTICS OF THE REPRESENTATIVE REGION FOR ACCELERATED TESTING*. The process described in the paper overcomes a significant issue with accelerated life testing that has hampered its acceptance in the design and development community, namely the ability to simulate the multitude of real life influences and their variance under field conditions. If all the proper influences are not simulated, there may be errors in the simulation process, especially with regard to the influences that have not been correlated to real data.

Dr. Klyatis's seven (7) step process provides a way of determining the representative field region under which testing or simulation may be conducted to assure the adequacy of the input variables. Such an approach would be invaluable from the standpoint of resource and time savings.

Here at DaimlerChrysler we strive not to conduct excessive testing while at the same time to be able to assure that our designs meet our customer's expectations and wants. We are constantly open to new methods of simulation and testing that will simultaneously reduce valuable resources and demonstrate that we meet design objectives. The work of Dr. Klyatis employs probabilistic methods and concepts to help us achieve this result. We feel that his work is deserving of the Probabilistic Methods Award from your committee.

Please feel free to contact me if you have any questions regarding this nomination.

Respectfully submitted,

Richard J. Rudy
Senior Manager – Product & Process Integrity
Corporate Quality
DaimlerChrysler Corporation
Phone: 248.576.2832
FAX: 248.576.2017
rjr11@dcx.com

** TOTAL PAGE.01 **

FIGURE 6.54 A letter from DaimlerChrysler Corporation endorsing the new concepts and methods in accelerated testing.

TRENGS IN DEVELOPMENT OF ACCELERATED RELIABILITY TESTING

(Seminar for FORD Motor Company, April 14, 2010).
Instructor Dr. Lev Klyatis (Director of Quality & Reliability ERS Corporation, Head of Reliability Department ECCOL, Inc.)

• Development techniques and equipment for more accurate simulation of real life input influences.

• Step-by-step development less expensive equipment for simultaneous combination of real life basic input influences.

•Development Accelerated Reliability Testing, which offers the possibility to obtain directly the

information for accurate prediction of reliability.

• Rapid obtaining accurate information for analysis the reasons of degradation mechanism and failures.

• Development the product quality through Accelerated Reliability Testing. Its specific.

• Development of accelerated analysis of the climate influence on the new product reliability.

What is ART development? It is when you have more opportunities for:

a) rapid finding of product elements that limit the product quality and reliability;
b) rapid finding of the reasons for the limitations;
c) rapid elimination of these reasons;
d) rapid elimination of product over-design (cost saving) to improve the product quality and reliability.
e) increasing product quality and reliability, therefore resulting in a longer warranty period.

This way is less expensive and offers more opportunities for rapid product reliability improvement.

It is essential to simulate real life input influences on the product accurately for ART. If we cannot simulate real life influences accurately, we cannot perform ART and rapidly improve our product reliability.

The author working in this direction and describes below how one can use it.

One can use the above for development technology of ART

1. Determination the failures that limited the product reliability and quality.
2. Finding the location and dynamic of mechanism development of the above failures (degradation).
3. Finding the reasons of the above failures.
4. Elimination of these reasons.
5. Increasing of reliability and product quality.
6. Increasing the warranty period of the product.

For the above ART implementation one needs accurate simulation real life influences on the actual product.

FIGURE 6.55   The Program of author's seminar for Ford Motor Company professionals who involved in testing area.

- The book *Accelerated Quality and Reliability Solutions* by Lev M Klyatis. 16 editions English and held by 356 WorldCat member libraries worldwide (Elsevier published in 2006);
- The book *Reliability Prediction and Testing Textbook* by Lev M Klyatis and Edward Anderson (Wiley published in 2018) five editions published in 2018 in English and held by 158 WorldCat member libraries worldwide, for several months after publishing"
- The book *Successful Prediction of Product Performance: quality, reliability, durability, safety, maintainability, life-cycle cost, profit, and other*

Information for the meeting with Nissan and Jatco Ltd. representatives.

A. ART/ADT implementation for accurate prediction of product quality, reliability, durability, safety, and human factors needs System of Systems approach which can be successful with using only interdisciplinary team. This is long term work for planning our collaboration. We have team of high level experts in advanced testing, prediction, medical, software, physical, design, and others who can provide consultant work for any industrial company.

B. Regarding Nissan's representatives questions:

   1. Short answer see in slides 31 and 34 of "Chat with the Expert" (Keynote Speaker Dr. Lev Klyatis) at the SAE 2012 World Congress. Wide answer sees in Dr. Klyatis's whole Chat... and books published by Elsevier (Oxford, UK, 2006) and Wiley (US, 2012). These books include overview of 241 references over the word in the first book and 257 references in the second book
   The above slides can be described in more detail during our meeting April 25, 2012.

   2. This question relates to our expert who is Dr. of Medical Science, professor who works together with Dr. Lev Klyatis. He said that any specific problem solution in human factors and safety can be considered during detail program of our consultant work for implementation for Nissan of ART/ADT as a key factor for accurate prediction of product quality, reliability, safety, and durability.

   3. Correlation of failure mode and failure mechanism is old criteria that cannot be useful for accurate prediction of the product life cycle cost, quality, reliability, and durability. One can see the results of use these criteria, for example, in a lot of recalls (relates to Nissan also) that leads to higher cost, lower safety, lost time than was predicted during design and manufacturing. For more detail understanding this problem, one needs to study real problems in automotive industry (and not this industry only) of inaccurate prediction, read carefully both mentioned Dr. Lev Klyatis's books, as well as his other dozens publications.
   We can show, as examples, that many companies want to save money and time for product testing, therefore use simple vibration testing (sometimes call them "durability testing"), without calculation how much money will be loss for subsequent processes after this testing. We can show also, as examples, that many companies use vibration testing for mobile product in one degree of freedom, instead of six degrees of freedom as in real world, then loss a lot of money for improvement the low quality of their product. One more example from vibration testing: one uses sinusoidal load inputs instead of random load inputs as in real field. It is cheaper, but one does not want to understand that the companies will loss much more money for subsequent procedures. The recall is one from basic result of inaccurate simulation of real life for testing during design and manufacturing, inaccurate evaluation then prediction. This is only one from many problems from poor testing.
   Finally, company has several times less profit that was predicted during design and manufacturing (see examples about above in Dr. Klyatis's books).

FIGURE 6.56   Information from NISSAN and Jatko Ltd. top management consultant (by author) meeting.

components by Lev M Klyatis (SAE International published); five editions published in 2016 in English and held by 522 WorldCat member libraries worldwide.

Figs. 6.58—6.66 demonstrates seminar programs on the implementation of accelerated reliability and durability testing through ASQ seminars, and the SAE International World Congresses, where, beginning in 2012, they were provided special technical sessions entitled "Trends in Development Accelerated Reliability and Durability Testing Technology", at which the author was invited to be the session chairman.

The author's experience as session organizer and session manager for many world congresses, world conferences, international symposiums, as well as

Dear Dr.Lev Klyatis

From Nissan, 2 persons are going to attend the meeting.

Mr. Kaoru Onogawa
Expert Leader , Powertrain Engineering Division

Mr. Shigao Murata
Expert Leader , Powertrain Engineering Division

Expert leaders are the most high position as engineers in Nissan.
They report to directly to board Member , repot to Fellow (Similar to CTO ,Chief Technology Officer))

Mr Onogawa has grown in Engine Department, and Mr. Murata has grown in Drive train Department.

Best Regards

Takashi Shibayama

FIGURE 6.57 Information Takashi Shibayama, Vice President Jatco Ltd. (Japan) for Lev Klyatis's consultant meeting with top managers NISSAN Motor Co., Ltd. (Japan) Kaoury Onogawa and Shigao Murata.

serving as a member of boards of reviewers for book and journal publishers, session chairs, and presenter (see Figs. 6.61 and 6.62 in this book, and other books [6.1,6.2]) by this author further demonstrates that the reviewing processes play an important role in the development and implementation of both the negative and the positive aspects of accelerated testing in automotive and aerospace engineering. Unfortunately, these experiences also demonstrated that, during these processes, the basic role played in the review process is most often the quantitative aspects only, with much less attention paid to the quality of the reviews and the organization of these events. Therefore, these events continue bringing less benefits than were possible for the professionals in attendance.

This weakness in the process has a negative influence on the implementation process of new solutions in automotive and aerospace engineering, including those in accelerated testing.

For example:

- Typically, each paper for a congress needs two or three reviewers. But session organizers often send the papers out to 11 to over 45 reviewers. This is a negative influence to the development and implementation of new achievements, because from this large number of reviewers, there

FIGURE 6.58 Title page of the Brochure of organized by American Society for Quality (ASQ) the author's seminar accelerated testing of products.

will be many who are not adequately familiar with the topic contained in the paper. Moreover, if some of these reviewers do not recommend the paper for the event's program, it will be rejected and the publication of the paper is blocked although it may contain important developments of new techniques or equipment, and the Congress attendees will not read or hear about these important new innovations.

- Another aspect is if the paper could help in the development of a new technology or methodology, and the paper was not accepted for the

ACCELERATED TESTING OF P

This one-day course will provide the foundation for designing a successful accelerated testing program (AT) so as to achieve high quality and reliability in existing products and future designs.

Seminar participants will learn:
- why current technique and equipment for AT give no more than 20-30 % of possible benefits;
- why the simulation of real life influences is usually not accurate for a high level of correlation between AT results and field results;
- how one can obtain maximum correlation;
- how engineers and managers can find and eliminate causes for failures and degradation of product quickly and at lower costs.

Course participants will also learn to apply AT to:
- shorten product time to the market;
- reduce design and product development cycle time , warranty costs, and minimize customer returns.

Accelerated testing is applicable to: mechanical, electro-mechanical, electronic, hydraulic and other devices used in Automotive, Railroad, Aerospace, Marine, etc.

Benefits of Attending:

By completing this seminar, you will know:

- Trends in development of AT.
- How to choose the successful AT method for your application.
- Proper test equipment selection for your process.
- How to utilize your existing test equipment.
- How to evaluate accelerated testing results.

- How to implement advanced and less expensive techniques to increase the product warranty period.
- Participants will receive Dr. Klyatis's new book "Successful Accelerated Testing".

Seminar Basic Content

- Introduction to successful accelerated testing (AT).
- The strategy for creating successful AT.
- Physical simulation of real life input influences on the product.
- Technology of step-by step accelerated testing.
- Conditions for accelerated multiple environmental testing.
- Accurate accelerated corrosion testing.
- Accurate accelerated vibration testing.
- Accelerated reliability testing.
- More accurate prediction of reliability.

Who should attend?

Corporate executives, quality engineers and managers, design & test engineers and technicians, reliability engineers and managers, supervisors, quality assurance managers, quality control engineers and managers, manufacturing engineers and others.

Course Schedule:

Date: 05/27/2003 - Tuesday

7:30 am – 8: 30 am	Registration
8:30 am – 12:00 pm	Course
12:00 pm – 1: 00 pm	Lunch
1:00 pm – 4:00 pm	Course

FIGURE 6.59 From brochure for seminar Accelerated Testing of the Products, which organized American Society for Quality.

Congress or symposium program, it never provides an opportunity for the improvement and instead is actually an obstacle to its implementation especially by professionals, who are not familiar with the new solution.
- If a paper, journal article, book, etc., includes positive development trends, but receives negative recommendation from reviewers, who may not be truly qualified in this area, evolving knowledge, will not get promulgated.
- Reviewer, who may not have adequate knowledge in a particular area, will review the paper, article, or book, and may write a formal review that concentrates on secondary areas, while failing to analyze the truly important aspects of the development. Such reviews could actually result in conclusions opposite from that presented in the document.

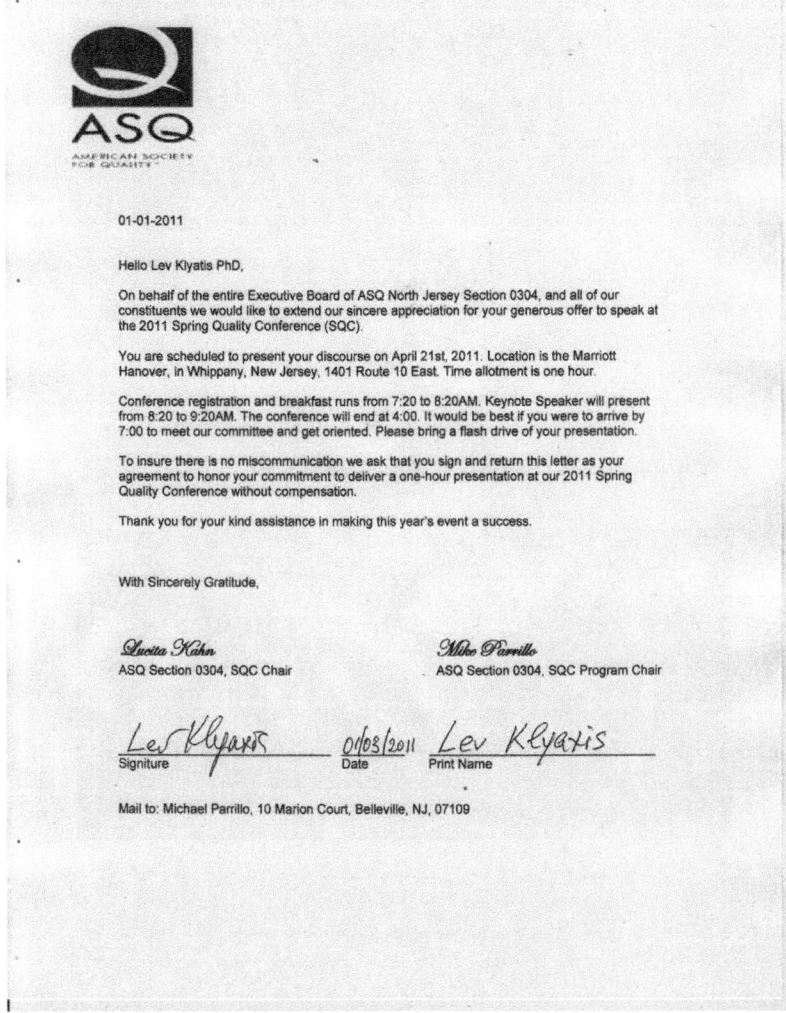

FIGURE 6.60 Information about the American Society for Quality (ASQ) presentation at the Spring Quality Conference.

- These events also have an influence on the implementation process, as they all contribute to failures in obtaining new knowledge from reading what are considered to be the important publications on this topic.
- From this author's experience, reviewers seldom "I am not familiar with this paper's technical area", when sending back the paper. But the author respects these reviewers and understands how this aspect can also be a positive in testing these new ideas and their implementation process.

FIGURE 6.61   Lev Klyatis, chairman of SAE World Congress (Detroit) Technical Session "Trends in Development Accelerated Reliability and Durability Testing". Speaker — Dr. Bryan Dodson, SKF.

SoHaR, Inc.

FIGURE 6.62   Title page from the author's presentation at the SAE 2019 World Congress, Detroit.

But with this said, it is important to realize that some of the publications with their analysis of the current situation and with their descriptions of the current methods and equipment are actually hindering progress.

It must also be recognized that some professionals who have the potential to improve testing in their organizations do not champion improvements, because of fear. They may be afraid of diminishing their stature or relationships with other professionals or companies. They do not understand that their actions are a negative influence on slowing the technical progress in the world. Many professionals know that there are instances where negative trends in

INTERNATIONAL.

SAE 2016 World Congress and Exhibition

Technical Session Schedule

Tuesday, April 12

Trends in Development of Accelerated Reliability and Durability Testing Technology
(Session Code: IDM300)

Room 331 C

This session presents the theory, practices and technology used in development of trends in reliability and durability testing (ART/ADT) technology and accurate physical simulation for successful performance predicting. The purpose is covering a new ideas and unique approaches to simulation interaction of full field inputs, safety, and human factors, improvement the ART/ADT steps-components, implementation that leads to development dependability, reduce recalls, life cycle cost, time, etc.

Organizers - *Bryan Dodson, SKF; Lev Klyatis, Sohar Inc.; Efstratios Nikolaidis, University Of Toledo*
Chairpersons - *Lev Klyatis, Sohar Inc.*

Time	Paper No.	Title
9:30 a.m.	2016-01-0318	**Improving Engineering Culture For Reliability, Quality & Testing in Automotive Engineering (Written Only -- No Oral Presentation)** *Lev Klyatis, Sohar Inc.*
	2016-01-0320	**Multivariate Analysis to Assess the Repeatability of Real World Tests (Written Only -- No Oral Presentation)** *Tejas Janardan Sarang, VJTI, Tata Motors Ltd; Mandar Tendolkar, VJTI; Sivakumar Balakrishnan, Gurudatta Purandare, Tata Motors Ltd*
	2016-01-0319	**Systemic Root Cause Early Failure Analysis during Accelerated Reliability Testing of Mass Produced Mobility Electronics (Written Only -- No Oral Presentation)** *David E. Verbitsky*

Planned by Global Supply Chain Committee / Integrated Design and Manufacturing Activity

---

FIGURE 6.63 Technical session IDM300 Trends in Development Accelerated Reliability and Durability Testing schedule at the SAE 2016 World Congress.

testing development are being used by professionals who should be promoting strategies that will move their organization to more positive testing effectiveness. Some of these are described in Chapter 4 of this book, as well as in other publications. Therefore, being well grounded in the critical aspects and approaches of these trends is very useful for the development improved testing, but not having enough support from the professionals who must be willing to advocate for their development remains a problem.

The author, as well as many readers, know there are many inventions that could be used to improve our lives, but unfortunately, they have not been advanced in real life.

SAE Metropolitan Section Governing Board Meeting – Draft Minutes

Meeting Date: May 7, 2018   **Prepared By**: Bob Santora

1. **CALL TO ORDER**
   1.1 Monthly Meeting called to order at 5:47 PM by Jeff Trilling.
   1.2 Members present: Jeff Trilling, Bob Santora, Dan Buckley, Ed Anderson, Steven Resch, Art Vatsky, John Anagnostos, Emil Beyer, and Lev Klyatis.
   1.3 Jeff announced that the Met Section earned another *Gold Status Award* from Warrendale for the 2016-17 Section Year. Congratulations to the SAE Met Section Board.
   1.4 Jeff reported that John Anagnostos received a *60 Years of Service* award plaque from SAE for his years of work and dedication to SAE, the teaching profession, and the local Board. Congratulations and great work, John!
   1.5 Before beginning the *Minutes* review Jeff announced that Lev would be making a presentation at this Board meeting, following business, which would count as an Activity Meeting, our third and final for the year. The meeting presentation would be based on Lev's production that he performed at this year's SAE World Congress 2018. Jeff led off the introduction of Lev as SAE Met Section's Crown Jewell! See Section 9 New Business Items 9.1 and 9.2 for details.

2. **MINUTES OF PREVIOUS MEETING**
   2.1 Motion to accept *April 2nd* Minutes: Ed: Item 8.2 should read Panama Canal *Expansion* Project, not "Experience".
   2.2 Motion to accept April 2nd, 2018 Minutes *as amended*: Motion: Ed; Second: Dan.

3. **TREASURER'S REPORT**
   3.1 Mark absent: Jeff reviewed the Treasurer's Report with the Board.
   3.2 Motion to accept Treasurer's Report: Ed; Second: John.

4. **ACTIVITY MEETINGS**
   4.1 Recent and Upcoming Meetings:
   4.1.1 *2018 Baja Vehicle Unveiling*: April 16th at NJIT Newark: CANCELLED due to weather. Rescheduled for **May 9th**. Participation encouraged by the Board.
   4.1.2 *NJIT Electric Vehicle Day*: April 25th: CANCELLED: All possibilities of getting the purported vehicles to the venue fell through. Our apologies to the NJIT crew for this disappointment. We will continue to try for another possibility for the future. Good luck with the rest of the competition schedule.
   4.1.3 *Rutgers Day*: April 28th. Jeff and Bob reported that despite all of our efforts to get through to electric vehicle sources to supply vehicles for this event, there were many examples on display that day, provided by individuals and clubs. Also a good display of vintage and sports production, and race cars present. Several past SAE Formula cars were on display, and last year's car and this year's product were on display running hot laps to shake down the new car for this year's competition. Several of the crew were on hand as pit

FIGURE 6.64 SAE Metropolitan Section Governing Meeting Minutes (first page). Announcement that the author will make presentation during this meeting.

There are also people who have positions of power in the implementation process, but do not use their positions to advance process improvement.

Successful implementation of new processes, especially in aerospace and automotive engineering also requires the support of the organization's top and middle management. The author described how Bekh N. N., General

6.1 This year's Engineering EXPO '18 event will again be held at White Plains High School in White Plains, NY on Sunday April 15ᵗʰ. Go to www.Beanengineer.org for more details. In light of the fact that we are light on give-away items with no time to get them, and a personnel shortage due in part to the SAE Design Competitions taking the students time and family activities for the Board members, it was decided to forfeit the event this year, and plan next year for a stronger showing.

6.2 Bob and Jim will try to continue to expand on Jim's earlier poster idea.

7 **MEMBERSHIP**

7.1 Dan reports that AOL would not send bulk e-mails at a certain time. Was asking for help from others to see if they had the same experience or success.

8 **OLD BUSINESS**

8.1 Stan absent; no update on *The Federal/State DOT Traffic Incident Management and Unified Command Program* activity meeting/tour of the Command Center in Woodbridge.

8.2 New slate of Board officers is due. New Section Year starts June 1, 2018.

8.3 The prior *Annual Reports* are now performed under the new format of *Central Banking* procedures.

9 **NEW BUSINESS**

9.1 Lev's presentation that he performed at World Congress 2018 was also reproduced at Vitale's at the Governing Board Meeting in May as an Activity Meeting: *Analysis of Current Practices with Reliability Prediction* commenced at 7:20 PM with a projector and laptop supplied by PANYNJ. Thanks to PA personnel for providing the tools and Lev for sharing his very interesting and important experience with these subjects with the Board as a presentation.

9.2 Lev brought a WCX program to the meeting to pass around to the Board for display. His presentation dealt mainly with automotive recalls and the focus was how to prevent them. Major points: recalls reflect quality; recalls and deaths due to accidents; manufacturers rely more on computer model testing than actual physical (real-life) testing. Basic reason for recalls: manufacturers can't predict how the product has quality built in, and why they can't predict it. Lev sited a top-three automobile manufacturer, also third on the recall list: instead of investing 100-200 million dollars in his program to improve quality, they lose billions per year due to recalls and law suits. The presentation lasted a little over an hour and Lev answered questions. It was an excellent meeting. Thanks again, Lev.

10 **FUTURE GOVERNING BOARD MEETINGS**

10.2 **June 11 at Vitale's at 5:30 PM** – New Section Year.

10.3 July: No Meeting – Summer Vacation.

10.4 August 13ᵗʰ at Vitale's at 5:30 PM.

10.5 September 10ᵗʰ at Vitale's at 5:30 PM.

10.6 October 1ˢᵗ at Vitale's at 5:30 PM.

FIGURE 6.65 SAE Metropolitan section Governing Board Minutes (3-rd page). About Lev Klyatis presentation.

Director KAMAZ, Inc., advanced the implementation of the new technology of accelerated testing in this large automotive industrial company. From this author's experience, many professionals, like Mr. Richard Rudy from DaimlerChrysler, Dr. Russel Vacante from RMS Partnership (DoD), Mr. Edward Anderson from the Port Authority of New York and New Jersey, Professor Zissimos Mourelatos from Oakland University, and many others learned about and worked to implementing accelerated reliability and durability testing, based on this author's processes, as well as through discussions with these experts.

In implementing this author's research in accelerated testing, which is a component of successful prediction of product's efficiency, he received many letters, such as the following e-mail dated.

February 5th, 2019:

"Dear Klyatis, Lev
Greetings!
It is learnt that you have published a paper titled *Analysis of the Current Practices with Reliability Prediction* in *SAE Technical Papers*, and the topic of the paper has impressed us a lot.
It has drawn widespread attention and interest from researchers and scholars in related fields.

Started with the aim to advance the development of scientific community, specialists and professionals in various fields can get the cutting-edge scientific research results from *American Journal of Applied Scientific Research*. Given the advance, novelty, and potential extensive use of your research results, **we** invite you with sincerity to contribute other unpublished papers of related fields to the journal. Your latest research of this article is also welcomed.
Click the link below to learn more information:
http://www.ajasr.org/submission

On behalf of the Editorial Board of the journal, we feel very honored to invite you to join our team as the editorial board member or reviewer of *American Journal of Applied Scientific Research*. Taking your academic background and rich experience in this field into account, the Editorial Board believe that you may be the most suitable candidate for this position. We believe that this opportunity will promote international research collaborations.
If you have any interest to join us, please click the following link:
http://www.ajasr.org/joinus.

Please kindly let us know your idea.
Yours respectfully,
Jessie Wright
Editorial Assistant of *American Journal of Applied Scientific Research*

And, some situations are important at the local or regional level, and not just in the world, international, or national level. For example, SAE International's Metropolitan Section organized a local presentation of the one this author prepared for the World Congress. This presentation was done during Metropolitan Section Governing Board meeting (see Figs. 6.64 and 6.65), as well as at an activity meeting in New York for the two professional societies, SAE and the National Association of Fleet Administrator (NAFA) [6.1].

FIGURE 6.66 Lev Klyatis—Chairman of Technical Session "Trends in Development Accelerated Reliability and Durability Testing" and Alexander Klyatis Co-organizer of this session, SAE 2018 World Congress & Exhibition, Detroit.

ELMER A. SPERRY BOARD OF AWARD

MEETING MINUTES
October 17, 2018

ASME Headquarters
6th Floor Boardroom
Two Park Avenue
New York, NY 10016-5990

Attendance:

AIAA	IEEE	
Richard Miles	Thomas Hopkins	
	Harvey Glickenstein	
ASCE	SNAME	ASME STAFF
Joseph Englot	James Dolan	David Soukup
	Naresh Maniar	
	Eugene Sanders	
	George Williams	
ASME	SAE	
George Hud	Ed Anderson	
Elizabeth Lawrence	Lev Klyatis	
	Art Vatsky	

I. CALL TO ORDER: Chair Hopkins called the meeting to order at 11:00 a.m.

II. HOUSEKEEPING

    a. The Minutes of the March 21, 2018 meeting were approved.

    b. The current Board membership was reviewed.

    c. Soukup discussed the financial statement, shown in Appendix I.

    d. Maniar described the forthcoming presentation of the 2018 Sperry Award to the Panama Canal Authority, that was scheduled for October 25, 2018.

III. OLD BUSINESS

    a. Active Candidates

        i. Shaft Bearing

        This is a candidate for the 2019 award. Maniar is the lead on this.

        ii. Tesla

FIGURE 6.67 First page from Elmer Sperry Board of Awards meeting minutes in 2018.

As was mentioned in Ref. [6.1], working as a SAE International Representative to the Elmer A. Sperry Board of Award provides the opportunity to research and recognize outstanding contributions, including those in the areas of accelerated reliability and durability testing technology implementation. The Board consists of representatives from six of the largest mobility related engineering societies. Dr. Klyatis made several presentations for this award. Fig. 6.67 demonstrates the first page of this Board's meeting minutes for October 2018.

Bibliography

[6.1]   Klyatis LM, Anderson EL. Reliability prediction and testing textbook. Wiley; 2018.

[6.2]   Klyatis L. Successful Prediction of Product Performance: quality, reliability, durability, safety, maintainability, life cycle cost, profit, and other components. SAE International; 2016.

[6.3]   Klyatis LM. Accelerated reliability and durability testing technology. Wiley (John Wiley & Sons, Inc.); 2012.

[6.4]   Klyatis LM, Klyatis EL. Accelerated quality and reliability solutions. Elsevier; 2006.

[6.5]   Klyatis LM, Klyatis EL. Successful accelerated testing. New York: Mir Collection; 2002.

[6.6]   L. Klyatis. The reliability/maintainability integrated with quality. 2007 RAMS (reliability and maintainability symposium.

[6.7]   Kononenko AF, Klyatis LM. Accelerated testing agricultural machinery. In: Paper for the committee for the agricultural problems economical commission for Europe. Geneva: United Nations; 1969.

[6.8]   Klyatis L. Why separate simulation of input influences for accelerated reliability and durability testing is not effective?. In: SAE 2017 World Congress. Paper # 2017-01-0276. Detroit; April 2017.

[6.9]   Klyatis L. Successful prediction of product quality, durability, maintainability, supportability, safety, life cycle cost, recalls and other performance components. The Journal of Reliability, Maintainability, and Supportability in Systems Engineering. RMS Partnership 2016:14−26.

[6.10] Klyatis L. Improving engineering culture for reliability, quality & testing in automotive engineering. In: SAE 2016 World Congress and Exhibition. Paper 2016-01-0318. April 12−14, 2016. Detroit; 2016.

[6.11] L. Klyatis. The role of accurate simulation of real world conditions and ART/ADT technology for accurate efficiency predicting of the product/process. SAE 2014 world congress and exhibition. Paper 2014-01-0746. Detroit.

[6.12] Klyatis L. Non-traditional solutions for current reliability, maintainability, and supportability problems. The Journal of Reliability, Maintainability, and Supportability in Systems Engineering RMS Partnership 2013:6−12.

[6.13] Klyatis L. Development standardization "glossary" and "strategy" for reliability testing as a component of trends in development of ART/ADT. Detroit, MI. In: SAE 2013 World Congress and Exhibition Paper 2013-01-0152; April 16−18, 2013.

[6.14] Klyatis LM. Why current types of accelerated stress testing cannot help to accurately predict reliability and durability?. In: SAE 2011 world congress and exhibition. Paper 2011-01-0800. Also in book reliability and Robust design in automotive engineering (in the book SP-2306). Detroit, MI, April 12−14, 2011; 2011.

[6.15]   Klyatis L. Accelerated reliability testing as a key factor for accelerated development of product/process reliability. In: IEEE Workshop Accelerated Stress Testing. Reliability (ASTR 2009). Proceedings on CD. October 7 − 9, 2009. Jersey City; 2009.

[6.16]   Klyatis L. Specifics of accelerated reliability testing. In: IEEE Workshop Accelerated Stress Testing. Reliability (ASTR 2009). Proceedings on CD. October 7 − 9, 2009. Jersey City; 2009.

[6.17]   Klyatis L, Vaysman A. Accurate simulation of human factors and reliability, maintainability, and supportability solutions. The journal of Reliability, Maintainability, Supportability in Systems Engineering. RMS Partnership 2007/2008.

[6.18]   Klyatis L. Reliability testing standardization. In: 2007 SAE AeroTech Congress & Exhibition. Reliability, Maintainability, and Probabilistic Technology G-11 Division Fall 2007 Meeting. Los Angeles, CA: SAE International; September 19, 2007.

[6.19]   Klyatis L. A new approach to physical simulation and accelerated reliability testing in avionics. In: Development Forum. Aerospace Testing Expo 2006 North America. Anaheim, California; November 14−16, 2006.

[6.20]   Klyatis L. Elimination of the basic reasons for inaccurate RMS predictions. In: A governmental-industry conference "RMS in A systems engineering environment". San Diego, CA: DAU-West; October 11−12, 2006.

[6.21]   Lev K. Introduction to integrated quality and reliability solutions for industrial companies. In: ASQ World Conference on Quality and Improvement Proceedings. May 1−3, 2006, Milwaukee, WI; 2006.

[6.22]   Klyatis L, Walls L. A methodology for selecting representative input regions for accelerated testing. Quality Engineering 2004;16(3):369−75. ASQ & Marcel Dekker.

[6.23]   Klyatis LM, Klyatis E. Accelerated reliability testing problems solving. SAE Transactions 2004;113:684−91. Section 6: Journal of Passenger Cars: Mechanical Systems Journal.

[6.24]   Klyatis LM. Climate and reliability. In: ASQ 56th Annual Quality Congress Proceedings. Denver, CO; May 20−22, 2002. p. 131−40.

[6.25]   Klyatis LM. Establishment of accelerated corrosion testing conditions. In: Reliability and Maintainability Symposium (RAMS) Proceedings. Seattle, WA; January 28−31, 2002. p. 636−41.

[6.26]   Klyatis LM, Klyatis E. Vibration test trends and shortcomings. Part 2. Reliability review. The R & M Engineering Journal (ASQ). Part 2 2001;21(4):19−27.

[6.27]   Klyatis LM, Klyatis E. Successful correlation between accelerated testing results and field results. In: ASQ 55th Annual Quality Congress Proceedings. Charlotte, NC; May 7−9, 2001. p. 88−97.

[6.28]   Klyatis LM, Teskin OI, Fulton JW. Multi-variate Weibull model for predicting system reliability, from testing results of the components. In: The International Symposium of Product Quality and Integrity (RAMS) Proceedings. Los Angeles, CA; January 24−27, 2000. p. 144−9.

[6.29]   Klyatis LM. A better control system for testing of mobile product − Part 2. Reliability review. The R & M Engineering Journal (ASQ) 1999;19(3):25−9.

[6.30]   Klyatis LM. Step-by-step accelerated testing. In: The International Symposium of Product Quality and Integrity (RAMS) Proceedings. Washington, DC; January 18−21, 1999. p. 57−61.

[6.31]   Klyatis LM. Conditions of environmental accelerated testing. In: The International Symposium of Product Quality and Integrity (RAMS) Proceedings. Anaheim, CA; January 19−22, 1998. p. 372−7.

[6.32]   Klyatis LM. One strategy of accelerated testing technique. In: Annual Reliability and Maintainability Symposium (RAMS) Proceedings. Philadelphia, January 13—16, 1997; 1997. p. 249—53.
[6.33]   Klyatis LM. WorldCat Identities. worldcat.org/identities/lccn-no2003091937.

Exercises

1. What are some specific positive trends in the development of accelerated testing (AT).
2. What are some areas that have seen implementation of accelerated reliability and durability testing (ART/ADT).
3. Describe the basic contents of the published reviews of ART/ADT, as described in the books.
4. Describe the strategic aspects and the basic steps for implementation of ART/ADT.
5. Describe some of the results of ART/ADT implementation in Iscar Ltd., and in other companies.
6. Describe some examples of what and where there were successful implementations of the new positive concepts of accelerated testing.
7. How were some American and Germany companies involved in the implementation of ART/ADT and their components.
8. How was the American Society for Quality involved in the development and implementation of the author's approaches to accelerated testing.
9. How was Dr. Klyatis' approach in the development of accelerated testing implemented and recognized in aerospace engineering.
10. Describe how the positive trends in accelerated testing development were implemented through international standardization.
11. Describe the basic content of Dr. Klyatis's seminars for Ford Motor Co. and Nissan in accelerated testing development.
12. Describe the basic content of the seminar "Accelerated Testing of Products", which was organized by the American Society for Quality.
13. Describe how DaimlerChrysler evaluated work in the development of accelerated testing.
14. Describe how Nissan, Toyota, Jatko Ltd., and other companies demonstrated their interest in the implementation ART/ADT?
15. As listed in data by WorldCat Identities, approximately how many libraries contain the author's books on accelerated reliability and durability testing.

Chapter 7

Trends in the development of equipment for accelerated testing

Abstract

Chapter 7 discusses trends in the development of testing equipment, including:

- Development of simpler testing equipment mostly for materials and details used in automotive and aerospace engineering;
- Demonstration of how the marketing of testing equipment tends to relate primarily to the electronics industry. This includes apparatus, instruments, materials, and machinery details;
- Demonstration of the global nature of the general-purpose test equipment market, largely consisting of the communications sector, the oscilloscopes segment, and the market in the Americas;
- General trends in testing equipment development;
- The growth and the development of the test and measurement industry;
- Key trends for the electronic test and measurement equipment;
- An overview of the automotive testing equipment market;
- Analysis of the trends in the Asia—Pacific region (China, India, and others);
- Trends in the development of equipment for aerospace simulation and aerospace materials testing.

7.1 Introduction

As was discussed in Chapter 4 the virtual approaches to accelerated testing are increasing, and the physical approaches to accelerated testing are decreasing. As the basic technology of testing consists of methods and equipment, the methods and speed of physical testing development decreases, and the speed of development of the equipment needed for physical accelerated testing is also decreasing.

Trends in Development of Accelerated Testing for Automotive and Aerospace Engineering.
https://doi.org/10.1016/B978-0-12-818841-5.00007-6
263

To see this, it is only necessary to do a comparison of Test Expos for the automotive industry now and 15–20 years ago.

The same companies that tried to develop equipment for more accurate physical simulation of field conditions by developing more complicated equipment can now often be seen to be moving back to simpler equipment and less complicated simulation. But this trend also produces less accuracy of the real-life simulation.

If you look at the accelerated reliability and durability testing equipment produced by companies, such as WEISS Technik, Instron Schenck, Advanced Test Equipment Corporation, WestTest, LDS, RENK, HORIBA, and others, you will see little in the way of new equipment for physical testing, but you will see a lot of equipment for materials and details testing and programming for virtual testing.

And while you will frequently see the term "durability testing," in fact, most of this is proving ground testing or fatigue testing, or some other types of testing, but it is not accurate to term it as "durability testing."

The reason for this situation is not with the companies that design and manufacture testing equipment, but with their clients whose interest too often is primarily in saving money in the testing area. Because virtual testing is less expensive, many companies in the automotive and aerospace field want to transition to virtual testing. While they believe that simple and less expensive testing will lead to accurate simulations of field conditions, they fail to take into account the results that frequently include more expenses to the subsequent processes of design and manufacturing, and increased product recalls that create losses. Too often the final result of simpler testing is decreased quality, reliability, durability, maintainability, profit, and increasing life-cycle cost.

As can be seen below, while there is an overall trend of an increased market for testing equipment, but upon closer analysis, this increase relates mostly to:

- Increasing development in the electronics industry;
- Increasing development of apparatus and instruments;
- Increasing development of the materials and details of the product.

But what about the market for equipment for testing entire cars, trucks, buses, airplanes, space research stations, and other machinery?

The author presents a partial analysis of this in Chapter 4 "Equipment for Accelerated Reliability Durability Testing Technology" (74 pages in the book [7.2]). From that time and continuing on to today, most changes have been in the area of control testing equipment, and such equipment is primarily connected with development in electronics.

Many of these electronic components are parts of systems of control and information for automotive and aerospace engineering.

What is being done for the other aspects of engineering technology testing development? What is being done to develop equipment for accelerated testing of units and complete products, such as cars, trucks, satellites, and many others?

There is a real need for more effective and successful testing equipment in these new technologies, especially successful prediction in these new technologies.

How does one obtain the initial information needed for the successful prediction of a new product's quality, reliability, safety, durability, maintainability, life-cycle cost, and many other performance components and characteristics? This author considered these problems in detail in the books [7.1−7.4], and other publications.

Unfortunately, the management of many industrial companies do not appreciate the simple fact of life that new product development for more complicated products needs correspondingly more complicated testing equipment.

7.2 General trends in development testing equipment

This section of the chapter will summarize the author's findings of general trends in the development of testing equipment in the world market.

Technavio's latest report on the global general purpose test equipment market provides an analysis of the most important trends expected to impact the market outlook from 2017 to 2021. Technavio [7.5] is a leading global technology research and advisory company. The company develops over 2000 pieces of research every year, covering more than 500 technologies across 80 countries. Technavio has about 300 analysts globally who specialize in customized consulting and business research assignments across the latest leading edge technologies.

Technavio analysts employ primary as well as secondary research techniques to ascertain the size and vendor landscape in a range of markets. Analysts obtain information using a combination of bottom-up and top-down approaches, besides using in-house market modeling tools and proprietary databases.

They corroborate this data with the data obtained from various market participants and stakeholders across the value chain, including vendors, service providers, distributors, resellers, and end-users.

General-purpose test equipment (GPTE) includes different testing and measuring (T&M) equipment, such as oscilloscopes, spectrum analyzers, signal generators, power meters, logic analyzers, electronic counters, and multimeters. Technavio forecasts the global general purpose test equipment market to grow to 6.58 billion USD by 2021, at a CAGR of nearly 5% over the forecast period.

7.2.1 Global general trend test equipment market

Top three Emerging Trends Impacting the Global General Purpose Test Equipment Market From 2017 to 2021:

● Communication Sector

In 2016, the communication sector dominated the global GPTE market with a 27.84% market share.

● Oscilloscopes Segment

The global oscilloscopes market is expected to reach $2, 114.11 million by 2021.

● Market in the Americas

The GPTE market in the Americas was valued at $2034.48 million in 2016.

	Global Market Growth	
2016 VALUE	$5.19 billion	$1.39 billion.
2021 VALUE	$6.58 billion	incremental growth.

Technavio has published a report on its forecast of the global general purpose test equipment (GPTE) market from 2017 to 2021. This report provides an analysis of the most important trends expected to impact the market from 2017 to 2021.

The research study by Technavio on the global general purpose test equipment market for 2017—21 provided not only a detailed industry analysis based on the product (oscilloscopes, spectrum analyzers, and signal generators), end-users (communication, aerospace and defense, and mechanical sectors), but also included the geographical area (the Americas, APAC, and EMEA). In this analysis, Technavio identifies emerging trends that they consider as a factor that has the potential to significantly impact the market and contribute to its growth or decline. These trends include:

Increase in outsourcing testing activities

The progress in technologies from 2G to more advanced technologies such as LTE and LTE-A has created the demand for continuous maintenance and quality testing of frequency and spectrum. Various vendors outsource the quality testing to external entities to ensure enhanced quality for customers.

Interoperability era for T&M equipment

T&M equipment vendors supply equipment that is designed to work accurately with a variety of products. Customers often install the products themselves and show a preference toward devices with greater interoperability. T&M products made for home gateways, video monitors, home networking/automation, and smoke alarms are all products that make use of GPTE equipment to ensure interoperability. Technavio analysts forecast a boost in the GPTE market, with set standards being adopted to ensure interoperability.

Anju Ajaykumar, a lead analyst from Technavio, specializing in the research test and measurement sector, says, "The construction industry is expanding at a large scale across the world. The construction sector in developed countries and emerging markets is expected to grow exponentially during the forecast period, particularly in countries in APAC. China is the largest consumer and producer of steel" [7.5].

7.3 Trends in the testing and measurement industry

7.3.1 Transformational shifts ahead. 1.94K

7.3.1.1 Growth prospects and key trends for electronic testing and measurement (T&M) equipment

The electronic test and measurement (T&M) industry has been taken by storm lately, with trends such as the Internet of Things (IoT), millimeter-wave (mm-wave) frequencies, and 5G. These trends suggest significant changes in the requirements for test equipment, thus opening up opportunities for vendors, notably in the areas of RF/MW **test** and measurement **equipment,** such as signal generators and network test equipment. It appears that there will be a robust near-future market for electronic **test** and measurement equipment in research and development [7.7].

Test and measurement instruments are critical to ensure product performance and shorten time-to-market (TTM). They are used across the entire life cycle of a product, from research and development (R&D) to manufacturing and deployment, and even into aftermarket servicing although equipment requirements may differ widely across these segments.

By generating about $2 billion globally in 2015, the R&D segment of the new general purpose test market is expected to witness single-digit growth rates over the next 5 years that will steadily accelerate, resulting in the market reaching $2.4 billion by 2021. The R&D T&M market will benefit from high-speed digital trends and a revitalized bit error ratio tester (BERT) market. It will be driven by the data center's buildout with the emergence of 400 gigabit Ethernet (GbE) technologies and complex modulation signals, such as Pulse Amplitude Modulation (PAM). Although, the R&D T&M market has traditionally relied heavily on the oscilloscopes market, the contribution of radio-frequency/microwave (RF/MW) test equipment to revenues, especially that of signal analyzers, is expected to be much greater 5 years from now. RF/MW test and measurement equipment, such as signal generators, network analyzers, and power meters will also benefit from the move to higher frequencies and wider bandwidths due to the advent of disruptive technologies like 5G, but also by a general move toward higher frequencies in aerospace and defense (A&D), automotive radars, and other industries. Today, over half of the market revenues for test and measurement equipment used in the laboratory come from instruments with a frequency range of up to 6 GHz. By 2021, the share of revenue coming from instruments with a frequency range of over 26.5 GHz is expected to be about 5 percentage points higher than it was in 2015 [7.7a].

In manufacturing applications, test and measurement equipment vendors have faced a tougher environment, with stiff competition due to lower technical requirements in comparison to the R&D segment. While the market was challenged over the past few years, only generating about $1.6 billion in 2015, it is expected to follow a recovering trend and reach $1.8 billion by 2020. One of the greatest challenges is making the testing of mm-wave frequencies cost-effective in the mass production environment. This challenge will grow in importance over time and become more significant with the deployment of technologies such as 802.11ad (WiGig) and 5G. In the short term, IoT generates plenty of challenges for manufacturers. Many are struggling to add connectivity to devices that have never been connected before. They lack expertise in RF and related fields. There is also significant price pressure from the consumer that impacts the entire value chain. Manufacturers are, thus, looking for cost-effective solutions from T&M OEMs, who must provide such solutions while addressing more complex technologies that drive up costs. The wide variety of IoT manufacturers also requires solutions providing extensive capabilities and flexibility. Plagued by this host of challenges, the manufacturing environment presents significant opportunities to T&M for innovation.

Outside of providing faster and less expensive instruments of higher performance, one of the key opportunities for T&M vendors in the manufacturing environment is providing technical expertise.

There is a strong desire across industries for using the huge amount of data being generated by manufacturing operations to improve production yield, among other traditional metrics [7.7a].

7.3.2 Testing is more just new equipment

While 2016 was expected to be soft for the electronics test equipment market, it was largely due to the high concentration in the market on new test equipment. A plethora of opportunities related to the test equipment market exist that participants can and should tap into. Market participants must broaden their view of the test market and capture adjacent growth opportunities while addressing areas of need in their core business market.

7.3.3 WiseGuy reports forecast "electronic test and measurement market"

WiseGuy Reports adds in their report "**Electronic Test and Measurement Market 2018 Global Analysis, Growth, Trends, and Opportunities Research Report Forecasting to 2023**" in [7.7a].

This report provides an in-depth study of the "**Electronic Test and Measurement Market**" using what they term Strength, Weakness, Opportunities, and Threat (SWOT) analysis to the organization [7.7]. Their Electronic Test and Measurement Market report also provides an in-depth survey of key players in the market.

The report is based on the analysis of the various objectives of an organization, such as profiling, the product outline, the quantity of production, required raw material, and the financial health of the organization.

Test and measurement equipment are devices used for testing and measuring the various electronic and mechanical products throughout their life cycle. These devices are used throughout the product's life cycle, from the initial design through development, verification, maintenance, and repair of various electronic and mechanical products. Test and measurement equipment are used for testing and measuring a variety of electronic devices such as cellphones, digital cameras, MP3 players, and solar inverters. And, in addition to these electronic products, some mechanical products also use test and measurement equipment, including turbines, automotive car suspensions, and aircraft propulsion systems.

Some of the major factors contributing to this growth are their increasing production and consumption by the automotive sector, technological advancements pertaining to the 5G solutions in the IT and telecommunication sectors, the growing focus of stakeholders in the industry and smart applications.

Market segment by Application, split into Communications and Electronics Manufacturing, Aerospace and Military/Defense, Industrial Electronics, Automotive, and Other Industries.

7.3.4 Overview of the automotive test equipment market [7.8]

The automobile test equipment industry is one among the most challenging industries. New designs of automotive products, shorter vehicle time to market, and increasing regulations by the government, especially in the area of emission control, are all creating new challenges for the automotive industry.

In order to address such challenges, it is becoming increasingly necessary for the automotive manufacturers to introduce new and more sophisticated automotive test equipment, thereby driving the growth of the T&M market.

The automobiles equipment testing market is gaining traction in the areas of rheology fire mechanics, media resistance, and surface performance. Automobiles are increasingly being equipped with modern systems and complex electronic safety devices and systems, which are generating greater needs for the adoption of automotive test equipment. The percentages of electronic devices in automotive products are increasing at a very fast pace, with many mechanical components being replaced by electronic components.

This generates a corresponding increase in the demand for suitable technologies to ensure that these new automotive components are being adequately tested before being incorporated into automobiles.

Robert Bosch GmbH (Germany), Honeywell International Inc. (U.S.), Siemens AG (Germany), ABB Ltd. (Switzerland), Delphi Automotive PLC (U.K.), Actia S.A. (France), Advantest Corp (Japan), Horiba Ltd. (Japan),

Softing AG (Germany), ACTIA Group (France), EM TEST (Switzerland), Freese Enterprises Inc. (U.S.), Moog Inc. (U.S.), Sierra Instruments (U.S.), and Teradyne Inc. (U.S.) are some of the prominent players profiled in the MRFR Analysis and are at the forefront of competition in the Global Automotive Test Equipment Market. Included in these, Freese Enterprises Inc. is a specialist in the automotive testing equipment market. They provide equipment for testing applications, such as Air Bag Controller, Automotive Paint and Coatings, Lighting Systems, Bumper Stiffness, Electric Motors, Electronic Throttle, Multifunction Switches and Instrument Panel Cluster Gauges, among others.

7.3.5 Industry/innovation/related news

A noteworthy development in the market was in September 2017, when Advantest Corp sold their V93000 Platform with Universal Analog Pin Module to TDK-Micronas for testing the full range of automotive sensors. By acquiring this flexible system, the company gained an integrated solution for the final testing of sensors and controllers used in automotive and industrial applications.

Another noteworthy development was in May 2013when Moog Inc. showcased its electric and hydraulic simulation tables at the Automotive Testing Expo Europe 2013 [7.9]. Moog also offered a wide range of other automotive testing equipment and systems, including their electric simulation tables, tire coupled simulation systems, electric and hydraulic actuators, servo valves, test controllers and test software for multiaxis test systems.

And then, in December 2008, EM TEST launched its automotive test equipment that included new ISO control software for automotive testing. This software provides the user with features such as a structured user interface that allows for the easy setting up of tests, pulse windows, which tells the user where the selected test stands, which also integrates external measuring devices for DUT monitoring and automated pulse verification.

This acquisition helped the company in the rapid extension of its emission measurement and powertrain research and development activities enabling them to become a total system supplier to the automotive industry.

7.3.6 Automotive test equipment market – segmentation

The Automotive Test Equipment Market can be segmented into the following key dynamics for the convenience of analysis and an enhanced understanding of the market. These key segments are:

- Segmentation by product type, e.g., wheel alignment tester, engine dynamometer, chassis dynamometer, vehicle emission test system, etc.
- Segmentation by vehicle type, e.g., light commercial vehicle, heavy commercial vehicle, passenger cars, etc.

- Segmentation by application, e.g., mobile device-based scan tool, handheld scan tool, PC/laptop-based scan tool, and others.
- Segmentation by regions, e.g., North America, Europe, APAC, and other specified regions.

7.3.7 Automotive test equipment market − regional analysis

Looking briefly at one of these key segments, "Segmentation by regions," the Asia−Pacific region is expected to dominate the global automotive test equipment market. This is owing to the presence of developing countries, such as India and China, where the automotive industry is growing at a fast pace. The market in the Asia Pacific region is also growing owing to factors such as the increasing number of vehicle manufacturing facilities, which is largely due to the low cost of production, a resultant increase in production capacity, and the growing demand for light and heavy vehicles.

The North America region is a matured market for automotive test equipment. The original equipment manufacturers in the region are primarily focusing on efforts for the improvement of the production quality and for delivering quality products. Much of the demand for automotive test equipment in the North America region is being driven by the strict emission requirement, especially regarding CO_2 and NOx emissions. However, North American original equipment manufacturers are setting up new engine and assembly plants in Mexico and Canada, which is increasing the demand for new automotive test equipment.

7.3.8 A historical perspective of EMC & field strength test solution [7.9]

EMS and EMI test solutions from the world market leader Rohde & Schwarz offers an exceptional range of EMC and field strength test equipment, from stand-alone instruments to customized turnkey test chambers.

The EMI and EMS test instruments and systems are designed to determine the causes and effects of electromagnetic interference and to ensure compliance with the relevant EMC standard. The EMC test solutions support all relevant commercial, automotive, military and aerospace standards, as well as ETSI and FCC standards for radiated spurious emissions and audio breakthrough measurements.

Over the past decade, a number of technology trends have started to significantly impact the design and function of new test and measurement market products. In 2013, [7.10] indicated that it was expected that this trend would continue accelerating. Further FPGA and DSP advances will enable manufacturers to quickly develop more and more advanced products as their processing power continues to grow. This enables new products to address applications where more advanced and costly ASIC-based instruments were previously required. Over the past few years, this trend has coincided with a trend toward increased memory in instruments. That trend could be seen to change in 2013 [7.9] when the memory capabilities of oscilloscopes priced under $1000 was measured in the 10s of millions of points. And, as the data provided by these instruments increases, it will be coupled with improvements in the instruments, which will start trending toward more advanced analysis and capabilities included in the instrument. The types of calculations that were previously available only on Windows-based instruments will become simple and straightforward directly on the instruments without the need, cost, and overhead of a separate PC and operating system. Examples of these capabilities include more advanced statistical pass/fail analysis of large sets of data and customizable user formulas that will be created and executed directly on the instruments.

Evidence of this trend can be seen in the statements of some of the industry leaders.

Charles Sweetser, in [7.10], wrote, "The industry is always looking for better methods and techniques for confidently determining the condition of power transformers. Maintenance practices and philosophies are always being scrutinized and re-evaluated in hopes of maximizing diagnostic value and balancing economic efficiency. Traditionally, our industry has practiced conventional off-line tests which depend on a single measurement at a single frequency, constant voltage, or constant current.

Often, only having conventional test data for review has resulted in inconclusive analysis that often leads to more unanswered questions. The industry is demanding reliable diagnostic information that is representative of the best possible condition estimation. In 2013, the emerging trend is to extract as much additional diagnostic information as possible by applying smarter advanced methods and techniques to existing procedures. This will require using multifunctional test instruments with advanced features. The idea is not to create new tests and increase testing overhead. Varying parameters to conventional tests, such as frequency, provide a new avenue for analysis. Based on research, practical experience, and advancement in measurement instruments, it is now possible to extract in-depth information that was not available in the past. These advanced diagnostic methods or 'extensions' provide new and critical information about the transformer condition. This proliferation, along with modern instrumentation, has transformed diagnostic applications and the need for in-depth information for making decisions."

Mark Schrepferman wrote [7.11] that: "… Test & Measurement system designers will be challenged to provide faster, extremely repeatable, rugged, and best-in-class testing environments. Test equipment will be expected to last over multiple generations of product introductions, meaning that the performance requirements of the RFICs used in test equipment must be better than the device under test by a factor of several generations. Further to this, next-generation communication systems that use higher-order modulation schemes such as Orthogonal Frequency-Division Multiplexing (OFDM), with high peak to average ratios, are driving the need for the components used in the test equipment's signal chain to have higher linearity. Additionally, more frequency bands will be introduced, driving the need for broader bandwidths and higher operating frequencies. This new, crowded spectrum will require additional filtering, so filter-bank switching is expected to drive the need for lower-loss components.

Finally, despite the fact that test solutions are growing in complexity, end customers will continue to expect lower overall test costs per unit."

And, as Mike Fox wrote in [7.12]: "A fresh new way of thinking about test and measurement is emerging. As we look to 2013, we are transforming testing and diagnostics with technologies that accelerate and improve the efficiency of diagnostic/repair/approval workflows and processes into what is a true 'diagnostic ecosystem.' In addition to the use of touchscreens with intuitive menu controls that emulate today's personal electronics, several of thermal imagers use Wi-Fi technology and mobile apps to connect to Android or Apple iOS tablets and smartphones. Electrical readings can be stamped right on the image. And wireless datastreaming DMMs can share readings with PCs. Look for smartphone/tablet connectivity as well. The goal of the diagnostic ecosystem is not only to improve communication among diagnostic and communication devices but also among technicians and their customers and managers, by leveraging accurate and coordinated readings from related tools as well as rapid and actionable communication."

Another interesting historical reflection is the testing laboratories at the Schaltwerk Berlin of the Siemens AG [7.13]. These are testing laboratories for electrotechnical equipment of high and medium-voltage, which were started in 1928. Table 7.1 traces some of the highlights of their history.

As has been stated, automotive test equipment [7.14] is an important segment of the testing and measuring market. Automotive test equipment is primarily used for testing and evaluation of a vehicle's performance.

A wide range of automotive test equipment such as accelerometers, speedometers, and brake testers are used to quantify and adjust the performance characteristics of vehicles.

7.3.9 Market size and forecast

The global automotive test equipment market is projected to grow at a 5.2% CAGR over the forecast period. Factors such as the increasing sales of new vehicles and rising awareness among the population regarding the importance of and the need for preventive maintenance of vehicles are expected to foster the growth of the test equipment market in the upcoming years.

In terms of regional platforms, Global Automotive Test Equipment Market is segmented into North America, Asia Pacific, Europe, Latin America, the

TABLE 7.1 Milestones in the history of the testing laboratories in the Schaltwerk, Berlin (Germany). More than 80 years of testing expertise.

1928	Commissioning of the high-power and high-voltage testing laboratories
1940	Capacity upgrade of the high-power testing laboratory to 1200 MVA
1954	Commissioning of the testing laboratories that were reconstructed after the war
1960	Construction of new high-voltage testing halls
1961	Cofoundation of PEHLA
1975	Commissioning of a new high-power testing laboratory with a maximum capacity of 3200 MVA
1982	Upgrade of the open-air test area with a 5 MV impulse voltage generator
1985	Capacity upgrade of the high-power testing laboratory to a maximum of 6400 MVA
1992	Accreditation of the testing laboratories according to the standard previous to ISO/IEC 17025
1994	Construction of a new temperature rise test area for 50/60 Hz up to 6000 A
1995	Commissioning of the vibration test system
2005	Upgrade of the synthetic test circuit to a maximum voltage of 1150 kV
2011	Extension of the mechanical testing laboratory by a new, independent mechanical endurance test hall

Middle East, and African regions. Of these, the Asia Pacific automotive test equipment market is likely to dominate the global automotive test equipment market by the end of the forecast period. This is largely due to the increasing production of vehicles in Asian countries, such as India and China, owing to the low cost of production and the presence of a large base of automotive manufacturing plants, which is expected to bolster the growth of the automotive test equipment market in the Asia Pacific region in the upcoming years. According to the International Organization of Motor Vehicle Manufacturers, production of new vehicles in China grew by 14.5% in 2016 as compared to 3.3% in 2015.

And, in addition, the European automotive test equipment market is poised to display substantial growth over the same forecast period. This is due to the growing production of vehicles in the European countries, especially for electric vehicles that are forecast to be a major factor likely to augment the growth of the automotive test equipment market in the European region. One example is the production of new vehicles in Finland rose by 53.3% in 2016.

Two events that occurred in 2018 are also enlightening.

First, the brochure [7.16] previewing "Automotive Testing Expo 2018" includes:

- Digital and DC-operated torque transducers offered by S. Himmelstein, who designs and produces sensor technology testing equipment;
- Autonomous driving calibration solutions by the Burke Porter Group that supplies testing instrumentation and assembly systems;
- Automated X-ray inspection system by North Star Imaging;
- Ultrasmall Ethernet DAF system;
- Innovative oil shear technology by Force Control Industries;
- One-stop-shopping proving ground by Navistar Proving Grounds who provides garage facilities, equipment, instrumentation, computerized data acquisition systems, and personnel required to support full new vehicle development programs);
- New injury testing module (Head Acoustics has released the latest version of its Artemis Suite data acquisition and analysis software);
- Sensor signal conditioners;
- Reliable pressure measurement solutions using the trusted Druck product range from Baker Hughes at GE Measurement and Control Solutions company;
- Autonomous vehicle simulation hardware by dSpace;
- Test facility operation partner;
- Miniature high-speed cameras;
- Test benches for modern drive concepts (modern alternative drive concepts for the future generation of vehicles, manufacturers, and suppliers. Test bench solution for battery-powered and hybrid concepts from Kratzer Automation);
- Handheld vibroacoustic data analysis device;
- Advanced data logger;
- Measurement calibration and diagnostics solutions;
- Research vehicles for hire;
- E-mobility test rig;
- Winter tire test facility;
- Autonomous vehicle test systems (LaunchPad is a self-propelled platform from AB Dynamics that allows complex and accurate control of all VRU targets);
- Electronically friendly test chamber refrigerants (special chambers from Weiss Technik);
- High-performance force and torque sensor;
- Temperature and humidity test chamber from Tenney/Blue M;
- Shock and vibration test solutions from Team Corporation;
- Next-generation materials test software and platform;
- State-of-the-art vibration test hardware and software;
- Cyclic corrosion test chambers;

- Ergonomic car movers;
- Modular, high-performance drive and DC/DC converter series;
- Fast and responsive pressure imaging sensors;
- High-speed tire footprint analysis system;
- Infrared temperature sensor;
- Advanced vehicle radar test system;
- Portable emissions measurement system;
- Rugged in-line flow meters.

MTS test systems are bringing a range of engineering test solutions. First is the new AWIFT Evo family of wheels force transducers, which was introduced to the market following MTS's acquisition of PCB Piezotronics. Other components of this productivity-enhancing portfolio on display in Novi include the all-electric ePostTire-coupled Road Simulator, new long-lasting and efficient DuraGlide actuators, and echo Intelligent Lab connectivity solutions."

Second, a similar situation could be seen with the Automotive Testing Expo 2018 Europe, June fifth through the eighth, in Stuttgart, as well as in China and Korea.

At this second event, which was in conjunction with the Expo [7.16], there was the Autonomous Vehicle Test & Development Symposium.

The symposium included 34 presentations. Of these, the following eight are directly related to the testing area:

1. Tony Gloutsos, director, Siemens USA. Using advanced simulation to test and train artificial intelligence algorithms.
2. Ram Mirvani, director, global business development. ADAS, Konrad Technologies, USA. Customized sensor test for autonomous driving with sensor fusion HIL.
3. Yoshiguki Usami, associate professor, Kanagawa University, Japan. Autonomous driving test between the two biggest cities in Japan.
4. Dierk Arp, executive director, Messing Systembau MSG, Germany. ADAS testing advanced: 6D target mover.
5. Dr. Craig Shankwittz, principal R & D engineer, MTS Systems Corporation, USA. Hybrid simulation for AV/ADAS test and development.
6. Alexander Noack, head of automotive electronics, b-plus GmbH, Germany. Sensor HIL testing based on raw camera and radar data − the next challenge?
7. Alanna Quail, manager ADAS and cybersecurity, FEV, USA. Automated cybersecurity testing for automotive applications.
8. Edward Leslie. Senior electrical engineer, Leids, USA. Stakeholder consideration for CAV testing on a managed lane facility.

But from the above, it can be observed that there is not enough attention being paid by industrial companies and by academia to the area of testing development.

7.4 Equipment for aerospace simulation

Weiss Envirotronics, Inc. is now Weiss Technik North America, Inc. [7.18] that provides product and service solutions wherever the customers are located around the world. The company is planning [7.18] to continue to be a local partner by offering the product solutions, service support, and industry-leading knowledge for custom and preengineered environmental test chambers that fit customer's testing requirements. The example of this company test chamber can be seen in Fig. 7.1.

7.4.1 Aerospace/altitude/space simulation

Weiss Technik North America altitude test chambers allow life stress testing conditions at various simulated altitudes and temperatures as described below. These chambers are specifically designed for the Aerospace, Military, and Space Simulation industries.

Their WT/D and WK/D vacuum-temperature and vacuum climate test chambers provide a way for reproducible tests of aviation industry components, in a highly stressed environment. These chambers allow the simulation of extreme flight programs in accordance with the relevant standards. For many years they have been applied in the fields of research, development, production, and quality control:

- Vacuum system for continuous dilution of the test space atmosphere until it reaches the required vacuum;

FIGURE 7.1 Vacuum-temperature and Vacuum-climate test chamber (Weiss Technik North America, Inc.) [7.18].

- The outer casing is made of corrosion-resistant, galvanized sheet steel with an environmentally friendly coating;
- Test chamber door is hinged on the left for the whole cross-section of test space;
- Test chamber door has optimum pressure when closed;
- Maintenance-friendly position;
- The entire power electronics is located in a switch cabinet at the right side wall of the test chamber;
- Safety mechanism which switches off the relevant function circuit and/or the entire test chamber;
- The test space is made of high-quality stainless steel and is vacuum- and vapor-tight welded;
- Separate sensors to protect the specimens irrespective of the temperature control system;
- Additional ducts in the side panels are provided for the electrical connections of the specimens from the outside.

7.4.2 Standard version

- Combined temperature and vacuum test $\geq$400 mbar;
- Low/high-temperature safety cutout as per EN 60519-2 (1993) with a separate sensor, thermal safety class 2;
- Touch panel;
- Ethernet interface;
- 4 potential-free switching inputs and outputs;
- Contactless switching of the heating panel;
- 50 mm entry port in the right-side panel;
- Water-cooled condensers.

7.4.3 Options

- Software package Simpati* for Windows;
- Additional potential-free switching inputs and outputs;
- Measured data recording system for Pt 100 and voltage signals $\pm$10V;
- Temperature extension;
- Analog outputs for setpoint and actual values;
- Additional Pt 100 sensor/thermal elements;
- Door with window;
- Shelf, height-adjustable;
- Flange port;
- Temperature conditioning panels for extreme combination tests;
- Ports 50 mm $\varnothing$;
- Other main supplies and frequencies;

- Air-cooled condenser;
- Protection against condensation with a dehumidifier for prevention of condensate on the specimens;
- Sound insulation.

Element 14 are providers of materials and product qualification services to the global aerospace testing sector operating throughout the world. They work in partnership with all of the aerospace primes and their supply chain partners to develop better products, get them to market on time, save time and money, and minimize the risks to their businesses through their product development activities.

Element 14 support the sector through its entire product life cycle — from R&D into manufacturing, life extension, and onto disposal. When combined with 29 Nadcap accredited laboratories, with 41 different Nadcap accreditations, global and local quality credentials and many aerospace customers, they have an international capability, footprint, and geographic reach that is unrivaled in the Aerospace Testing, Inspection, and Certification industry.

They provide aerospace testing that helps to deliver certainty to the Aerospace Primes and their supply chain. Their aerospace testing expertise allows them to provide a comprehensive range of mechanical testing, chemical testing, wear properties testing, grain size testing, structural, failure analysis services, and nondestructive testing and inspection services for metals, polymers, composites, and ceramic materials. They provide as well a range of EMC testing (to the RTCA DO-160 standard) and testing for all major components and systems.

7.4.4 Aerospace materials testing

The global network of Nadcap accredited, and Aerospace supplier approved aerospace materials testing laboratories, together with an experienced team of Aerospace Materials Testing Experts, provide a comprehensive range of materials testing services. These include fatigue testing and fracture mechanics, tension and compression; impact and hardness testing; stress rupture and creep testing for metals, nickel and titanium alloys, aluminum, superalloys, ceramic matrix composites, polymer matrix composites, elastomers, plastics, and adhesives.

Commonly tested materials and components include:

- Metals and alloys;
- Ceramic and polymer matrix composites;
- Plastics and polymers;
- Fasteners;
- Pipes and tubing;
- Welded samples and structures;
- Industrial components, subassemblies, and equipment.

7.4.5 Fatigue testing capabilities

Element's fatigue testing labs can perform high cycle fatigue, low cycle fatigue, and specialized programs on a range of metals and alloys, polymers, and components.

Element's materials testing laboratories provide testing to ASTM E606, the standard test method for strain-controlled fatigue testing. This specification is typically used for low-cycle endurance testing, where components are subjected to mechanical cyclic plastic strains that cause fatigue failure within a short number of cycles.

Low cycle fatigue testing is performed using a tension-compression endurance testing machine. The subject material is first machined and longitudinally polished or ground into cylindrical or flat test specimens with a uniform-gage section. The samples are then loaded into the test frame and subjected to repeated stress under a constant strain rate in accordance with ASTM E606. Extensometers are used to control strain range by measuring deformation in the gage section of the specimen.

Typical low cycle fatigue tests are targeted to run no more than 100,000 cycles. Test frequency typically ranges between 1 and 5Hz, depending on material and requirements. By testing multiple specimens at varying strain levels, S/N or strain-life curves can be developed for the material. ASTM.

E606 recommends testing at least 10 samples to gain statistical confidence in strain-life curves.

ASTM E606 also outlines reporting requirements for strain-controlled fatigue testing, which includes:

- Objective of testing;
- Description of material and specimens;
- Test environment and conditions;
- Initial, stabilized, or half-life values;
- Cyclic strain range;
- Number of cycles to failure;
- Mode of failure.

7.4.6 Abrasion & wear testing

When friction is the predominant factor causing deterioration of the materials, abrasion and wear testing will give data to compare materials or coatings.

Abrasion testing is used to test the abrasive resistance of solid materials. Materials such as metals, composites, ceramics, and thick (weld overlays and thermal spray) coatings can be tested with these methods. The intent of abrasion testing is to produce data that will reproducibly rank materials in their resistance to scratching abrasion under a specified set of conditions.

Standard abrasion testing methods should not be used to evaluate the resistance of a given material in a specific environment. Rather, its value lies in ranking materials in a similar relative order of merit as would occur in an abrasive environment.

A customized wear testing program, on the other hand, can be configured to closely mimic actual operating conditions, including temperature fluids, and direction of wear. This custom approach will result in wear testing data that is much more relatable to the specific work environment in question.

Element's abrasion testing capabilities include Taber abrasion, rubber wheel abrasion, pin abrasion, RCA abrasion, and more.

Pin abrasion testing (ASTM G132) is performed using two pin specimens: the subject material and reference material. A pin is positioned perpendicular to an abrasive surface, which is mounted on and supported by a flat surface. The pin abrasion testing machine permits relative motion between the abrasive surface and the pin surface. The wear track of the pin is continuous and nonoverlapping. The pin rotates about its axis during testing. The amount of wear is determined by the weight loss of the pin. ASTM G132 calls for a reference specimen to be included in the calculation in order to correct for abrasivity variations.

Rubber wheel abrasion testing (ASTM G65) is performed by loading a rectangular test sample against a rotating rubber wheel and depositing sand of controlled grit size, composition, and flow rate between them. The mass of the test sample is recorded before and after conducting a test. To develop a comparison table for ranking different materials, it is necessary to convert this mass loss data to volume loss in order to account for the differences in material densities.

Taber abrasion testing (ASTM D1044 and ASTM D4060) is performed by mounting a flat specimen, either square or round, to a turntable platform that rotates two abrasive wheels over the specimen at a fixed speed and pressure. One wheel rubs the specimen outward toward the periphery and the other, inward toward the center. Specimen mass (ASTM D4060) or haze (ASTM D1044) is measured pretest and posttest to allow for material property comparisons. A wide variety of abrasive wheels are available for Taber abrasion testing, depending upon the project goal.

Blade-on-block wear testing typically utilizes an object (block) that articulates back and forth on a stationary specimen (blade) while being subjected to a constant normal load. Blade-on-Block testing is especially useful when a specimen needs nonstandard environmental conditions or a higher load force than Pin-on-Disk testing can achieve.

Pin-on-disk wear testing (ASTM G99, ASTM G133, and ASTM F732) involves abrading two materials — one material is machined into a pin, the other into a disk — to determine a variety of properties, including wear rates and frictional force coefficients. Pin-on-Disk can be conducted at elevated temperatures or in submerged environments to more accurately simulate "real life" wear conditions.

7.4.7 Electrostatic discharge testing

Satellite equipment needs to be immune to electrostatic discharges (ESD) which are likely to occur in flight.

Testing for the susceptibility to ESD on spacecraft equipment is, therefore, an important part of the EMC test program [7.20]. The test method according to ECSS-E-ST-20-07C [7.20], which is commonly used in space industry, leaves an undesirable number of degrees of freedom on how the test can be performed, which, in turn, has an influence on the injected current, and therefore on the test results.

Because of the indirect coupling of ESD to circuits, the current waveform applied during the test and its repeatability are of high importance. In the frame of an ESTEC research contract (4000109887/13/NL/GLC), a critical review of the ECSS-E-ST-20-07C is performed to identify the influence factors, which have an impact on the test's repeatability.

To comply with future pending gas regulations, environmental simulation chambers [7.21] come as standard models or customized for users' needs. The new Climates Excel line of environmental simulation test chambers offer performance and are compliant with the new F-gas regulations that are progressively being introduced, leading up to 2030 [7.21]. The range of test chambers uses more environmentally friendly gases and features a new, more compact compressor; and other features, such as extended temperature ranges.

7.4.8 F-gas compliance

The F-gas Regulation (BC) No 842/2006, which came into effect on January 1, 2015, prohibits the purchase of certain refrigerating machines that use gases or techniques that pose an environmental hazard from their global warming potential (GWP). GWP is an indication of the harmfulness of a gas relative to the greenhouse gas effect that would be equivalent to one molecule of carbon CO over a given period of time. In real terms, 1 kg (2.3 lb) of CO_2 represents a GWP of 1. The higher this number, the more harmful the fluid or the gas's GWP.

While the implementation of this European standard is not yet complete, but it is the plan that by 2030, the legislation will fully prohibit the maintenance of machines using fluids with GWP higher than 2500 CO_2 equivalent. FUTURE-PROOFING, NOW.

Before being confronted with this deadline, it is prudent to invest in devices that are already compliant with, or even exceed, the new regulations in the field of extreme environment simulation. The Climate has already positioned itself with new Excel machines, which use R449A gas. In equal quantities, this gas has over 99% of R404 specifications, but with a 65% reduction in GWP.

For users, the consequences of this are twofold. The first is the use of the gas makes it possible to comply with the new regulations. Second, with a GWP of 1397, R449A gas gives an advantage in terms of maintenance. Under regulations, machines that have a maximum charge threshold below 50 metric tons equivalent CO_2 require only an annual inspection. Machines in the Excel line can, therefore, keep the same load of calorific without generating any extra inspections compared with the use of R404A refrigerant gas.

Safety, reliability, and accuracy of conditions for testing are among the critical criteria that can be met by optimizing climate chamber designs with finite element analysis [7.22].

Hardy Technology's department of technical design in Chongqing received an order in 2015 for an altitude climate test chamber with a volume of 55 m^3 from a third-party testing organization of the Chinese government. While celebrating this winning bid, the designers started to think about how to produce an optimized design. Based on the concerns for safety and reliability, it was decided that the equipment should use an enhanced internal frame design. However, the required thicker steel plates and stronger ribs would also increase the thermal load on the equipment, such that the heating and refrigeration rates would be affected, and thereby increasing the inefficiencies in energy consumption in operation.

Additionally, the thin air at high altitudes reduces the circulating air in the test space, which would extend the time to reach the required uniform temperatures and reduce the available time for the actual test.

Following the suggested solution, Hardy's design department finished the design, and the equipment was delivered to the customers and passed the associated tests successfully. The inner wall of the equipment uses 304 stainless steel with a 6 mm thickness. When starting with an internal pressure of 1 sun, the pressure reduction rate is precise, taking 25 min to bring the chamber down to 50 kPa (0.5 bar) while the system deviation is less than ± 1kP2.

7.5 Combined testing equipment

7.5.1 Aerospace testing equipment

Flight hardware. When building a spacecraft and associated space flight hardware, the design process most often requires a comprehensive set of testing. The objective of the testing is to verify the ability of the design to survive its design loads and to qualify the design for flight. Such tests frequently include static loads, model, vibration, acoustic, and shock tests. But one impact of this testing is the resulting extension of the time and increased cost required for putting the hardware into space. The idea of combined testing has been developed over recent years with some real success, but it has received little attention in the technical media.

One such example is the Burke E. Porter Machinery Company (BEP) [7.23], which has worked in combination with Vehicle Test Systems. As the manufacturer of the first fully electric DVT Roll/Brake machine, BEP, has paved the way for innovation in automotive and heavy-duty vehicle testing. Supplying End-of-Line equipment, Dynamometers, and Tire and Wheel Assembly testing systems to all major OEM's throughout the world.

Making of the Solar Impulse 1 (HB−SIB): Wind Tunnel Test Slowly Coming to Life

The Solar Impulse 1 (HB-SIA) solar-powered prototype aircraft has proved itself in demonstration flights in Europe, as well as with a successful solar-powered flight across the USA.

These accomplishments have paved the way for the next milestone in the innovative Solar Impulse project − the development and manufacture of the long-range and larger Solar impulse 2, designed to fly around the world. [7.24]. Although originally designed to fly around the world in 2015, the testing phase of the Solar Impulse 2 is now projected to be completed in 2019. The airplane was completed at the end of 2013, and flight tests were started in early 2014 [7.25].

Extensive tests were required to ensure the new design's soundness, including wind tunnel. The testing Campaign is conducted in RUAG Aviation's Large Subsonic Wind Tunnel Emmen (LWTE).

The RUAG Large Subsonic Wind Tunnel in Emmen (Switzerland) is one of the largest structures in Europe and is used for a variety of tests, including aircraft development, automotive research, rain tests, and boat aerodynamics. For the Solar Impulse 2 testing, a wooden mock cockpit will be used, enclosed in a polyurethane foam shell with covering material on the external surface. These tests are meant to verify the airworthiness of the fairing, and the cabin door, and to simulate the behavior of the materials during flight.

One key element of this test is to confirm the ability to jettison the cockpit's door during an emergency. Another key element is to evaluate the overall behavior of the structure with winds blowing from various angles. At the time of the writing of this book, this testing was estimated to be completed in 2019.

7.5.2 Environmental testing and test facilities

European Test Services (ETS) [7.22] is maintaining and providing test facility services to European industries by managing and operating the environmental test center of the European Space Agency(ESA), located in Noordwijk, The Netherlands. ETS is active in providing mechanical testing, EMC testing, thermal vacuum testing, altitude simulation, and more. In addition to the testing of spacecraft and space applications, ETS has also become a major supplier of testing services to the railway, marine, and the electric power industry [7.22,7.23].

7.5.3 ETS mechanical data handling facilities

ETS's Mechanical Data Handling facilities provide data acquisition, reduction, and presentation for vibration, acoustic, and shock testing. Their acquisition system can handle various analog inputs and consists of four mobile/modular 128 channel racks interconnected to the test article via mobile patch panels.

In addition to this acquisition hardware, ETS can also provide acceleration sensors and a Force Measurement Device (FMD) to their customers. This FMD consists of load cells connected via a summation unit to the data acquisition system. In this way, the interface forces between the shaker and test articles are available during the test for control/notching and immediately available after the test for response analysis and model identification.

7.5.4 ETS's maintenance, management and test facility services

ETS is contracted to maintain the facilities of the ESA ESTEC Test Center in the Netherlands.

The facilities are divided into mechanical, thermal, EMC, data handling, and infrastructure facilities:

- The mechanical test facilities/services comprise vibration/shock, acoustic noise, balancing, and physical properties measurement.
- The thermal facilities consist of thermal vacuum facilities with which the space environment can be simulated (incl. sunlight and cold), as well as altitude simulation/decompression.
- The EMC facility is equipped to provide MIL standard EMC testing.
- The data handling facilities comprise a 500 + channels mobile/modular mechanical acquisition system and a 1000 channels mobile thermal acquisition system.
- The main parameters of the mechanical data handling facilities are listed below:
- Available acquisition channels:
  - Piezoelectric transducers   3 x 128#
  - Voltage signals   128#
  - Strain gauge channels   60#
- Charge amplifier characteristics:
  - Input range ±25 to ±51200 pC
  - Frequency bandwidth   96 Hz
  - Dynamic range   >80 dB
  - Amplitude accuracy of   0.2%
- Voltage amplifier characteristics:
  - Voltage input range ±100 to ±10 V
  - Frequency bandwidth   96 Hz
  - Dynamic range   >92 dB
  - Amplitude accuracy of   0.2%
  - ICP supply   4 mA

- Bridge amplifier characteristics:
  Full bridge, half-bridge, and quarter bridge
  - Frequency bandwidth   20 kHz
  - Dynamic range   >90 dB
  - Amplitude accuracy   0.2%
  - Bridge supply voltage   0.5, 2 and 5 V
  - Bridge max, supply current   30 mA
- Connectors:
  - Piezoelectric sensors   microdot 10/32
  - Voltage sensors   BNC
  - Strain gauges   LEMO/soldered

ESCO Technologies Company. ETS Lindgren (Subsidiary company). Chamber, Enclosure and Test Cell Solutions for Test and Measurement Applications. ETS-Lindgren [7.28,7.29] is working on systems and components for the detection, measurement, and management of electromagnetic, magnetic, and acoustic energy. The company adapts technologies and applies proven engineering principles to create value-added solutions.

This company began in 1995 when EMCO, Rantec, and Ray Proof combined their resources to create a new entity called EMC Test Systems. Later, after this acquisition, the company changed its name to ETS-Lindgren.

In 2000, Holiday Industries was also acquired. The company's testing technology is used in today's electromagnetic field sensing (EMF) systems for test and measurement, and health and safety applications.

Continuing their growth in 2002, ETS-Lindgren purchased Acoustic Systems, a supplier of acoustic test and measurement, audiology, and broadcast applications.

ETS Provides Customized Solutions

Examples of how ETS-Lindgren allows their customers to achieve a custom solution can be seen in the following:

- A wireless service provider needed to be able to test the operability of wireless phones, which were brought into local customer service centers. ETS developed a bench-top test cell with an integral antenna that enabled the provider to diagnose the phones quickly and accurately at the customer's service centers.
- A large automotive manufacturer needed a chamber that could test fully operational vehicles at speeds of up to 112 km and needed special antennas to perform the test. ETS-Lindgren designed a facility with a combined dynamometer, air cooling, and exhaust systems for the vehicles while their RF experts designed the one-of-a-kind antennas and positioning system.
- A well-known integrated circuit manufacturer needed to test its products for both EMC emissions and acoustic properties. ETS-Lindgren built one chamber that could do both without compromising either EMC or acoustic measurement results.

As can be seen in these examples, ETS-Lindgren can provide engineering and manufacturing resources to create custom solutions.

Accessories and Equipment

ETS-Lindgren manufactures most of the equipment or accessories required for RF testing, including:

- Antennas
- EMC
- Microwave
- Wireless test
- EMF sensors
- Broadband E-field
- Laser powered
- Battery powered
- Shaped Response
- FCC requirements
- ICNIRP requirements
- Positioning equipment
- Antenna towers
- Equipment (DUT) turntables
- Multiaxis DUT positioners
- RF hardened CCTV cameras
- Power line and RF signal line filters
- RF shielded lighting
- RF honeycomb vents
- RF bulkhead feedthroughs
- Specialized equipment.

7.5.5 Chambers, enclosures, and test cell product gallery: EMC chambers

7.5.5.1 Free-space Anechoic Chamber Test (FACT)

Free-space Anechoic Chamber Test (FACT) represents the state-of-the-art technology for EMC measurement using demountable modular panels, anechoic absorber, and sliding, swing, or hinged doors. FACT chambers provide the test environment for meeting most international emissions and susceptibility standards, such as CISPR, IEC, VCCI, ANSI, FCC, and SAE.

7.5.6 Statistical mode averaging reverberation chambers (SMART)

Statistical Mode Averaging Reverberation Test (SMART) chambers provide an electromagnetic environment for performing both radiated susceptibility and emissions testing. IEC 61000-4-21 draft standard addresses the reverberation chamber as a test environment.

7.5.6.1 SMART™ 80

- 80 MHz–18 GHz frequency range
- Continuous or Stepped Tuner rotation
- Typical size for SMART 80 chamber: 13.44 x 6.09 × 4.87 m (interior)

7.5.7 Environmental combined testing equipment for military vehicles

The GSPEL, located at the U.S. Army Tank Automotive Research, Development and Engineering Center (TARDEC) in Warren, Michigan, provides the ability to develop, test, and troubleshoot vehicle systems and components under a variety of conditions, leading to more efficient and mobile ground vehicles.

The centerpiece of the GSPEL is the Power and Energy Vehicle Environmental Lab (PEVEL), which provides specialized test chambers. In 2007, the Army contracted Jacobs Engineering (Jacobs) to perform the design/build of the PEVEL facility. This environmental laboratory provides full mission profile testing under various environmental conditions, including temperatures ranging from 60°F to 160°F, relative humidity up to 95%, and wind speeds up to 60 miles per hour [7.30,7.31].

The humidity in the test chamber is controlled to within 0.5 grains of moisture per pound of dry air, while the surrounding ambient air can contain upwards of 100 grains of moisture per pound of dry air. The test chamber has also been designed to simulate a high-humidity environment containing upward of 250 grains of moisture per pound of dry air.

In addition to humidity simulation, the test chamber is able to simulate various solar loads, with the ability to provide a simulation of the sun for the prescribed solar intensity setting of the chamber.

The PEVEL also has the ability to simulate road loads for wheeled and tracked vehicles. This can be applied for any situation from a single drive wheel up to 10 drive wheels. Each wheel load and speed can be controlled independently, enabling full road-load simulation.

Along with environment and load testing, the PEVEL can accommodate hybrid-electric and fuel-cell vehicles, enabling the Army to test vehicles with alternative fuel sources for the ground power systems of the future.

7.5.7.1 Test equipment for commercial vehicles

RENK Systems Corporation [7.32] has built numerous test stands for heavy-duty and off-road vehicles: trucks, buses, forklifts, other material-handling equipment, and agricultural equipment, including tractors and harvesters. This company has designed two-axle dynamometers for combined testing in climate test chambers (Fig. 7.2).

FIGURE 7.2 Two axle dynamometer in climate test chamber (orange MAN semi-tractor) [7.32].

Fig. 7.2 Two-axle dynamometer in climate test chamber (MAN semi-tractor). [7.32] ESPEC design and manufacturing of equipment for combined testing [7.33] along with mounted equipment, usually electronics. While the company inaccurately advertises this as "reliability testing," their work is interesting as it demonstrates a step to accelerated reliability testing. This company designed and produced equipment for combined multiaxial vibration, temperature, and humidity simulation and testing.

MIRA's [7.34] combined environment facilities enable tests combining cyclic temperature, humidity, solar loading, and shock/vibration to be run according to a customer's exact requirements. By combining these conditions, real-world environments and damage mechanisms can be reproduced in the laboratory.

A number of facilities are fitted with state-of-the-art fire suppression systems and flexible interfaces designed to provide a safe shutdown of equipment based on configurable trigger conditions. These systems facilitate the safe testing of equipment requiring immediate power down, for example lithium-ion batteries and their management systems.

All of these facilities have access ports to allow power feeds, including electrical, hydraulic, pneumatic, etc.

Their available combined environmental testing capabilities include:
Climatic capability:

- Max temperature ranges: 70°C to +180°C
- Max chamber size: 3 m × 3 m x 4m
- Humidity range: 10%−95% RH
- Solar loading programmable up to 1200W/m^2

Vibration capabilities

- Five EM shaker systems and two multiaxis EH systems
- Maximum thrust: 62 kN
- Frequency ranges: 1 Hz to 3 kHz

Vibration types

- Sine (including resonance search, track, and dwell)
- Random
- Shock
- Sine on Random, Random on Random and gunfire (Mixed-mode testing)
- Time History Replication (Road Load data, etc.)

E-Labs, Inc. [7.35] provides equipment for combined environment testing that includes any of three vibration systems into combined environments, combining temperature/humidity chambers with electro-dynamic vibration shakers.

These chambers can perform rapid rates of temperature change, and the vibration facilities can produce 18,000 force-pounds, with frequencies up to 2000 Hz.

The items tested range from test samples, parts, and components, to finished products. The industries they have served for climatic testing include Aerospace/Avionics, Appliances, Automotive, HVAC, and Industrial Machinery.

Weiss-Umwelttechnik GmbH is among the most famous worldwide producers of standard testing chambers and systems for environmental simulation [7.36]. Their test systems combine a climate test chamber with variable temperature and air humidity, and with salt spray tests complying with the standard DIN EN ISO 9227 (DIN 50021).

The system is suitable for cycling salt mist and climate tests in accordance with relevant standards for the automotive industry, such as VDA 621-415 B.

Their corrosion climate test chamber for cycling tests type SC 1000/15−60 IU has a temperature range between +10 °C and +60 °C, with a temperature accuracy of ±0.5 °C at air humidity between 10% and 95% relative humidity with an accuracy of ±3% relative humidity for climate testing.

Their chamber is suitable for simulation of the values specified in the test standards:

- Typical settings are +50 °C/10% r.h. (dry climate) and +50 °C/95% r. h. (humid atmosphere).
- Dewpoint between +6 °C and +59 °C.
- Temperature tests (without controlled humidity) can be performed to −15 °C.
- These working ranges can be extended with controlled climate and freezing according to further corrosion test standards as an option.
- The test space volume is more than 1000 L.

- The interior test space can be almost completely used for the tested specimens. Therefore, large-dimensioned components, such as complete car body parts, can be placed in the test space.
- The test chamber floor is designed to carry a maximum load of 100 kg.

Their product range comprises temperature and climate testing systems, as well as test systems for simulated exposure to weather, temperature shock, corrosion, and for long-time testing in various test chamber volumes. Walk-in/drive-in chambers and process-integrated plants for environmental simulation and biology are designed, produced, and installed in accordance with customer specifications. Weiss Umwelttechnik is certified according to DIN EN ISO 9001.

Their test systems combine climate test chambers with variable temperature and air humidity and with salt spray tests complying with standard DIN EN ISO 9227 (DIN 50021).

Their systems are suitable for cycling salt mist and climate tests in accordance with the relevant standards for the automotive industry, such as VDA 621-415 B.

The Weiss [7.36] product range comprises test systems of various test chamber volumes for temperature and climatic testing, simulation of exposure to weather, temperature shock, corrosion, and long-time testing — in research, development, quality control, and production.

Walk-in systems and in-line plants are designed, produced, and installed in accordance with customer specifications. Key features include:

- Dust and spray water testing
- Explosion protected test systems according to ATEX
- Fitotron plant growth chambers and rooms
- Lithium-Ion test chambers
- Mobile AC systems
- Noxious gas
- Radiation and weathering
- Solar technology testing

Associated Environmental Systems (AES) [7.37] offers a range of pre-fabricated, tight-tolerance environmental walk-in rooms that are easy to ship and install. Major features include:

- Forced air circulation systems that distribute air evenly and constantly recirculate it, maintaining uniform temperatures and humidity.
- Temperature ranges are from $-73°C$ ($-100°F$) to $+85°C$.

A refrigeration system that meets the room's load demand and operates continuously in cooling mode. Single-stage mechanical units operate at $-35°C$ and above and two-stage cascade units at below $-35°C$.

- The electric heating system provides very accurate and straight-line control of temperatures. The system is reliable and has a long life.
- Humidity System choices are the conditioned fine Mist System or the Steam Generator System. Different ranges are available, including special dehumidity ranges to 5% RH.
- Prefabricated room construction consists of modular, sandwich-type panels with a metal skin inside and outside. Panels are easy to ship and assemble.
- Multiple viewing windows can be ordered in customized sizes.
- Incandescent or fluorescent lighting # is vapor-proofed and sealed.
- Instrumentation consists of single or two-channel microprocessor programmers with an accuracy of $\pm0.25°C$ or better at the sensor. Other options include: RS-232C, RS-422A, IEEE- 488 interface.
- Safety features include a redundant control circuit with its own controller and failsafe contactor.
- Many other options are available.

7.5.8 Refrigeration systems

AES supplies a refrigeration package that supports the load demands of the walk-in environmental room. It includes:

- Single-stage mechanical refrigeration systems for operations at $-35°C$ and above.
- Two-stage cascade refrigeration systems (air cooled or water cooled) for operations below $73°C$.

Scientific Climate Systems, Ltd. (SCS) [7.38] designs and builds drive-in chambers for testing any type of vehicle, such as automobiles and tractors, with the capability to simulate weather environments from extreme heat and humidity to subcold temperatures.

Environmental Tectonics Corporation (ETC) [7.39] Testing and Simulation Systems (TSS) group has been designing and manufacturing environmental simulation systems for the Automotive and HVAC Industries since 1969. ETC's experience in the design, manufacture, installation, and maintenance of Environmental Simulation Systems include expertise in the development of the following.

7.5.9 Other automotive test equipment

- Automotive Air Conditioning Systems Test Bench
- HVAC Component Calorimeters and Durability Test Stands
- Component Wind Tunnels
- Altitude Simulation Systems

- Conditional Air Supply Systems
- Sealed Housing for Evaporative Determination (SHED)
- Drive-in Vehicle Test Chambers
- Climatic Wind Tunnels
- Full Vehicle Corrosion Chambers

Climatic Wind Tunnel includes:

- Adjustable nozzle 7−13m2 and a long test section to accommodate a wide range of vehicle size and type, from small cars to Class 8 trucks and buses, with wind speeds in excess of 240kph;
- Temperature from −40°C to +60°C and humidity from 5% to 95% RH;
- Exceptional flow quality for advanced aerodynamic simulation and thermodynamic testing;
- Low background noise level (64dBA at 50kph) for the detection of vehicle drive-away anomalies, such as misfires, transmission hesitation, etc.;
- Independent power rollers on the chassis dynamometer with a turntable to enable cross-wind testing;
- Solar simulation system up to 1100W/m2 intensity with sunrise-sunset simulation capabilities;
- Blowing rain, falling and blowing snow simulation;
- A complete suite of ancillary systems for customer vehicle operation, including hydrogen and electric vehicle compatibility.
- HVAC/Industrial Test Equipment
- Environmentally controlled chamber
- Airflow measurement tunnels
- Compressor and other subcomponent calorimeter
- Balanced ambient and calibrated calorimeters.

During the last years, the design of environmental simulation systems has undergone considerable improvement due to the availability of technically upgraded measurement devices, PC-based computer data acquisition systems, and real-time data evaluation systems. The impact of these changes allows for more accurate testing and permits fuller automation of the existing test facilities.

Additionally, the TSS group has made substantial design and operational revisions, enhancing the equipment's compatibility with recent "green" initiatives. Various Vehicle Test Applications have been developed that address varied test environments, including simulating hot road and sun exposure conditions.

Soufflerie Climatique Ile de France (SCIDF) [7.40] is a French company that offers climatic wind tunnel testing for the automotive industry, and they now have installed a Froude Hofmann chassis dynamometer system.

Their test system is of the customized 4x4 configuration designed specifically to enable the company to test a wide variety of vehicles from small cars to large trucks. Froude Hofmann claims that the control system provides a unique control mode of 4x2 independent configuration, which enables two different vehicles (following each other) in the wind tunnel, to be tested simultaneously.

The wind tunnel has the capacity of testing temperatures from $-35°C$ to $+55°C$, humidity from 10% to 98%, solar radiation intensity from 0 to $1.200W/m^2$, and wind speed from 0 to 250 km/h.

The 1.6 m roll 4x4 chassis dynamometer was customized to suit SCIDF's exacting requirements.

In order to accommodate the wide range of vehicles tested at SCIDF, the 4x4 configuration featured a wheelbase adjustment from 2 to 7 m, an axle load of up to 6500 kg per axle, speeds of up to 250 km/h and a floor plate designed to carry 18-ton trucks.

Real-time data supervision is available on charts and tables supported for each measured parameter. An alarm or warning is also available for each channel.

7.6 Combined testing for vehicle components

Link Engineering Company [7.41] is a multinational company dedicated to the design and manufacture of test equipment and is a provider of comprehensive testing services for a wide variety of vehicle components.

Their particular focus is on providing test solutions for components behind the engine, including transmissions, brakes, clutches, wet and dry friction materials, wheels, tires, hubs, springs, steering systems, axles, and related subsystems. The link also provides a range of electric motor test systems, including alternators, starter motors, and steering system testers.

The link serves the transportation industry in segments, such as passenger cars, trucks, trailers, buses, motorcycles, aircraft, railway carriages, and off-road vehicles.

7.6.1 Wheel & hub test systems

Wheel and bearing test stands are available in several system configurations and are capable of testing performance and characteristics through various load inputs. Additionally, auxiliary systems are available for creating different extreme conditions of temperature, humidity, mud, and salt slurry test parameters.

Biaxial Wheel Test System: Designed for advanced design and development testing, full verification, product verification, and gristmill testing of automotive and light truck wheels and their camber angle control is designed to simulate actual road loads. Their capabilities include:

- Radial Fatigue Machine: Tests the performance and characteristics of automotive wheels through a tire interface and road wheel system.

- Wheel Impact Test Stand: Provides a means of simulating aggressive curb impacts on wheel and hub assemblies.
- Bearing Spalling Test System: Provides both radial and axial loaded testing of hub bearings.

With the increased emphasis on vehicle fuel efficiency, vehicle manufacturers are currently investigating methods to reduce hub drag more than ever before.

7.6.2 Transmission & driveline test systems

As a multifaceted engineering developer and comprehensive manufacturer, Link provides fully integrated test systems for the evaluation of transmissions and driveline assemblies and components.

Link test equipment provides measurement from simple components to full vehicle systems.

Their equipment works with transmission and driveline systems, including; oil and friction, reaction plates, gear interface, belt and chain, housing, torque converter, piston, spline shaft, output shafts, axles, differentials, wheel ends, mounting hardware, full and partial systems, and assemblies.

An example of the transmission and driveline test systems that Link Engineering develops and manufactures, include:

- High-speed Automatic Transmission Test System
- SAE No.2 Wet Friction Test Stand
- Manual Clutch Durability Test System
- Transmission Torque Cycling Durability System
- T0-4 Test System
- Four Square Test System

SERVOTEST Company [7.42] designs and manufactures multiaxis multistation systems for the testing of automotive ball joints. SERVOTEST has experience in the design and manufacture of testing systems used to provide a repeatable laboratory simulation of the loading experienced by automotive ball joints throughout their service life.

Automotive ball joints are typically expected to last the lifetime of the vehicle.

The multiaxis motion combined with the complex loading patterns resulting from both continuous and transient road-load conditions are just two of the challenges faced by the designers of these components. Frequently these joints are also subjected to high- and low-temperature thermal loading, as well as other challenging environmental conditions.

SERVOTEST has developed a number of configurations of multiaxis ball joint test rigs to meet the varied needs of the ball joint testing community. Depending upon the size of the ball joint to be tested, these can be either single- or multistation test rigs.

The most common configuration is a dual-station three-axis rig which combines vertical and horizontal loading with ball joint rotation and rocking motions. Variations on this design include a single-station design for larger joints, incorporation of a thermal chamber for temperature cycling, and a single-axis rig for lateral load testing only.

An alternative heavy-duty configuration offers simultaneous independent three-axis loading combined with both rocking and rotation for a pair of ball-joint specimens. This configuration also provides for thermal conditioning if required.

Cincinnati Sub-Zero's AV/CV-Series vibration chambers offer rapid temperature change rates with combined temperature, humidity, and/or vibration [7.43]. All models are designed for compatibility with the consumer's choice of electrodynamic or mechanical vibration systems.

These chambers may be used with existing vibration shaker tables and can also be utilized as a separate temperature/humidity cycling chamber providing a greater return on their investment.

7.6.3 AV-series agree vibration chambers. temperature/humidity/ vibration chambers

The AV-Series AGREE Vibration chambers offer vibration testing combined with temperature change rates.

These chambers are available for temperature and vibration testing or temperature, humidity, and vibration testing. All models are designed for compatibility with electrodynamic or mechanical vibration systems. Dual-purpose chambers may also be used as a vibration chamber or as a temperature cycling chamber without vibration utilizing a solid floor plug.

The AV-Series AGREE models have the optional capability to interface with vibration systems in both the horizontal and vertical modes of operation. While each model is manufactured to standard designs, they may also be customized with added features, such as two-sided doors, double doors, vertical lift with height adjustment, rear sliding doors, and more.

7.6.3.1 CV-series vibration chambers. temperature − humidity − vibration chambers

CSZ provides drive-in chambers for combined environmental testing to simulate a variety of road conditions and climates. These chambers are designed for full vehicle testing. Drive-in chambers may be constructed using a modular/panelized box or fully welded walk-in box depending on customer requirements.

Their drive-in chambers simulate vibration and a variety of climate conditions, including solar simulation or infrared lighting. They may be used for different types of tests. Every chamber may be tailor-designed to meet various requirements and specifications, including integration with vibration systems for four-post road simulation.

7.6.3.2 Drive-in chamber features

Instrumentation

- The CSZ EZT-570i Touch Screen Controller offers a "7 or optional 10" touch screen and the latest in test chamber programming for ease of use. It comes standard with data logging, data file access via memory stick or PC, Ethernet control and monitoring, alarm notification via email or phone text message, data file backup, full system security, online help, and voice assistance in multiple languages and more.
- A Temperature Limit and alarm are included to protect the test subject.

Communications

- RS-232/485 serial communications, ethernet control, and monitoring offer a selection of communications options;
- EZ-View optional software allows the user to control and monitor up to 20 chambers from any location.

Cabinet

- Reinforced floor to support heavy loads;
- Fog-free viewing window and interior lighting make viewing the workspace hassle free;
- Stainless Steel Interior liners are constructed from Type 304 brushed stainless steel that is easy to clean and thoroughly sealed to prevent moisture migration;
- White embossed or aluminum scratch-resistant exterior.

Refrigeration

- Pressure gauges allow operating pressures to be continuously monitored and provide early warning indicators;
- CSZ test chambers utilize refrigerants, which are environmentally safe, nonflammable, nonexplosive, and have a zero ozone depletion potential (ODP).

CSZ CV-Series vibration chambers offer rapid temperature change rates with combined temperature, humidity, and/or vibration environments. All models are designed for compatibility with electrodynamic or mechanical vibration systems. While each model is manufactured to standard designs, they may be custom engineered to meet a wide range of testing requirements.

Their CV-Series temperature/humidity and vibration chambers are designed to integrate with vibration systems only in the vertical mode of operation. Chambers may also be used as temperature cycling chambers without vibration utilizing a solid floor plug.

Cincinnati Sub-zero has also developed a drive-in test chamber for testing complete vehicles.

The BMW Climatic Test Complex [7.45] consists of three large climatic wind tunnels, two smaller test chambers, nine soak rooms, and the support infrastructure. The capabilities of the wind tunnels and chambers are varied, and on the whole, give BMW the ability to test at practically all conditions experienced by their vehicles, worldwide. The wind tunnel test section was designed to meet demanding aerodynamic specifications, including a limit on the axial static pressure gradient and low-frequency static pressure fluctuations.

7.7 Equipment for accelerated reliability and durability testing

Fig. 7.3 illustrates a test chamber with computer-controlled universal equipment for different types of engine reliability/durability testing that includes [7.46,7.47]:

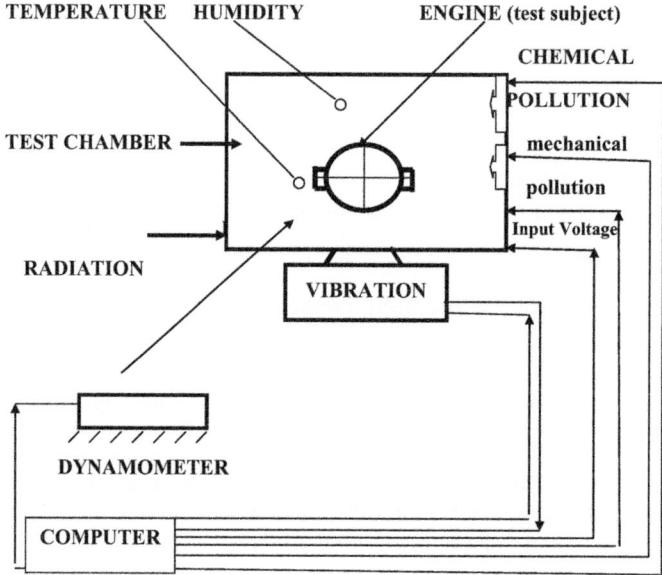

FIGURE 7.3 The schema of equipment for engine reliability/durability testing.

- Temperature;
- Humidity;
- Vibration;
- Dynamometer;
- Pollution chemical;
- Pollution mechanical;
- Input voltage.

One specific difference of this testing equipment, as compared with other equipment that is used for engine testing by companies that design and manufacture the engines, is this equipment uses vibration testing in conjunction with other loadings. This is a more accurate real-world simulation for reliability and durability accelerated testing.

A similar test chamber to that shown in the above schema was developed by the State Enterprise TESTMASH (Russia) for transmissions reliability/durability testing. This author participated in the development of a group of test chambers for reliability/durability test equipment, which combined specific vibration equipment, equipment for technological process simulation, drive simulation, and equipment for mechanical and corrosion process simulation. All of the loadings applied to the test subject were random processes so as to be close to real-world conditions.

A drive-in test chamber with universal equipment for accelerated reliability and durability testing was developed by State Enterprise TESTMASH) [7.46,7.47].

As depicted in Fig. 7.4, this chamber simulates the following fully-integrated input influences:

- Vibration testing equipment;
- Dynamometer testing equipment;
- Equipment for simulating chemical air pollution;
- Equipment for simulating mechanical (dust) air pollution;
- Equipment for simulating solar radiation;
- Equipment for simulating input voltage;
- Equipment for simulating various temperatures;
- Equipment for simulating humidity simulation.

By using multiple chambers in one system (Fig. 7.5), the cost of accelerated reliability/durability testing can be significantly reduced. In comparison with using chambers for testing one test subject (vehicle or its component), the advantages of such a system, include:

- Ability to test simultaneously multiple test subjects in one system;
- A single power system for N test chambers;
- A single computerized data acquisition and control systems;
- A single exhausts system for testing multiple test subjects;
- A single system of input influences the simulations for N test chambers.

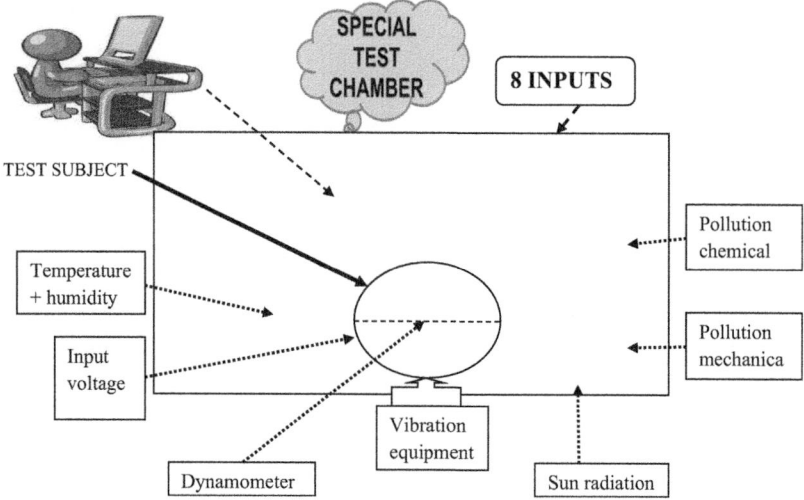

FIGURE 7.4   Scheme of the test chamber for accelerated reliability/durability testing (ART/ADT) (TESTMASH) (8 inputs).

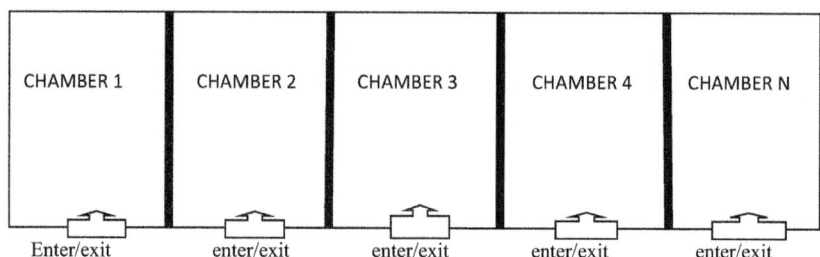

FIGURE 7.5   A scheme for multiple test chambers in one system.

The Automotive Center of Excellence (ACE) Institute of Technology at the University of Ontario at the beginning of the 21st Century created, acquired, and operated a climatic wind tunnel. This tunnel was designed to provide automotive manufacturers, tier, suppliers, and other industries of all sizes with testing capabilities to validate prototype products under a range of simulated field conditions, and that could be used in the future, after necessary development, for accelerated reliability/durability testing [7.46,7.47].

The project in which ACE developed a testing facility in partnership with the government and a university provided solutions to many interrelated problems, such as cost, utilization, and availability of testing facilities for accelerated reliability and durability testing. However, such a project requires strategic long-range thinking, the cooperation of varied partners, and the commitment of resources by all concerned parties.

The solution developed by ACE followed the ideas presented in this book and can be useful for future implementations of these ideas by others.

To better understand this interesting venture, a brief history of Automotive Centre of Excellence (ACE) will be provided. The ACE was developed in partnership with the University of Ontario Institute of Technology (UOIT), General Motors of Canada Ltd. (GMCL), the Government of Ontario, the Government of Canada, and the Partners for the Advancement of Collaborative Engineering Education (PACE). This innovative government-industry-university partnership is extraordinary in its scope and vision and remarkable for its inclusiveness. Its founding partners have created a welcoming environment where the dynamic pursuit of automotive and manufacturing innovation can involve everyone with a great idea, a desire to learn and a commitment to sustainable progress. ACE is the first testing and research center of its kind in Canada and in many respects, the world.

The initial funding for ACE was part of the General Motors Beacon Project announced on March 2, 2005 in partnership with the Canadian and Ontario Governments. The $2.5 billion CDN Beacon Project included strategic investments in GM's Canadian operations and the establishment of ACE at UOIT to drive automotive innovation and engineering by better-linking companies, suppliers, universities, researchers, and students.

In 2005, $58 million of the Beacon Project funds was earmarked for ACE. Based on a market analysis undertaken by UOIT in 2006, the university opted to expand plans for the climatic test facilities to include a larger fan, more-flexible test cell facilities, and a larger Integrated Research and Training Facility (IRTF). These changes would allow the wind tunnel to be utilized for trucks and buses, as well as light automotive applications, thereby contributing to the potential broad market appeal of the ACE and its unique facilities.

As the ACE project evolved, the university partners committed over $100 million, including contributions from the Government of Ontario, Government of Canada, UOIT, and General Motors of Canada through its PACE partnership.

The three main contractors were Diamond and Schmitt Architects (Toronto, Ontario), the Aiolos Engineering Corporation (Toronto, Ontario) and Vanbots Construction Inc. (Concord, Ontario).

The design and engineering work started in 2007, with construction beginning on June 5, 2008.

Five different countries were involved in the construction of the ACE facility.

The CRF offers full-size chambers that allow for full climatic and life cycle, including one of the largest and most sophisticated climatic wind tunnels. In this test chamber, wind speeds can exceed 240 km per hour, temperatures range from -40 to $+60°C$, and relative humidity can range from 5 to 95%. The climatic wind tunnel has a unique variable nozzle that can optimize the airflow from 7 to 13 square meters (and larger), allowing for an unprecedented range of vehicle and test property sizes (See components of this tunnel in Figs. 7.6–7.8). Coupled with this feature is a large flexible chassis dynamometer that is integrated into an 11.5-meter turntable.

Now, for the first time anywhere, vehicles and test properties can be turned into the airstream under full operating conditions to facilitate analysis in crosswind conditions. The large open chamber has a readily reconfigurable solar array that will replicate the effects of the sun. It is also hydrogen-capable, allowing for alternative fuels and fuel cell development.

Features of the climatic wind tunnel will include a 7–14.5 m^2 variable nozzle concept to enable a wide range of vehicle sizes and wind speed combinations. A turntable with a chassis dynamometer and vibration equipment

FIGURE 7.6 Climatic wind tunnel [7.45].

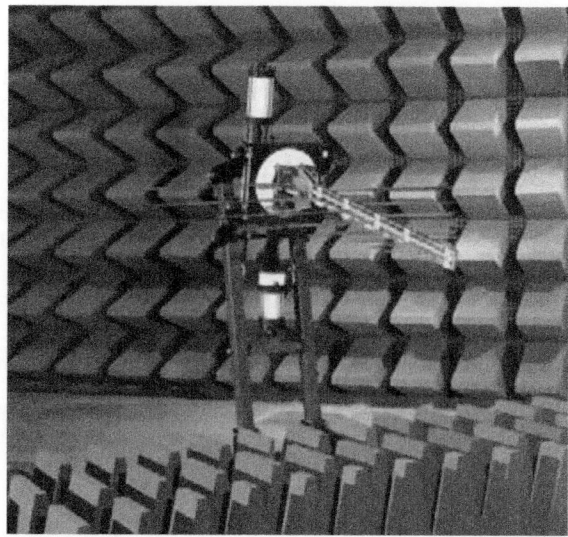

FIGURE 7.7 Test chamber for testing product for both EMC emissions and acoustic properties [7.45].

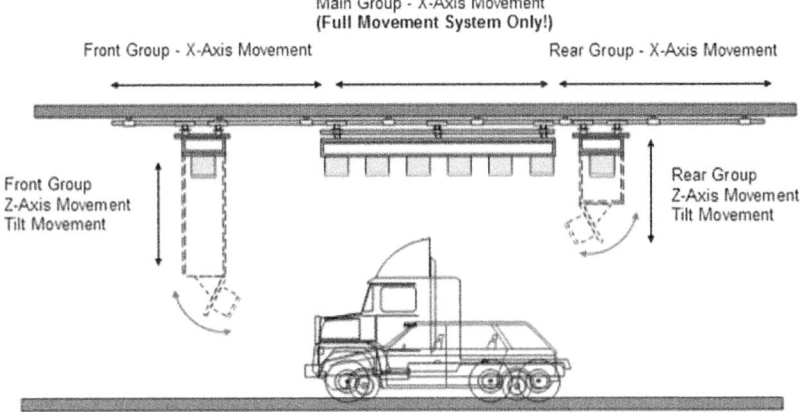

FIGURE 7.8 Test section with solar simulation system (A & B) [7.45].

enables vehicle yaw tests. The dynamometer assembly has been designed for removal from the test chamber and to include an elevator assembly and air bearing transport system.

ACE Climatic wind tunnel highlights:

- Climatic control;
- Dimensions: L20.1 m x W13.5 m x H7.5 m;
- Diurnal solar array;
- Fuel dispensing (hydrogen capable);
- Snow, rain and arid desert;
- Temperature and humidity (60 C to −40 C, five to 95% humidity);
- Turntable with a chassis dynamometer to enable vehicle tests at yaw;
- Variable nozzle (7−14.5 m^2) to enable a wide range of vehicle sizes and wind speed combinations;
- Wind speeds exceeding 240 km per hour (km/h).

Fan and motor system:

- Air flow − aero 888 m^3/s;
- Air flow − large/small nozzle 293/470 m^3/s;
- Expansion/vibration isolation;
- Motor speed 610 r/min;
- Nominal power 2500 kW (kW);
- Rotor/blade mass 6000 kg (kg);
- 0 to 40% in 6 s, 40 to 100% in 15 s, 100 to 0% in 15 s.

Air circuit viewing platform (ground floor)
Over 5.5 floors building envelope, top to bottom.
Final common conclusion: equipment for accelerated testing is increasing in quantity but decreasing in quality.

For accelerated testing, especially in the area of space, the problem of sensors remains very important.

Table 7.2 demonstrates the list of a number of sensors that are used routinely in the laboratory to perform various testing tasks. The first step in developing a program directed to solving this problem is deciding what needs to be exactly monitored and measured. Chapter 3 of this book demonstrated the common approach and examples of how to do this. One example presented in [7.1a] shows how the operating characteristics of interest are included.

Table 7.3 lists a number of sensors that are used routinely in the laboratory to perform various testing tasks.

TABLE 7.2 The characteristics and typical sensors that could be used [7.1a].

Characteristics	Typical sensors used
Sound and vibration	Accelerometers or capacitance probes
Temperature	Thermocouples or thermistors
Speed	Magnetic or optical pickups
Torque	Strain gages or piezoelectric sensors
Sliding or rolling	Electrical contact resistance or sound

TABLE 7.3 Characteristics to be measured and typical sensors used in the laboratory.

1. Temperature sensors	Thermocouples, thermistors, and infrared detectors.
2. Position sensors	Eddy current or capacitance probes, LVDT sensors, lasers and ultrasonic range detectors
3. Vibration sensors	Accelerometers, strain gages, capacitance probes
4. Speed sensors	Magnetic pickups, optical encoders, tachometers
5. Force or pressure	Piezoelectric or piezorestrictive sensors
6. Metal-to-metal contact detectors	Electrical resistance or voltage drop
7. Chemical sensors	Dielectric constant
8. Optical sensors	Densitometers, IR, UV fluorescence, spectroscope, photoconductivity cells, photodiodes
9. Acoustic emission	Ultrasonic and sonic detectors

Bibliography

[7.1a] Frank Murray S, Hestmat H, Fusaro R. Accelerated testing of space Mechanisms. April 1995. MTI Report 95TR29.

[7.1] Klyatis LM, Klyatis EL. Accelerated quality and reliability solutions. Elsevier; 2006.

[7.2] Klyatis Lev M. Accelerated reliability and durability testing technology. Wiley; 2012.

[7.3] Klyatis L. Successful prediction of product performance: quality, reliability, durability, safety, maintainability, life-cycle cost, profit, and other components. SAE International; 2016.

[7.4] Klyatis LM, Anderson EL. Reliability prediction and testing Textbook. Wiley; 2018.

[7.5] J. Cavazos, Industry Director. Frost & Sullivan. www.frost.com.

[7.6] Business Wire, A Berkshire Hathaway Company. Global general purpose test equipment market to grow to USD 6.58 billion by 2021, at a CAGR of nearly 5%.Global general purpose test equipment market to grow to USD 6.58 billion by 2021, at a CAGR of nearly 5%. Graphic: Business Wire. May 17, 2017. https://www.businesswire.com/...Top-3-Emerging-Trends-Impacting-Global-General.

[7.7] Electronic test and measurement market 2018 global trends, market share, industry size, growth, opportunities and forecast to 2023. "WiseGuyReports.com adds "electronic test and measurement market 2018 global analysis, growth, trends and opportunities research report forecasting to 2023". October 16, 2018.

[7.8] Automotive test equipment market 2018 global emerging technologies, top key leaders, recent trends, industry growth, size and segments by forecast to 2022. September 12, 2018.

[7.9] C. Armstrong, North America for Rigol Technologies USA. What trend or new technology will drive the test instrument market in 2013?. www.rigolna.com.

[7.10] C. Sweetser, Omicron. www.omicron.at.

[7.11] M. Schrepferman,Peregrine semiconductor corporation. www.psemi.com.

[7.12] M. Fox, FLIR test & measurement. http://www.flir.com.

[7.13] SIEMENS.The testing laboratories at the Schaltwerk Berlin.

[7.14] Automotive test equipment market overview. New York, United States — October 18, 2018. MarketersMedia.

[7.15] Show Preview. Automotive testing expo 2018. Novi, Michigan. October 23—25, 2018.

[7.16] Novi, Michigan. In: Autonomous vehicle test & development symposium; October 23—25, 2018.

[7.17] WT-D/WK-D. Aerospace testing industry leader Weiss Technik weiss-na.com Adwww.weiss-na.com.

[7.18] Element. Aerospace_testing-exova.com.Adwww.exova.com.

[7.19] Pelissou P, Daout B, Romero C. Critical review of the ECSS-E-ST-20-07C ESD test set-up for testing spacecraft equipment. 2016. ESA Workshop on Aerospace EMC (Aerospace EMC).

[7.20] Perraux R. Environmental test chambers: extreme environments testing with F-gas compliance. Aerospace Testing International 2018. Showcase.

[7.21] Chen L, Zhao F, Wang L. Accurate test chambers. Aerospace TestingInternational 2018. Showcase.

[7.22] .

[7.23] Burke Porter Machinery Company. www.bepco.com.

[7.24] Aerospace Testing International Solar impulse into the tunnel. April 2014.

[7.25] Solar impulse — building a solar airplane. www.solarimpulse.com/en/our/building-a-solar-airplane/.

[7.26] Test Center — European Test Services (ETS). www.european-services.net.

[7.27] European Space Agency. PepiColombo. Mercury composite Spacecraft. Sci.esa.int/ … /50,547 — sunshield-being-omstalled-on.

[7.28] ESCO technologies. ETS-Lindgren. Automotive chambers. www.ets-lindgren.com.

[7.29] ESPEC Technology Report. Special Issue: Evaluating Reliability.No. 3. 1997.

[7.30] Ground Systems Power and Energy Laboratory. U.S. Army. www.army.mil/standto/archive/issue.php?issue.

[7.31] Jacobs. Environmental testing for military vehicles. Automotive Testing Technology International 2012:78—9.

[7.32] RENK Systems Corporation. www.renksystems.com/.

[7.33] ESPEC Technology Report. Special Issue: Evaluating Reliability.No. 3. 1997.

[7.34] Combined Environment Testing. MIRA. www.mira.co.uk. Defense Vehicle Engineer.

[7.35] E— labs. Industrial testing laboratory. Frederickburg. Virginia.

[7.36] Weiss Umwelttechnik GmbH www.compositesworld.com/suppliers/weissumw/DIR.

[7.37] Associated environmental systems (AES). www.associatedenvironmentalsystems.com/.

[7.38] Scientific Environmental Testing, Ltd. www.scs-usa.com.

[7.39] Environmental Tectonics Corporation (ETC). https//www.etcusa.com/.

[7.40] Chassis dynamometer system. Automotive Test Technology International June 2013. La Soufflerie Climatique Ile de France, www.soufflerie-climatique.fr.

[7.41] Link Engineering Co. www.linkeng.com.

[7.42] Servotest — Test and Motion Simulation. www.servotestsystems.com/.

[7.43] Cincinnati Sub-Zero Chambers. All types and sizes made for you. www.cszindustrial.com/Environmental.

[7.44] Klyatis LM. Accelerated evaluation of farm machinery. Moscow: AGROPROMISDAT; 1985.

[7.45] UOIT — ACE - Automotive centre of excellence. ace.uoit.ca/.

[7.46] Gary M. Elfstrom and Greg L. Rohrauer. 8 design and construction of the UOIT Climatic Wind Tunnel. University of Ontario. Institute of Technology.

[7.47] L.Klyatis. Test Centers "Testmash". Journal Automotive Industry. 9/1992. Moscow, Russia.

[7.48] High quality equipment for test service of Farm machinery. Interview with Chairman of engineering center TESTMASH Dr. Lev Klyatis. Journal tractors and agricultural machines 1990;(11). Moscow, Soviet Union.

[7.49] Bender T, Hoff P, Kleemann R. The new BMW climatic testing complex — the energy and environment test Centre. Paper 2011-01-0167. 2011. Detroit.

Exercises

1. What is the specific key finding taken from the referenced TestExpo regarding the current state of testing equipment?

2. Describe the specific status of the testing equipment market as taken from "Technavio."

3. Describe three test equipment global markets.

4. Describe the expected global market growth of test equipment from 2016 to 2021.

5. Describe some of the specific testing and measurement equipment for electronic product testing.
6. What are some key trends for electronic test and measurement?
7. Describe some of the basic aspects of the Automotive Testing Equipment Market.
8. Describe specific aspects of the testing equipment market in the Asia Pacific Region.
9. Describe some examples of combined aerospace testing equipment.
10. Discuss some features of the combined environmental testing equipment for military vehicles.
11. Describe some examples of combined environmental testing.
12. Describe combined testing equipment for vehicle components.
13. Describe some of the basic components of test chambers designed with universal equipment for testing of engines.
14. How many and what types the inputs are simulated in the test chamber for accelerated reliability and durability testing, which was developed by TESTMASH?
15. Describe some specific capabilities of the Automotive Center of Excellence (ACE −Canada).

Chapter 8

How to use the positive trends in the development of accelerated testing and avoid the negative aspects and misconceptions prevalent in the industry

Abstract

This chapter is written as an aid to teachers and students in engineering education, training courses, seminars, lectures, tutorials, and other aspects of the use of ART/ADT. It includes two major areas of interest.

First, the analysis of the current situation in accelerated testing. This provides the necessary understanding of what needs to be done in using and developing the positive trends in accelerated testing, and in recognizing and minimizing the negative trends prevalent in testing;

Second, how to advance in a step-by-step process the positive trends in the developments in accelerated reliability and durability testing technology (ART/ADT).

8.1 Introduction

The idea for writing this chapter developed from a reviewer's comments on this author's paper prepared for and delivered at the SAE 2018 World Congress. This reviewer observed the need for detailed written instructions and examples for professionals and students to improve their knowledge in the development of accelerated testing.

As a precursor to such teaching or learning, one must first know:

1. Most types of laboratory and proving ground testing, as well as intensive field/flight testing, are accelerated testing. This is because testing results are obtained quicker than in normal field/flight conditions.

Trends in Development of Accelerated Testing for Automotive and Aerospace Engineering.
https://doi.org/10.1016/B978-0-12-818841-5.00008-8

2. A product's life performance, including quality, reliability, durability, safety, maintainability, supportability, life cycle cost, profit, recalls, complaints, other interacted components and research results, **depends significantly on the testing's effectiveness.**

3. Effective testing requires **accurately** simulating the real field/flight conditions.

4. Product recalls have increased during the last dozen years. This is because while the products have become more complicated, and while the technical progress of the design process is increasing at a rapid rate, the advances in testing are increasing at a much slower rate. This is especially true in the electronic, automotive, and aerospace areas.

5. Product recalls have resulted in the loss of many billions of dollars. Examples were presented in Chapter 2 (Figs. 2.2, 2.3, and text).

6. Presently, much of the testing, whether in the laboratory, proving ground, and/or utilizing intensive field/flight conditions, fails to **successfully predict** the product's long-term effectiveness in real-world conditions. Too often this results in unforeseeable recalls, customer dissatisfaction, lower quality, poor reliability, safety issues, and/or higher life cycle costs than were predicted during the design and manufacturing phases.

The conclusion from all of the above is a product's effectiveness is highly dependent on how accurately the actual real-world field/flight conditions were simulated in the testing.

8.2 Analysis of the current situation in accelerated testing

In order to implement the positive trends in accelerated testing, it is, first, necessary to recognize the negative trends prevalent in much of today's testing environment.

Detailed descriptions of both accelerated testing's positive and negative aspects can be found in this author's books, including in the Preface, Introduction, and Chapter 1 of Ref. [8.1], in Chapter 2 Analysis of Current Approaches in Simulation and Testing of Ref. [8.2], and in Chapters 3 and 6 of the book [8.8].

While these books provide much in the way of in-depth information on the subject, they did not fully address the trends in the most recent development of accelerated testing. Therefore, the following will provide a brief review of some of the previously provided material, but with the addition of detailed information on the newer developments in the field of testing.

While in the real world, many field/flight inputs interact simultaneously and in combination, most frequently employed accelerated testing still uses either separate discrete or at most several of the many factors in testing. This does not truly represent the real-world product performance and produces inaccurate accelerated testing.

For example, even some professionals in the field incorrectly consider accelerated testing to be:

1. Separate simulation of the following inputs
   - Vibration in the laboratory;
   - Vibration from proving grounds;
   - Input voltage;
   - Chemical pollution;
   - Mechanical pollution;
   - Temperature;
   - Air fluctuations;
   - Explosive atmospheres;
   - Step stresses;
   - Constant stress;
   - Temperature shock;
   - Mechanical shock;
   - And others.
2. Simulation of the following combination, but not most of the real combined inputs:
   - Temperature + humidity;
   - Temperature + vibration (HALT, HAAS);
   - Temperature + humidity + vibration + input voltage;
   - And others.

This testing does not fully account for the product's real-world exposure, where many input influences act simultaneously and in combination, and not as separate or as a small number of combined influences.

As a result, the simulation provides inaccurate performance prediction. The basic problem is: the product not has been exposed to the actual interactions experienced when these varied and different inputs actually occur.

Moreover, too often, none of these actual input influences are accurately simulated, because the process of identifying the real-world influences is poorly developed or not fully understood.

Therefore, the results of much of the accelerated testing using these methodologies does not provide the accurate initial information necessary for the successful long-term prediction of the product's effectiveness, including both the product's technical aspects (quality, reliability, safety, durability, maintainability, supportability, and others) and the economic aspects (life-cycle cost, safety, profit, recalls, and others).

And, this issue is prevalent in both physical and virtual simulations of real-world conditions that are necessary for providing accurate accelerated testing.

As the aforementioned references demonstrate and analyze many of the negative and positive aspects present in current accelerated testing, they are not repeated in this book, but they should be reviewed by teachers, students, and professionals involved in testing to obtain a complete understanding of these problems.

Finally, this chapter is focused on the developments in accelerated testing that are related specifically to automotive and aerospace engineering.

8.3 Adopting the positive trends in accelerated reliability and durability testing technology

This section will focus on enabling the reader on **how** to take advantage of the important positive trends that are developing in the field of accelerated reliability and durability testing (ART/ADT). Other examples of the proper use of the concepts of ART/ADT can be found in section 8.2 of this book, and in other publications that are focused on this direction of accelerated testing development. While many companies are still using separate factors, or at most several combined factors in their testing, they seldom use fully integrated accelerated reliability testing technology from the first to the last step. Too often as a result of this incomplete or improper testing, the investment in the methods and equipment for ART/ADT is assessed as providing a poor return on investment or as being too expensive. But, an organization can help dramatically increase the benefits obtained from these testing investments by including the following components of ART/ADT technology.

The first step-by-step approach to accelerated reliability/durability testing technology, shown below, was published briefly in an article [8.3]. Building on this, the author then developed and formulated this to an 11 step process, as follows:

1) Collection of the Initial Information from the Field;
2) Analysis of the Field/Flight Initial Information as a Random Process;
3) Establishing Concepts for the Physical Simulation of the Input Influences on the Product;
4) The Development and Use of Test Equipment, which Simulates the Field/Flight Input Influences of the Actual Product in the Laboratory for ART/ADT;
5) Determining the Number and Types of Test Parameters for Analysis During Accelerated Reliability/Durability Testing;
6) Selecting a Representative Input Region for Accelerated Reliability Testing;
7) Preparation Procedures for the Actual Accelerated Reliability/Durability Testing;
8) Use of Statistical Criteria for Comparing Accelerated Reliability/Durability Testing Results and Field/Flight Results;
9) Collection, Calculation, and Statistical Analysis of Accelerated Reliability/Durability Testing Data;
10) Prediction of the Dynamics of the Test Subject's Reliability, Durability, and Maintainability During its Service Life;
11) Using Accelerated Reliability/Durability Testing Results for Rapid and Cost-Effective Test Subject Development and Improvement

The following provides greater detail on each of these 11 steps.

8.3.1 Step 1: Collection of the initial information from the field

This step formally defines the input influences that act on the specific product in the field/flight operations. It is critical to determine how the input influences act, and to select the appropriate input influences that must be simulated in the laboratory to provide accurate ART/ADT. Accurate ART/ADT parameters and regimes cannot be estimated without this information.

In order to accomplish this step, it is necessary to obtain:

- The actual input field influences, which affect the product's reliability, durability, maintainability, and supportability, i.e., the range, character, speed, and the limits of the changing values of temperature, air pollution, air fluctuations, humidity, complete solar radiation (ultraviolet, infrared, visible), input voltage, air pressure, full features of the road (type, surface, density, etc.), and other influences, which occur under various field conditions wherever the product is employed. In order to achieve this goal, one needs to study all the parameters of each input influence. Fig. 8.1 is an example of a diagram of the parameters for temperature influences.
- Note that the parameters as shown in Fig. 8.1 act on the product both in single steps and in simultaneous combinations, including their interactions, as would be experienced in the field.
- Fig. 8.2 depicts the input influences acting on the entire system in the full hierarchy of the equipment.

It needs to be recognized that the term "full hierarchy" means the complete product as a test subject (system) in real life, consisting of N units (subsystems) that act in a series of interactions. The subsystems (units) also consist of K details (subsubsystems) that also act in a series of interactions. Therefore, each subsystem and subsubsystem in real life acts in a series with connections to and interactions with all other subsystem and subsubsystems of the complete product.

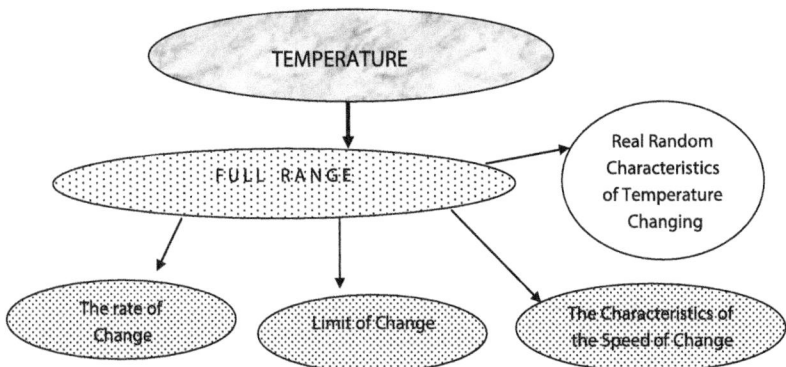

FIGURE 8.1 Schematic diagram for the study of the temperature-related parameters, as an example of studying the input influences on a test subject in the field/flight regime.

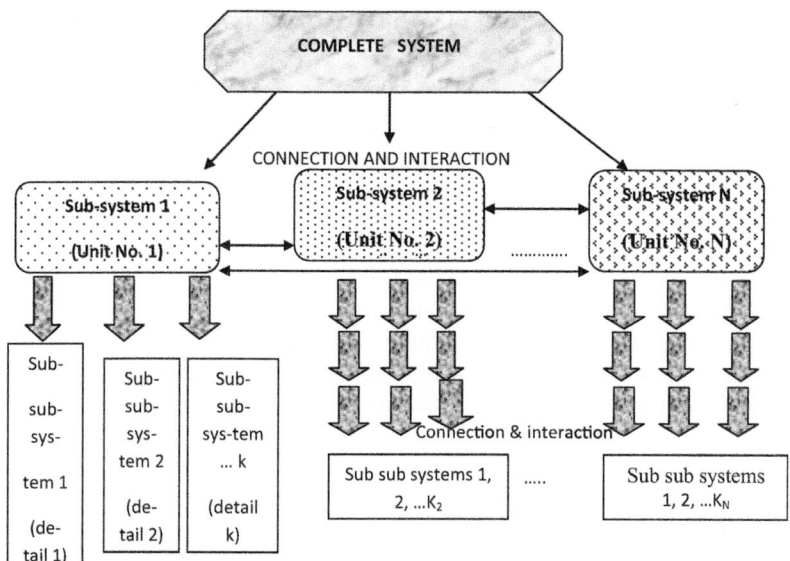

FIGURE 8.2 Depiction of the full hierarchy of the complete product and its components as a test subject.

As a result, the system (the complete product), its subsystems (units), and subsubsystems (details) must all provide their needed functions to make the system work. Therefore, in order to accurately simulate the real-life influences on the product in the laboratory during ART/ADT, one must accurately simulate the full hierarchy of these connections and interactions. If the simulation of the field input influences are only for the separate units (subsystems) or details (subsubsystems), and do not take into account the connections and interactions with other units (details), it cannot provide the accelerated reliability or durability testing as a component of the complete system's reliability/durability testing. Failing to treat the system as a hierarchy of its subsystems and subsubsystems is an instance of inaccurate simulation, which then leads to an inaccurate ART/ADT. The ART/ADT results will be different from the product's field/flight results and will lead to an inaccurate prediction of the test subject's life cycle cost, reliability, durability, maintainability, and quality in the actual world.

- Output variables are the direct result of the action of the input influences, such as loading, tension, wear, resistance, output voltage, the decrease of protection after deterioration, the range, character, speed, and the limits of values and rates of change under various field conditions, including climatic areas (see Chapter 3). With a developed product, these studies can be brief, as the data obtained from their previous generations of the products can be used. It must also be recognized that there will be variations depending on the facilities of the researchers, the goals of the research, the conditions of the experiments, and the subject of the study, etc.

- Mechanism of degradation for the components or the test subject, including the parameters of degradation, their value, speed, and the statistical characteristics of these parameters, any or all of which may change during the usage with time, etc. [8.4].
- Input influences and output parameters of the data collection and analysis. This includes the types of input influences that have a significant impact on the degradation and failure process. If it is known that one or several influences do not contribute to the product's degradation (failure), then they can be eliminated from consideration. But it must be positively known that those factors are of low enough significance that they may safely be ignored. For the electronic equipment, this was described in detail in Refs. [8.5,8.6].
- The use distribution of the test subject when it is working under different conditions. This is an important factor as most products are used in different usage conditions, with different loads, duty cycles, etc. These use conditions affect the values of the output parameters. Moreover, the usage results lead to the distributions that are needed for programming and understanding the results of the ART/ADT. Some examples of these distributions for car trailers and fertilizer applicators are shown in Ref. [8.7]. If the test subject is used in different climatic regions, this may also affect these distributions.
- The interactions of the specific test subject as a part of a system, which consists of subsystems (units) and subsubsystems (details) with their connections and interactions (see Fig. 8.2).

8.3.2 Step 2: Analysis of the field/flight initial information as a random process

For products, especially for mobile products, the field input influences and output parameters most likely have a random probability characteristic.

To account for the random nature of these influences it is necessary to evaluate the statistical characteristics of the studied parameters, including the mean, standard deviation, correlation (normalized correlation) or the power spectrum, and the distribution of input influences and outputs.

8.3.3 Step 3: Establishing concepts for the physical simulation of the input influences on the product

The field/flight input influences, which must include safety and human factors, must then be accurately simulated in the laboratory with periodical field testing in order to obtain a higher correlation of the accelerated reliability/

durability testing results with the field/flight results. Achieving this objective leads to the accurate prediction of reliability, durability, and maintainability in the real-world results of the ART/ADT. Practice has shown that, in general, the most accurate physical simulation of the input influences occurs when each statistical characteristic $[\mu, D, \rho(\tau), S(\omega)]$ of all of the input influences differs from the field conditions by no more than 10 (10%) percent.

As it was demonstrated in Figure 4.25 in Ref. [8.1] presents an example of ensembles of experimental normalized correlation and power spectrums of tension data, which was registered by sensor 1 for a car's frame point in different field (operating) conditions. This probability distribution may plausibly be used for a more detailed evaluation of the above processes.

For each specific situation, these statistical criteria should be used to compare the reliability (durability, maintainability, etc.) measured in the accelerated reliability/durability testing with the field/flight reliability (durability, maintainability, etc.).

An important concept in ART/ADT methods, as well as in other accelerated testing methods, is the "acceleration coefficient." The acceleration coefficient is the ratio of the predetermined time for the product's timeline in service (years, hours, cycles, etc.) as the numerator versus the time of testing as the denominator. Thus an accelerator coefficient of 1 would indicate the testing time would be equal to the real-world time. Thus, all accelerated testing yield acceleration coefficients are greater than 1 (>1). The similarity of the degradation processes in the field as compared to those in the laboratory is a factor in determining the practical limit of the acceleration (acceleration coefficient description), as shown in Figure 5.3, Chapter 5.

The most common method of ART/ADT is based on decreasing breaks in the working time to the minimum; but, while testing can be accelerated by testing up to 24 h a day every day, this method of acceleration reduces or does not include idle time or operating time at reduced or minimum loading. Therefore, this methodology is based on the principle of reproducing the complete range of operating conditions, which requires maintaining the correct proportion of heavy and light loads. And, for simulation, including environmental factors, the testing needs to be done inside of the test chambers. From this author's experience, this method of acceleration is especially effective for products with low usage cycles or with long periods of storage.

This method of acceleration has the following basic advantages:

- A high degree of correlation between field/flight results and the results of the ART/ADT.
- Each hour of pure work performed by the product in real-world operation is faithfully close to the testing stress schedule. Testing stresses are nearly identical in its destructive effect to that experienced in each hour of pure work under normal operating conditions.
- There is no need to increase the testing facility's size and proportion of stress applied in this method of acceleration.

Typically this method of ART/ADT results in 10−18 times faster than would be experienced by the product in the field. Moreover, there is often a good correlation of the ART/ADT results with the actual field/flight results. Calculation the accelerated coefficient is accomplished by dividing the Accelerated Testing Time Scale by the actual Field Use Time Scale, i.e., t_1/t_1^1, etc. Storage time acceleration, although more difficult to simulate, is generally performed by using accelerated stresses of the environmental parameters (temperature, humidity, pollution, etc.) in the test chambers. However, calculating the correlation between these laboratory testing results and the field results is more complicated. Therefore, it needs to be accompanied by periodic field testing as a component of the ART/ADT, especially if we take into account human factors and safety problems.

As a general concept, increasing the level of testing stress to a level greater than the maximum real field/flight conditions results in a lowered correlation with the real-world operation. For this reason, this reduced correlation also means reduced accuracy of the simulation and lowered probability of obtaining an accurate prediction of the product's reliability and durability in the field/flight regime. Therefore, when utilizing this type of testing, limiting the acceleration coefficient is very important. This is shown pictorially in Figure 5.4 (Chapter 5).

This reduced correlation also makes it more difficult to find the root causes and actual reasons for failures (degradation) and the correct solution for resolving them. Therefore, in determining the appropriate degree of acceleration (acceleration coefficient), it is necessary to specifically decide the approach that will be most effective for each product.

Accelerated laboratory testing for ART/ADT must always be provided as a simultaneous combination of the different pertinent groups of testing (mechanical, multienvironmental, electrical, etc.). For example, if performing vibration testing of a product, this testing will not accurately evaluate the reliability or durability of the product unless the other appropriate stress factors are applied simultaneously. The same is true for the simulation of temperature-humidity conditions. Testing for only one or two parameters is not true environmental testing, because it does not accurately simulate the full environmental influences experienced by the product in real-world operations.

Acceleration (stress) factors are factors that hasten the product's degradation process as compared to that experienced in normal usage. There are many methods of acceleration field/flight conditions, including:

- Higher concentration of chemical pollution and gases;
- Higher air pressure;
- Higher temperature, fog, and dew, and other forms of precipitation;
- Higher rates of change of the input influences.

There is also a widely used testing methodology that simulates with only a minimum number of combinations of field input influences. For example, a temperature/humidity environmental chamber may be used for environmental testing that only uses two simultaneous influences even though it is known that these are only two of the many environmental influences acting on the product. Care must be exercised whenever a high level of acceleration is combined with the field input influences used in ART/ADT, as these tend to reduce the correlation between the testing's result and field performance.

For more details and in-depth coverage of ART/ADT concepts, the reader is referred to the book [8.1].

8.3.4 Step 4: The development and use of test equipment, which simulates the field/flight input influences on the actual product in the laboratory for ART/ADT

This step discusses some of the special designs of testing equipment that are a result of the need for the analysis of the combination of field/flight input influences on the product $(X_1^1 \dots X_M^1)$, so they can be simulated in the laboratory.

Therefore, there is a substantial need for vibration testing, and this need results in the vibration testing equipment. Many of the testing equipment available in the market are either universal test equipment or influence specific test equipment. Universal test equipment is designed so it can be used on different types of products. These types of test equipment are usually used by many companies in the design and manufacturing of a wide range of products, as well as for specific applications. For example, virtually all types of mobile products and many components of stationary machinery vibrate.

The selection of appropriate test equipment is very important as it has a major influence on the accuracy of the ART/ADT results. For example, users of test equipment can buy single-axis and multiaxis vibration test equipment (VTE). For many types of stationary and for all mobile products, the single-axis VTE cannot accurately simulate the real-life vibration. The modern solution to this area is the multiaxis VTE.

VTE vibration inputs can be generated from three up to 6 degree of freedom. Vibration can be simultaneously generated for one to three linear motions (vertical, lateral, and longitudinal) and for one to three angular motions rotational (pitch, roll, and yaw).

A similar choice is available for test chambers, which simulate the environmental influences and the combinations of environmental influences for all types of products. There are many types of environmental test chambers available in the marketplace, but for reliable ART/ADT, test chambers providing multiple environmental influences are recommended. Some of the different types of test chambers that are presently available in the market and some other types of universal testing equipment can be seen in Ref. [8.1, Chapter 5].

Companies, especially suppliers that design and manufacture units or entire machines, typically have many types of specific testing equipment.

In the quest to provide effective ART/ADT these companies invest in both universal and specific testing equipment, which are typically used in the development of their products. Depending of the type of product, specific testing equipment is required. Much of this testing equipment is self-designed by the companies for their own use.

But when self-designing testing equipment, it should be remembered that the fundamental reason for developing testing equipment is to provide an accurate simulation of the field/flight situation.

8.3.5 Step 5: Determining the number and types of test parameters for analysis during accelerated reliability/durability testing

The objective of this step is to determine the minimum number of test parameters that will enable the comparison of the ART/ADT and field/flight testing results to accurately predict reliability, durability, maintainability, and other performance components of the marketed product.

In order to determine the optimal number of test conditions, it is, first, necessary to establish the basic areas of each influence that can be introduced into all varieties of operating conditions. This can be expressed as:

$$E > N,$$

where:

E is the number of field input influences $X_1 \ldots X_a$ (Fig. 3.1) [8.1];
N is the number of simulated input influences $X_1^I \ldots X_b^I$ (Fig. 3.1) [8.1].

And the allowable error for simulation input influences is $M_1(t)$:

$$M_1(t) = X_1(t) - X_1^1(t)$$

where:

$X_1(t)$ are input influences of the field/flight;
$X_1^I(t)$ are simulated input influences.

The basic steps for selecting the areas of influence, which will assure introducing all of the basic influences present in the field/flight situation are:

1. Determine the type of random processes to be studied. For example, the stationary random process is determined by the dependence of the correlation from the difference of the variables only.

   Establish the basic characteristics of this process. For example, in a stationary random process these characteristics would include the mean, the standard deviation, the normalized correlation, and the power spectrum.
2. Define each area's ergodicity, i.e., the possibility to make judgments about the process from one realization. This occurs when the correlation approaches zero as the time $\tau \to \infty$.
3. Check the hypothesis that the process is normal. For this goal, use the Pearson or another criterion.

4. Calculate the size of the area of influence.
5. Select the size of the divergence between the basic characteristics in different areas.
6. Minimize the selected measures of divergence and find the area of influence, which introduces all of the possibilities of the field operation.

Following these steps will provide the number and types of field/flight input influences that need to be simulated. The result of the analysis of these influences' actions on the product's degradation (failure) mechanism shows how these influences provide the information on how each of these influences' actions connect with the product's degradation.

8.3.6 Step 6: Selecting a representative input region for accelerated reliability testing

The content of this step presented in detail in the previous subchapter. Briefly, to accomplish this step, it is, first, necessary to identify one representative region from the multitude of input influences (or output variables) under specified field conditions with a minimum of divergence. Next, simulate this representative region's characteristics to provide the required accelerated reliability and durability testing. The solution for normal processes can be determined by the use of the Pearson criterion, but for other types of processes, random process analysis can be used. A complete description of this step can be found in the article [8.11].

8.3.7 Step 7: Preparation procedures for the actual accelerated reliability/durability testing

Preparation for ART/ADT should include the following:

- Scheduling of the product's technological processes, including all areas and conditions of use (time of work, storage, maintenance, etc.) for each condition.
- Determining the sensor's requirements and disposition.
- Determining the nature of the input influences and the output variables, which occur under each of the field conditions that are to be analyzed.
- Determining the needed measurement regimes of speed, productivity, output rate, etc.
- Determining the typical measurements for the conditions and the regimes of the field/flight conditions, and their simulation for accelerated reliability/ durability testing, and by comparison to the actual field/flight values.
- Establishing the value of the proposed testing.
- Executing the testing.
- Maintaining and updating the scheduling of the testing regimes.
- Obtaining the test results and analysing them.

8.3.8 Step 8: Use statistical criteria for comparing accelerated reliability/durability testing results with field/flight results

The determination of the similarity of the degradation mechanism is the basic objective criteria for comparing the field/flight results to the ART/ADT results. Calculating this similarity is accomplished by using statistical criteria that compare the accelerated reliability/durability testing results to the real-world operational results.

Using these criteria can help in deciding whether to use the current ART/ADT technique or whether it is necessary to develop this technique further until the difference between the reliability/durability and other necessary components functions distribution during ART/ADT and that occurring during real-world usage is not more than the desired fixed limit.

The statistical criteria presented in this book should be used in these three stages of testing:

- During ART/ADT providing a comparison of the output variables or in providing a comparison of the degradation process with the degradation process in the real world.
- During ART/ADT to provide a comparison of the appropriate time, cost, frequency, and other factors, such as the requirements and intervals for maintenance processes with those experiences during normal field/flight conditions.
- After ART/ADT for a comparison of the predicted reliability/durability indices (time to failure, failure intensity, etc.), as well as maintainability and other components indices with actual field/flight corresponding indices. Additional information may be generated by this step through analysis of failed and not failed test subjects.

The difference between the real world and ART/ADT results for the above situations should not be more than that established by the predefined fixed limit. The limit for the fixed-parameter differences between the real world and the ART/ADT should be determined by the desired level of accuracies, such as maximum permission difference of 3%, 5%, or 10%.

Another analysis that may be used for correlating the results of ART/ADT with field/flight testing is:

$$\left[(C/N)_{ART/ADT} - (C/N)_{f/f} \right] \leq \Delta_1$$

$$\left[(D/N)_{ART/ADT} - (D/N)_{f/f} \right] \leq \Delta_2$$

$$\left[(V/N)_{ART/ADT} - (V/N)_{f/f} \right] \leq \Delta_3$$

$$\cdots\cdots\cdots$$

$$\left[(F/N)_{ART/ADT} - (F/N)_{f/f} \right] \leq \Delta_i$$

where Δ_1, Δ_2, Δ_3 ... Δ_i—are divergences calculated from the results of the ART/ADT and field/flight testing;

ART/ADT and *f/f*—are ART/ADT and field/flight conditions;

C, D, V, and *F*—are measures of output parameters, such as corrosion; destruction of polymers, rubbers, wood, etc. (*D*); vibration (*V*) or tension; etc., and failures (*F*).

N is the number of equivalent years (months, hours, cycles, etc.) of exposure in the field/flight (f/f) or ART/ADT.

Fig. 8.3 shows an example of the normalized correlation data for $\rho(\tau)$ and the power spectrum S(w) of a car's trailer frame tension data in the field/flight and ART/ADT after using the above criteria.

The implementation of criteria for correlating ART/ADT results and field/flight results are described in greater detail by this author in Ref. [8.1].

The results of the implementation of these criteria for a car's trailer are also presented in Tables 8.1 and 8.2, and they can also be seen in Chapter 6 of this book.

Correlation can also be found by using experimental results. Fig. 8.5 shows the experimental distribution of tension amplitudes or frequencies on a car

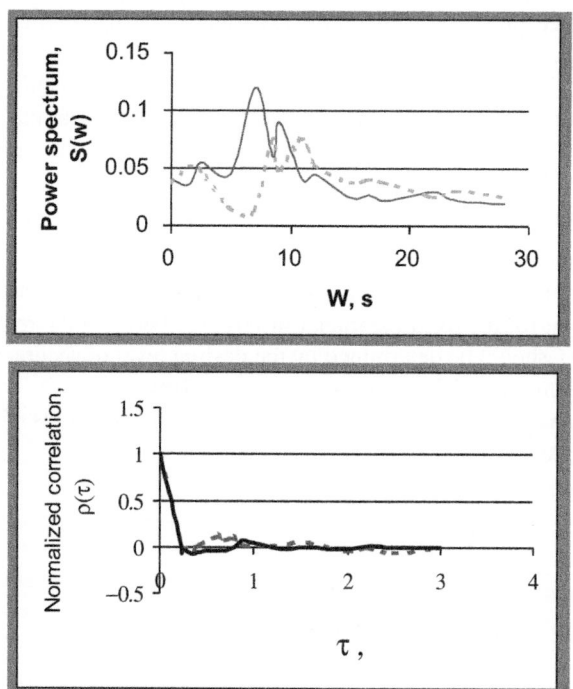

FIGURE 8.3 Normalized correlation $\rho(\tau)$ and power spectrum S(w) of the car's trailer frame tension data [8.3]. Straight line represents field/flight and dashed line represents ART/ADT.

TABLE 8.1 Results of the calculation of tensions on the wheel axles of fertilizer applicator [8.10].

Class number	Class frequency N_i	N_j	Class accumulated frequency $\sum_{Ni}$	$\sum_{Nj}$	Accumulated relative frequency $P_i = \dfrac{\sum_{Ni}}{N_i}$;	$P_j = \dfrac{\sum_{Nj}}{N_j}$	Modulus of congruence accumulated relative frequencies difference Max |D|
1	10	16	10	16	0.020	0.040	0.020
2	18	10	28	26	0.070	0.065	0.05
3	20	14	48	40	0.121	0.100	0.021
4	32	32	80	72	0.202	0.180	0.022
5	39	50	119	122	0.303	0.305	0.002
6	84	68	203	190	0.512	0.476	0.037
7	68	72	271	262	0.680	0.655	0.025
8	49	63	320	325	0.808	0.812	0.004
9	30	36	350	361	0.884	0.900	0.016
10	24	16	374	377	0.994	0.940	0.004
11	10	12	394	389	0.969	0.971	0.002
12	12	11	396	400	1000	1000	0.000

Where: i, during the field/flight; j, during the ART/ADT.

trailer's axle (sensor 1), the frame (sensor 2), and the carrier system (sensor 3) during both working in field conditions and during the ART/ADT as it was demonstrated in Figure 3.12 in Ref. [8.1].

The results of the calculation based on sensor 1 data, are presented in Table 8.1. This comparison of the data shows that the deviation is a very small.

$$\lambda = \max|D| \cdot \sqrt{\frac{\sum N_i \cdot \sum N_j}{\sum_i + N_j}}$$

If λ is less than 1.36, then one can adopt the hypothesis that both samples belong to one statistical population, i.e., the loading regimes in the field and in the ART/ADT are closely related. Here $\lambda = 1.36$ is the value of Smirnov's criterion at the 5% level.

TABLE 8.2 Part of assembly of fertilizer Applicator's tension data [8.7].

Digit number	Field/ flight n_i	ART/ADT testing n_j	Difference $n_i - n_j$	Definition from average $[(n_i - n_j) x]$	Definition from average in square $[(n_i - n_j) x]^2$
1	1	1	0	−32.5	1056.25
2	5	4	−1	−33.5	1122.25
3	18	8	−10	−42.5	1808.25
4	20	25	−5	−27.5	756.25
5	121	66	−55	−87.5	7658.25
6	223	174	−49	−81.5	6642.25
7	270	201	−59	−101.5	10,302.25
8	217	107	−110	−142.5	20,302.25
9	105	40	−55	−97.5	9506.25
10	37	11	−26	−58.5	3422.25
11	13	4	−9	−41.5	1722.25
12	3	1	−2	−34.5	1190.25

$$\sum_{ni} = 1033 \quad \sum_{nj} = 642 \quad \sum_{ni-nj} = 391 \qquad \sum_{[(ni-nj) - x]^2} = 65489.5$$

Table 8.2 shows an example of the use of the Student's distribution for the evaluation of the mean.

The following results were obtained by sensor 2. In this example the average is equal to 1.4.

Therefore, $1.4 < 1.8$, where 1.8 is the value taken from the tables of the Student's criterion at 5% for the degrees of freedom level (additional information on this theory is available in probability publications).

Therefore, in this example the hypothesis is proven true.

The results of the comparison of random tension data in the field and in ART/ADT are illustrated in Fig. 8.4 where the normalized correlation functions and power spectrums of a car trailer's frame tension data were collected by sensor 1. The time of correlation is between 0.09 and 0.12 s, the time of attenuation is the same, the maximum of the corresponding frequencies of the power spectrum equals 8−12 s, and the interval of frequencies was

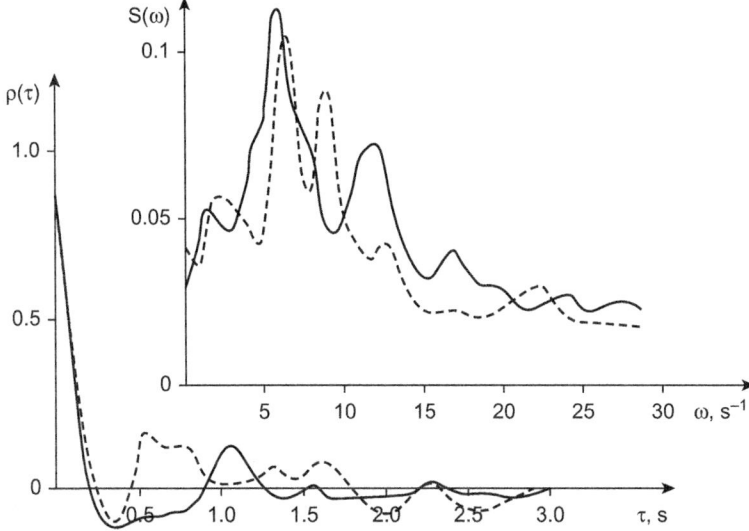

FIGURE 8.4 Normalized correlations $\rho(\tau)$ and power spectrum $S(\omega)$ of car trailer's frame tension data. Straight line represents field and dashed line represents accelerated (laboratory) testing.

substantially the same (from 0 to 16−18 s.). The maximum values of the power spectrum differences are of small velocities.

In this example, the conclusion is that overall the regimen of testing the carrier and the running gear systems of the car's trailer in the laboratory is closely related to the test regimen in the field.

8.3.9 Step 9: Collection, calculation, and statistical analysis of the accelerated reliability/durability testing data

Data is the foundation of accurate product reliability, durability, quality, life cycle cost, safety, profit, and maintainability analyses, and simulation.

Data include factors, such as total operating time, number of failure and the chronological time of each subsystem failure, component failures, and times to each component's failure, time for maintenance (repair), how failures display themselves, the reasons for failures, the mechanism of failures and degradation, and so forth. The system for collecting and using degradation and failure data from testing is often called a Failure Reporting and Corrective Action System (FRACAS). This step includes the test data collection during the test time, the statistical analysis of this data based on the failure (degradation) type and the test regimens, the reasons for deterioration leading to the ultimate failure of the test subject, and accounting for the acceleration coefficients. Fig. 8.5 shows an example of the accelerated loss of paint protection in the test chamber.

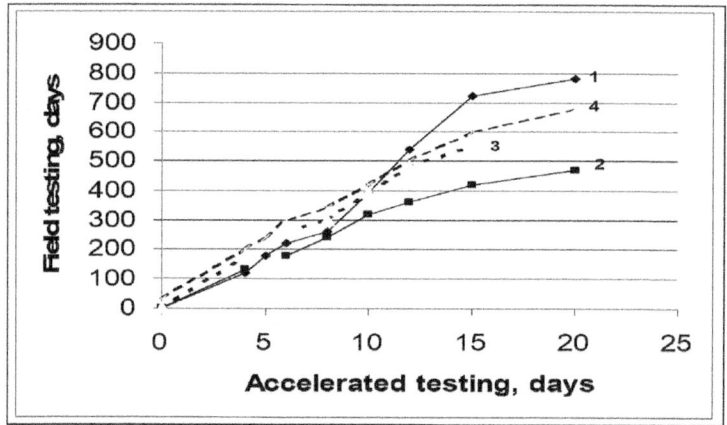

FIGURE 8.5   Accelerated deterioration of paint protection in test chambers (two types of paint). First type of protection (paint A): 1—protection quality; 2—impact strength; 3—bending strength. Second type of protection (paint B): 4—impact of strength.

In this case, the acceleration coefficient can be easily determined as shows below.

The loss of the protection quality of paint A (curve 1) for 5 days of accelerated testing is the same as that for 90 days of exposure in the field. The acceleration coefficient in this case is 18.

As with most accelerated stress testing (AST), the paint deterioration data involves collecting and analyzing many data points, and so the data are monitored by computer and the data is collected automatically. An additional important benefit from using the computer data collection is that, it can also be used to ensure that the test conditions are maintained during testing, and can monitor and alert to any deviations from the desired conditions. The importance of this data collection step is discussed in greater detail in Ref. [8.6].

8.3.10 Step 10: Prediction of the dynamics of the test subject's reliability, durability, and maintainability during its service life

Accelerated reliability and durability testing is not a final objective. It should be a source of initial information for determining quality, reliability, maintainability prediction and solutions, and for solving of other product problems. This initial information can be used in the evaluation of problems in the testing conditions or in predicting problems that will develop in field conditions.

References [8.2,8.8] provide further information on predicting these conditions.

8.3.11 Step 11: Using accelerated reliability/durability testing results for rapid and cost-effective test subject development and improvement

If the accelerated reliability/durability testing results have sufficient correlation with the field results, it is possible to quickly find the true reasons for test subject degradation and failure. These reasons can be determined by analyzing the test subject's degradation during the time of use, and by determining the location of the initial degradation and the development of the degradation process. Knowing this, it is possible to quickly remedy the reasons for the degradation. This is the approach that is used by the author, and recommended as this methodology is both time and cost-effective. Fig. 8.6 illustrates the strategic process of using this approach to accelerated reliability/durability/maintainability product improvement.

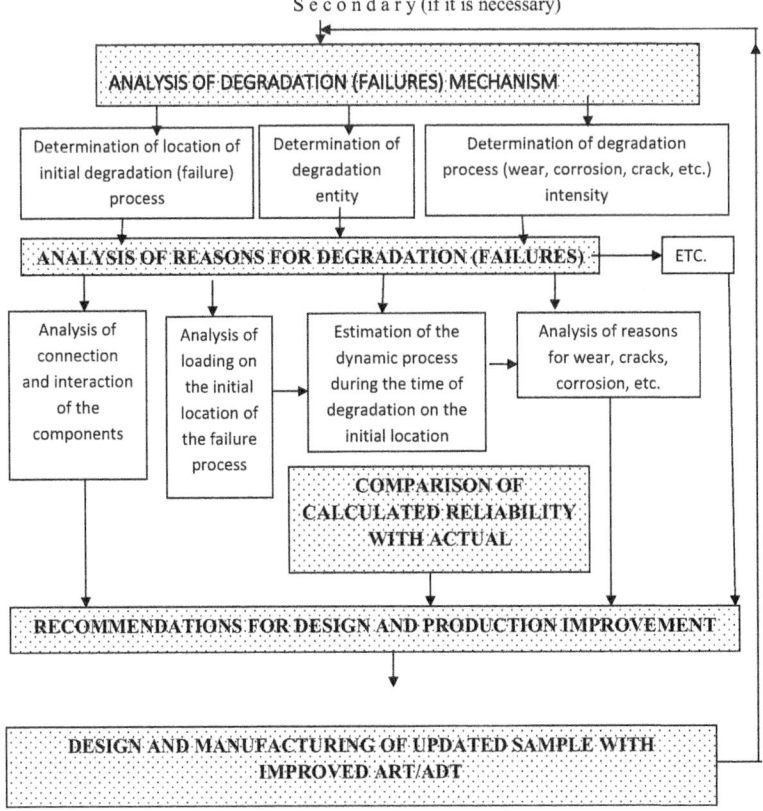

FIGURE 8.6 The scheme of updated process of rapid reliability/durability improvement during and after ART/ADT of systems, sub systems or components.

If there is insufficient correlation the real reasons for test subject degradation and the character of this degradation during accelerated reliability testing usually do not correspond to the degradation that occurs in the field. The conclusions achieved from this accelerated reliability or durability testing may be incorrect and will lead to increased cost and time for improvement and development of the test subject, rather than providing the hoped for accelerated development.

In such case, the designers and reliability engineers often thought they understood the reason for failure (degradation), and then changed the design or manufacturing process. But the changes were based on faulty information. Only after did they learn that their solution was incorrect when the "improvement" was demonstrated not to be working from field use. Then they had to revisit the problem and look for other reasons for the degradation, or other failures.

This type of situation does not allow for rapid product improvement or for the speedy development of test subject reliability (quality).

Moreover, its overall effect is to increase the cost and time of the test subject's development and improvement. This is a situation that is all too familiar in practice.

The following are two practical examples illustrating the possibilities of utilizing the author's approach to ART/ADT for the rapid improvement of product quality, reliability, and durability. The first example involved a problem with the harvester's reliability and durability, which was not resolved by the designers even after several years of field testing and data collection. At the author's direction a special complex was developed for ART/ADT to provide accurate prediction of the harvester's reliability/durability/maintainability. The methodology involved testing for 6 months as detailed in Ref. [8.1].

The approach and results are described below:

- Two of the harvester's specimens were subjected to evaluation for the equivalent of 11 years of operation.
- Three variations of one specimen unit and two variations of another unit were tested and the resulting information demonstrated satisfactory performance based on a service life of 8 years of equivalent operation.

As a result of the use of the new accelerated testing methodology elimination of the rapid degradation was accomplished by:

- Changing the harvester's design based on the conclusions and recommendations developed from the testing results.
- Incorporating these design changes on the tested specimens and then field testing the modified units.
- Reducing the cost (3.2 times) and the time (2.4 times) required for the harvester's development.

This resulted in a reliability increase of 2.1 times, which was validated by field operations. Also, the design changes for the basic components of the harvesters that had previously limited their reliability were confirmed. Usually, this work requires a minimum of two years to assure accurate comparison, but in this case, the above improvement was accomplished four times quicker.

The second example involved a problem with the reliability of the belts in a machine working heads.

For several years, beginning with the design stage and continuing through the manufacturing stage, it was recognized that the low reliability of the belts was limiting the reliability of the machine.

The designers tried increasing the strength of the belt, but even after several years they could not increase the reliability by more than 7%, and in doing so, they doubled the cost of the belts.

Using ART/ADT, special test equipment was developed to accurately simulate the belt's field input influences. After several months of testing and studying the reason for the low reliability the cause was found not to be with the belts, but in the rollers. The designers then improved the design of the belt/roller system, which resolved the problem.

Field testing of the modified machine demonstrated that the durability of the belts in the updated machine was increased by 2.2 times. The cost of implementing these changes only increased the machine's cost by 1%, which included the cost of the testing equipment and all of the work that was involved in finding the actual reasons for the problem and thereby increasing both the belts and the machine's reliability.

Without accurate initial information from ART/ADT, the best attempts at reliability prediction will not be useful.

Before closing this chapter, it should be mentioned that there is another approach to reliability prediction. This is the use of mathematical models that can relate the dependence of the product's reliability to the stress factors in the manufacturing process or the relationship of the product's reliability to the operating conditions in the field. It is also important to not forget to consider other factors in periodic field testing, such as the operator's influences and management's influences that may have a significant effect on the product's reliability. These influences can also be evaluated using mathematical models.

Unfortunately, these models as so yet cannot accurately simulate the complicated real field conditions.

The acceleration coefficients must also be changed to fit the desired testing objectives. The author recommends using the known formulas for the prediction of time to failures and time to maintenance.

Finally, it should also be mentioned that another approach is using a multivariant Weibull model utilizing ART results of components to predict system reliability. This approach can result in reduced testing time and reduced testing cost [8.9].

These results are especially important for new products. With new products, the testing results should show no more than a small number of failures, and they are particularly important when it is difficult or impossible to estimate the product's reliability.

Bibliography

[8.1] Klyatis LM. Accelerated reliability and durability testing technology. Wiley; 2012.

[8.2] Klyatis L. Successful prediction of product performance. Quality, reliability, durability, safety, maintainability, life-cycle cost, profit, and other components. SAE International; 2016.

[8.3] Klyatis LM. Step-by-step accelerated testing. In: Annual reliability and maintainability symposium (RAMS); 1999. Washington DC.

[8.4] Klyatis LM, Klyatis EL. Accelerated quality and reliability solutions. Elsevier; 2006.

[8.5] Alion Science and Technology. Accelerated reliability test plans and procedures. System Reliability Center. http://alionscience.com/consulting.

[8.6] Chan HA, Parker PT. Product reliability through stress testing. In: Annual reliability and maintainability symposium (RAMS). Tutorial notes. Washington, DC; 1999.

[8.7] Klyatis LM. Accelerated evaluation of farm machinery. Moscow: Agropromisdat; 1985.

[8.8] Klyatis LM, Anderson EL. Reliability prediction and testing textbook. Wiley; 2018.

[8.9] Klyatis LM, Teskin OI, Fulton W. Multi variety Weibull model for predicting systems reliability from testing results of the components. In: The international symposium of product quality and integrity (RAMS) Proceedings. Los Angeles; 2000.

[8.10] Klyatis LM, Klyatis EL. Successful accelerated testing. NewYork: Mir Collection; 2002.

[8.11] Klyatis L, Walls L. A methodology for selecting representative input regions for accelerated testing. Quality Engineering 2004;16(3):369–75. ASQ & Marcel Dekker.

Exercises

1. Describe what must first know in order to teach and/or learn how to improve their knowledge in accelerated testing.
2. Why is it incorrect to consider separate simulation for accelerated testing?
3. Describe some examples of simulation input combinations that do not reflect the real inputs of the field/flight conditions.
4. List the eleven steps provided by the author to provide ART/ADT.
5. What information need to be obtained in the collection of initial information from the field/flight to provide ART/ADT?
6. Why is there a need to analyze the field/flight information as a random process?
7. What concepts need to be established for determining the physical simulation of the input influences on the product?
8. What are some of the methods of acceleration of field/flight conditions that need to be used for the physical simulation of field/flight inputs?
9. Describe some of the types of test equipment that are needed for laboratory testing as component of ART/ADT.

10. Describe the process that can be used to establish the optimal number of test conditions.
11. Describe the basic steps provided by the author for selecting the appropriate areas of influence, which will assure introducing all of the basic influences in the field/flight situation.
12. What should be included in the preparation for performing ART/ADT?
13. List the three stages of testing that should use the statistical criteria presented in this book.
14. What kinds of analysis may be used for correlating the results of ART/ADT with field/flight results?
15. What data factors are a necessary foundation for accurate product reliability, quality, durability, and maintainability analysis?
16. Describe some of the methods of collection, calculation, and statistical analysis of ART/ADT data.

Index